AMINO ACIDS, PEPTIDES AND PROTEINS

— An Introduction —

AMINO ACIDS, PEPTIDES AND PROTEINS

— An Introduction —

H.-D. JAKUBKE
and
H. JESCHKEIT

Martin-Luther-Universität Halle, Wittenberg

Translated by
G. P. Cotterrell and J. H. Jones
in cooperation with R. Ulbrich
and the authors

A HALSTED PRESS BOOK

JOHN WILEY & SONS
NEW YORK

First published in Great Britain 1977 by
The Macmillan Press Ltd

Published in the U.S.A. by
Halsted Press, a Division of
John Wiley & Sons, Inc.
New York

Printed in the German Democratic Republic

Library of Congress Cataloging in Publication Data

Jakubke, Hans-Dieter.
Amino acids, peptides, and proteins.

A rev. translation of Aminosäuren, Peptide, Proteine.
"A Halsted Press book."
Includes bibliographies.
1. Amino acids. 2. Peptides. 3. Proteins.
I. Jeschkeit, Hans, joint author. II. Title. [DNLM:
1. Amino acids. 2. Peptides. 3. Proteins. QU55J25a]
QD431.J2513 1977 547'.75 77-23945
ISBN 0 470-99279-4

Contents

Foreword to the First German Edition

In writing this book we have made no attempt to describe the three classes of compound in comprehensive detail; such a task would require several volumes. The purpose of the book is to document important developments in the field of amino acids, peptides and proteins and to make them accessible to students of natural sciences and medicine.

It is our conviction that a representation based on chemical differentiation is of prime importance, since it ensures a more complete understanding of the subject. An important section of the book deals with peptide synthesis including modern methods allowing some degree of automation — for example, the MERRIFIELD synthesis, which has been considered in detail. Such methods are important because they allow the industrial production of polypeptide active constituents.

The past two decades have witnessed an explosive growth in our understanding of the chemistry of proteins, and so classical protein chemistry is only touched on: it was more important to discuss recent developments of sequence analysis and structure elucidation.

At the end of each section references are given to important reviews, monographs and original papers.

The book is designed as a textbook for university students in chemistry, biochemistry, biology, pharmacy and medicine, and as a reference book for all beginning research work in this field.

We wish to express our sincere thanks to all who have helped in the preparation of this book. We are aware of the imperfection of our efforts, and would welcome critical comment from readers.

H.-D. Jakubke

H. Jeschkeit, 1968

Foreword to the Second German Edition

Our book has been carefully revised for this second edition. Improvements in the arrangement of chapters have been made, and rapid advances in peptide and protein research have necessitated some rewriting in certain sections — for example, peptide synthesis on polymeric supports.

In view of the multitude of publications in the field of solid-phase synthesis, we cannot claim a comprehensive coverage of the area, but we have tried to give a critical analysis of solid-phase techniques alongside our discussion of conventional stepwise synthesis.

Amendments and additions were also necessary in the chapter on naturally occurring peptides, to reflect new work on discoveries, total synthesis and structure–activity relationships. Advances in methods of structure analysis of proteins have also been included. Many of the figures in the protein sections have been improved, and we have also mentioned attempts at the total synthesis of ribonuclease A, S and T_1.

Grateful acknowledgement is made to numerous colleagues who provided information for this edition. Particular thanks are due to our publishers for their co-operation.

H.-D. Jakubke

H. Jeschkeit, 1971

1. Amino Acids[1-5]

Investigations of fossil microorganisms have shown that amino acids existed three billion years ago, and their extraterrestrial occurrence has been proved by chromatographic analysis of the organic constituents of meteorites. Furthermore, traces of glycine and alanine have been detected in samples of lunar material.

Amino acids are simple organic compounds. Their physical and chemical properties can be explained by the presence of both acidic and basic groups in the same molecule. We can distinguish α-amino acids (in which the amino and carboxylic functions are both linked to the same carbon atom) from $\beta,\gamma,\delta,\varepsilon$-amino acids, which have a greater distance between the functional groups.

$\cdots CH_2-CH(NH_2)-COOH$	α-Amino acid
$\cdots CH_2-CH(NH_2)-CH_2-COOH$	β-Amino acid
$\cdots CH(NH_2)-CH_2-CH_2-CH_2-CH_2-COOH$	ε-Amino acid

Naturally occurring amino acids are usually α-aminocarboxylic acids. They are the fundamental building blocks of proteins which—like the nucleic acids—take part in all living processes. In addition to the protein-bound amino acids, the cellular tissue and fluid of living organisms contains a permanent reservoir of free amino acids ('amino acid pool'). These free amino acids take part in numerous metabolic reactions. The amino acid distribution patterns of blood plasma and urine are of great diagnostic significance in medicine. Specific metabolic roles of amino acids include the biosynthesis of polypeptides and proteins, and the synthesis of phosphatides, porphyrins and nucleotides. Particular amino acids are required

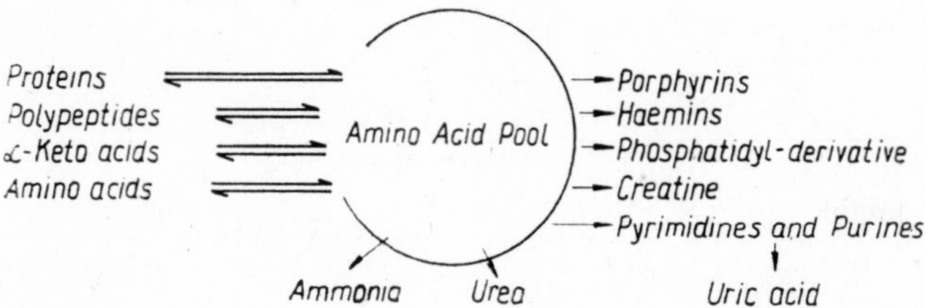

Figure 1. Schematic representation of the amino acid pool

Table 1.

	Name	IUPAC–IUB abbreviation		Chemical structure
Aliphatic neutral amino acids	Glycine	Gly	G	H_2N-CH_2-COOH
	Alanine	Ala	A	$CH_3-CH(NH_2)-COOH$
	Valine	Val	V	H_3C $\rangle CH-CH(NH_2)-COOH$ H_3C
	Leucine	Leu	L	H_3C $\rangle CH-CH_2-CH(NH_2)-COOH$ H_3C
	Isoleucine	Ile	I	H_3C-CH_2 $\rangle CH-CH(NH_2)-COOH$ H_3C
Aliphatic hydroxy amino acids	Serine	Ser	S	$HOH_2C-CH(NH_2)-COOH$
	Threonine	Thr	T	$H_3C-CHOH-CH(NH_2)-COOH$
Sulphur-containing amino acids	Cysteine	Cys	C	$HS-CH_2-CH(NH_2)-COOH$
	Cystine	Cys Cys		$S-CH_2-CH(NH_2)-COOH$ | $S-CH_2-CH(NH_2)-COOH$
	Methionine	Met	M	$H_3C-S-CH_2-CH_2-CH(NH_2)-COOH$
Imino acids	Proline	Pro	P	H_2C-CH_2 | | $H_2C\quad CH-COOH$ \\ / N H
	Hydroxy-proline	Hyp		$HO-HC-CH_2$ | | $H_2C\quad CH-COOH$ \\ / N H

The proteinogenic amino acids

Discoverer and first isolation	Content (%) in	
H. Braconnot, from gelatin (1820)	Silk fibroin	43.6
	Gelatin	25.7
Th. Weyl, from silk fibroin (1888)	Silk fibroin	29.7
E. V. Gorup-Besanez, from pancreatic extracts (1856)	Elastin (bovine aorta)	17.6
J. L. Proust, from cheese (1819)	Serum albumin	12.8
	Zein (maize)	19.0
F. Ehrlich, from beet sugar molasses (1904)	Serum albumin	2.6
	Globulin (oats)	4.3
E. Cramer, from silk (1865)	Silk fibroin	16.2
W. C. Rose and co-workers, from fibrin (1935)	Keratin (human hair)	8.5
	Avidin	10.5
E. Baumann, from cystine by reduction (1884)	Keratin (hair)	14.4
	Keratin (feather)	8.2
	Keratin (wool)	11.9
W. H. Wollaston, from urinary calculus (1810)	Keratin (human hair)	18.0
J. H. Mueller, from casein (1921)	γ-Casein	4.1
	Ovalbumin	5.2
E. Fischer, from hydrolysates of casein (1901)	Salmine	6.9
	Casein	10.6
	Gelatin	16.3
E. Fischer, from hydrolysates of gelatin (1902)	Gelatin	13.6
	Collagen	12.8

Table 1 (continued)

	Name	IUPAC–IUB abbreviation		Chemical structure
Basic amino acids	Lysine	Lys	K	$H_2N{-}CH_2{-}(CH_2)_3{-}CH(NH_2){-}COOH$
	Hydroxy-lysine	Hyl		$H_2N{-}CH_2{-}CHOH{-}(CH_2)_2{-}CH(NH_2){-}COOH$
	Arginine	Arg	R	$(H_2N)(HN{=})C{-}NH{-}(CH_2)_3{-}CH(NH_2){-}COOH$
	Histidine	His	H	imidazole ring (HC, N, NH, C, CH) $-CH_2{-}CH(NH_2){-}COOH$
Acidic amino acids and mono-amides thereof	Aspartic acid	Asp	D	$HOOC{-}CH_2{-}CH(NH_2){-}COOH$
	Asparagine	Asn	N	$H_2N{-}CO{-}CH_2{-}CH(NH_2){-}COOH$
	Glutamic acid	Glu	E	$HOOC{-}CH_2{-}CH_2{-}CH(NH_2){-}COOH$
	Glutamine	Gln	Q	$H_2N{-}CO{-}CH_2{-}CH_2{-}CH(NH_2){-}COOH$
Aromatic and hetero-aromatic amino acids	Phenyl-alanine	Phe	F	$C_6H_5{-}CH_2{-}CH(NH_2){-}COOH$
	Tyrosine	Tyr	Y	$HO{-}C_6H_4{-}CH_2{-}CH(NH_2){-}COOH$
	Tryptophan	Trp	W	indole (N–H) $-CH_2{-}CH(NH_2){-}COOH$

Discoverer and first isolation	Content (%) in	
E. Drechsel, from casein (1889)	Myoglobin Serum albumin	15.5 12.8
D. D. Van Slyke, from gelatin (1938)	Gelatin Collagen	2.0 0.93
E. Schulze, from lupin seedlings (1886)	Salmine Gelatin Histone (rat liver)	86.4 8.3 15.9
A. Kossel, from hydrolysates of sturine (1896)	Haemoglobin	7.0
H. Ritthausen, from peas (1868)	Globulin (barley)	10.3
L. N. Vauquelin and P. J. Robiquet, from asparagus juice (1806)		
H. Ritthausen, from wheat gluten hydrolysates (1866)	Gliadin (wheat) Gliadin (rye) Zein (maize)	39.2 37.7 22.9
E. Schulze, from beet juice (1883)		
E. Schulze and J. Barbieri, from lupin seedlings (1879)	Serum albumin γ-Globulin	7.8 4.6
J. Liebig, from alkaline degradation of casein (1846)	Silk fibroin	12.8
F. G. Hopkins and S. W. Cole from a pancreatic digest of casein (1901)	Lysozyme (egg) α-Lactalbumin	10.6 7.0

for special functions, such as glutamic acid for transamination and methionine for transmethylation reactions.

The final products of amino acid degradation in general metabolic processes are ammonia, urea and uric acid. The replacement of amino acid losses takes place by protein degradation, by transamination of α-keto acids and by transformations of amino acids.

It is clear from the general formula $R-CH(NH_2)-COOH$ that the naturally occurring amino acids differ only in the nature of the side chain residue R. This residue may be acidic, basic or neutral, and either aliphatic, aromatic or heteroaromatic in nature. All members of the series, with the exception of glycine (where R is hydrogen), have an asymmetric carbon atom, so we have to distinguish between optically active D- and L- and optically inactive (racemic) DL-amino acids. The naturally occurring amino acids generally belong to the L-series. Only a few metabolic products of lower organisms and particular animals contain D-amino acids.

In the following sections the nomenclature, stereochemistry, chemical reactivity, synthesis and analysis of amino acids will all be discussed.

1.1. The Naturally Occurring Amino Acids[6–8]

To date more than 150 different amino acids have been discovered in nature, and the frequency of such discoveries has risen with advances in chromatographic techniques for the detection and isolation of small amounts of amino acids. The immense array of possible vegetable and animal sources is being investigated systematically. Very many amino acids have been detected in plant sources. They occur as metabolic products, particularly during germination, and are often stored in soluble forms. Unusual amino acids have been detected as hydrolysis products of antibiotics and other biologically active products of microorganisms. The first natural amino acid to be discovered was asparagine, which was isolated in 1806 by VAUQUELIN and ROBIQUET from the juice of the asparagus plant. Asparagine is one of about 20 amino acids which are the normal constituents of vegetable and animal proteins. A summary of these 'proteinogenic' amino acids is given in the following section.

1.1.1. Proteïnogenic Amino Acids

There are several possible ways of classifying the proteinogenic amino acids. From their isoelectric points we can distinguish acidic, basic and neutral amino acids, or alternatively we can consider in turn those with aliphatic or aromatic side chains, each of these classes being subdivided according to the functional groups present. Proline and hydroxyproline are strictly imino acids because here the secondary amino group —NH— is part of a pyrrolidine ring system. Biochemists classify the amino acids according to the metabolism of their carbon chains and divide them into glycogenic and ketogenic amino acids. Glycine, alanine, serine, threonine, valine, aspartic acid, glutamic acid, arginine, histidine, proline and hydroxyproline are glycogenic amino acids: they are converted into oxalacetic acid, phosphoenol pyruvic acid and carbohydrates (gluconeogenesis). Leucine, isoleucine, phenylalanine and tyrosine are termed ketogenic since they give rise to ketonic substances. Another classification divides amino acids into essential or inessential amino acids according to whether they can or cannot be synthesised in the body.

1.1.2. Essential Amino Acids[9–11]

Plants and some microorganisms are able to produce all the amino acids necessary for the synthesis of their cellular proteins. Animals can only synthesise about 15 amino acids. The remainder must be supplied by foodstuffs of suitable composition or deficiency symptoms occur (inhibition of growth and protein biosynthesis, negative nitrogen balance, etc.). These amino acids are therefore called 'essential amino acids'[12].

The amino acids essential for man and the daily requirements in grams are listed in Table 2. The values in parentheses are the recommended quantities.

Table 2. Daily requirements of essential amino acids

Arg	1.8	(3.0)*	Met	1.1	(2.2)
His	0.9	(1.5)*	Phe	1.1	(2.2)
Ile	0.7	(1.4)	Thr	0.5	(1.0)
Leu	1.1	(2.2)	Trp	0.25	(0.5)
Lys	0.8	(1.6)	Val	0.8	(1.6)

* Necessary only during growth.

Some amino acids may be supplied as their D-antipodes, since D-amino acid oxidases can transform them into keto acids which give the corresponding L-amino acids on stereospecific transamination.

The amount and nature of the amino acids required by an organism vary according to its physiological or pathological state. Thus some amino acids needed for growth in young mammals cease to be essential to adults. Tryptophan and lysine requirements increase during pregnancy and lactation: tryptophan and isoleucine are particularly important for infants. Especially heavy amino acid requirements follow blood loss and tissue repairs following physical injury.

Glycine is an essential amino acid for birds. In ruminants the microorganisms of the digestive tract take over the biosynthesis of all the essential amino acids if there is a sufficient nitrogen supply. In man the provision of a sufficient supply of essential amino acids is a major nutritional problem. The nutritional value of proteins depends upon their amino acid composition. Only a limited number of animal proteins are of high nutritional value; for example, the complete proteins of egg and milk. Here the essential amino acids occur in sufficient quantities and in suitable proportions for human nutrition. Plant proteins have lower nutritional value because they are deficient in several essential amino acids, especially lysine and methionine. This deficit may be made up by addition of the corresponding amino acids or by combinations with other suitable proteins. In Table 3 are listed the essential amino acid values of several foodstuff-proteins. The high lysine content of petroleum-derived protein is remarkable.

Table 3. Essential amino acid composition of proteins from different sources[13]

Protein content (dry basis %)	Wheat flour	Beef	Cow's milk	Torula yeast	BP protein concentrate
	13.2	59.4	33.1	44.4	43.6
	Essential amino acid content of the protein (%)				
Leu	7.0	8.0	11.0	7.6	7.0
Ile	4.2	6.0	7.8	5.5	3.1
Val	4.1	5.5	7.1	6.0	8.4
Thr	2.7	5.0	4.7	5.4	9.1
Met	1.5	3.2	3.2	0.8	1.2
Lys	1.9	10.0	8.7	6.8	11.6
Arg	4.2	7.7	4.2	4.1	8.0
His	2.2	3.3	2.6	1.7	8.1
Phe	5.5	5.0	5.5	3.9	7.9
Trp	0.8	1.4	1.5	1.6	1.2
Cys	1.9	1.2	1.0	1.0	0.1

1.1.3. Other Natural Amino Acids

There is a strong structural relationship between the proteinogenic amino acids and other naturally occurring amino acids. For instance, there are at least 20 compounds related to alanine, which differ only in the nature of the substituent on the methyl group. This substituent may be a simple amino group as in *α,β-diamino propionic acid*, H_2N-CH_2-$CH(NH_2)$-COOH, which occurs in Mimosaceae, or it may be the unusual cyclopropane residue as in *hypoglycine A* or in 1-*aminocyclopropane*-1-*carboxylic acid*. Both of these amino acids have been found in various fruits:

H_2C—CH—CH_2—$CH(NH_2)$—COOH (ring closed by CH_2)

Hypoglycine A

H_2C—CH_2 (ring closed by C, bearing H_2N and COOH)

1-Aminocyclopropane-1-carboxylic acid

Stizolibinic acid, a pyrone derivative of alanine, has been isolated from pea seedlings:

HOOC (pyrone ring, O) CH_2—$CH(NH_2)$—COOH

L-*Thyroxine* and other iodine-containing amino acids occur in thyroid tissue. *3,3′,5-Triiodo-*L*-thyronine* has been isolated from ox thyroid. It has about five times the biological activity of thyroxine:

HO—(I, I)—O—(I, I)—CH_2—$CH(NH_2)$—COOH

Thyroxine (tetraiodo-L-thyronine)

3,5-Dibromotyrosine and *monobromotyrosine* have been discovered in certain corals.

Additional 'unusual' amino acids are *β-alanine*, $H_2N-CH_2-CH_2-COOH$, an isomer of alanine and a constituent of carnosine, anserine, panthotenic acid and coenzyme A; *3,4-dihydroxyphenylalanine* (*DOPA*), an intermediate in the formation of melanin; and last but not least *β-pyrazolylalanine*, the first pyrazole derivative to be isolated from natural sources:

(pyrazole N)—CH_2—CH(NH_2)—COOH

DOPA occurs in the free state in beans and is thought to be responsible for the aphrodisiac side-effects said to occur on eating beans. Recently DOPA has been used in the treatment of Parkinson's disease.

With a few exceptions, the other rare amino acids may also be related to corresponding proteinogenic amino acids. However, the common structural elements in these amino acid families do not allow any firm deductions as to biogenetic relationships.

The sulphur-containing amino acids (cysteine–cystine family) are also represented among the rare amino acids by, for example, L-*djenkolic acid*,

$$H_2C\begin{cases} S-CH_2-CH(NH_2)-COOH \\ S-CH_2-CH(NH_2)-COOH \end{cases}$$

which occurs in the East asian djenkol bean, *lanthionine*, which has been isolated from hydrolysates of wool and hair,

$$S\begin{cases} CH_2-CH(NH_2)-COOH \\ CH_2-CH(NH_2)-COOH \end{cases}$$

and *alliine* (*S*-allylcysteine sulphoxide), which has been found in garlic oil,

$$CH_2{=}CH-CH_2-\underset{\underset{O}{\|}}{S}-CH_2CH(NH_2)-COOH$$

(it is enzymically converted to allicin, which gives rise to the characteristic odour of garlic). Other compounds in this series are *ethionine*, the homologue of methionine, $CH_3-CH_2-S-CH_2-CH_2-CH(NH_2)-COOH$, and *homocysteine*, $HS-CH_2-CH_2CH(NH_2)-COOH$, which has been isolated from several fungi.

In the *aminobutyric acid* series the α-isomer has not as yet been found as a protein constituent, but it occurs in animal and plant tissues and has been detected in normal human urine. γ-Aminobutyric acid occurs in brain and several plant tissues. *Homoserine*, $HOCH_2-CH_2-CH(NH_2)-COOH$, has been isolated from *Pisum sativum* as the corresponding lactone: it is an intermediate in the metabolism of several proteinogenic amino acids. Canavanin, α-amino-γ-guanidinooxybutyric acid,

$$H_2N-\underset{\underset{NH}{\|}}{C}-NH-O-CH_2-CH_2-CH(NH_2)-COOH$$

occurs in the free state in soybean and inhibits the growth of several microorganisms. L-α,γ-*Diaminobutyric acid*, $H_2N-CH_2-CH_2-CH(NH_2)-COOH$, has been isolated from hydrolysates of the polymyxin antibiotics. The naturally occurring imino acids are related to proline, pipecolic acid or azetidine-2-carboxylic acid.

Pipecolic acid, a metabolite of lysine, has been isolated from various plants and microorganisms.

NH COOH

Pipecolic acid

NH COOH

Azetidine-2-carboxylic acid

Azetidine-2-carboxylic acid is the toxic constituent of *Convallaria majalis* L. (lily of the valley). Its toxicity may be based on an ability to deceive the natural system of protein biosynthesis. Azetidine-2-carboxylic acid is therefore incorporated into proteins instead of proline. Lily of the valley itself is protected against this confusion by the presence of a highly specific prolyl-transfer-RNA-synthetase.

Ornithine, $H_2N-CH_2-CH_2-CH_2-CH(NH_2)-COOH$, a basic amino acid, has been found in the free state in plants and as a component of several antibiotics: it does not occur in proteins although it was once thought to. It has been isolated from hydrolysates of marine algae.

Among the 'unusual' aminodicarboxylic acids are L-*α-aminoadipidic acid*, $HOOC-CH_2-CH_2-CH_2-CH(NH_2)-COOH$, and L,L-*α,ε-diaminopimelic acid*, $HOOC-CH(NH_2)-CH_2-CH_2-CH_2-CH(NH_2)-COOH$. Aminoadipidic acid is an intermediate in lysine metabolism and has been found in penicillins. Diaminopimelic acid is a constituent of bacterial cell wall and was first isolated from *Corynebacterium diphtheriae*.

The first amino-tricarboxylic acid, γ-carboxy-glutamic acid (Gla), $HOOC-CH(NH_2)-CH_2-CH(COOH)_2$, was found in prothrombin and other homologous blood-clotting factors.

The D-antipodes of the proteinogenic amino acids also occur naturally. For example, D-*alanine* occurs in the blood of milkweed bugs and guinea-pigs; D-*ornithine* has been isolated from the liver of particular types of shark; D-*serine* is formed (together with D-2,3-diaminopropionic acid) during metamorphosis of silk moths (*Bombyx mori*); and D-*glutamic acid* is found in the capsular material of *B. anthracis*.

1.2. Nomenclature of the Amino Acids

Amino acids have often been named according to the sources from which they were first isolated: for example, *asparagine* was isolated from *asparagus* juice, *cystine* from urinary calculi (Greek *cystis* = bladder), *glutamine/glutamic acid* from wheat *gluten*, *serine* from silk (Greek *seros* = silkworm) and *tyrosine* from casein (Greek *tyros* = cheese).

Some names are based on peculiarities of the first isolation procedure. *Arginine* was obtained as its silver salt (Latin *argentum* = silver) and *tryptophan* was obtained by *tryptic* degradation of proteins.

Structural relationships to other natural compounds have, of course, also contributed to the naming of amino acids. For instance, *valine* is derived from *valeric* acid and *threonine* is so named because its structure is related to the tetrose sugar D-*threose*. *Proline* is related to the systematic name *pyrroli*dinic acid. With the exception of tryptophan and the aminodicarboxylic acids, the names of amino acids all end in *ine* (*cf.* amine).

The aminoacyl residues of the general formula R—CH(NH_2)—CO— are named by replacing the suffix *ine* of the amino acid name (the suffix *ic* in the case of Asp or Glu; the terminal *an* in the case of Trp) with the suffix *yl*. The optical configuration of an amino acid is indicated by means of the small capital letters D or L. Racemic compounds are given prefix DL. The subject of optical activity will be considered in greater detail in Section 1.3.

In order to avoid the labour of writing down the full names of formulae of amino acid residues, especially when describing peptide sequences, protein chemists have an agreement over the use of abbreviating symbols. These symbols are derived from the trivial or chemical names of the amino acids. Three letters represent the name and structural formula of the amino acids. The L-configuration is understood: for example, Ala for L-alanine and Met for L-methionine. Only with D- or DL- amino acids does a qualifying D- or DL- appear before the symbol, e.g. D-Ala and DL-Met.

Allo- amino acids are characterised by the prefix a, e.g. *a*Ile for *allo*-isoleucine.

Hydroxy amino acids are denoted by the prefix Hy (for hydroxy) and the initial letter of the amino acid — for example, *Hyl* for *hydroxylysine* and *3Hyp* or *4Hyp* for the isomeric *hydroxyprolines*. The prefix 'nor' is used to change the trivial name of a branched chain amino acid into that of the straight-chain isomer. The abbreviations for *norvaline* and *norleucine* are *Nva* and *Nle*.

In the higher unbranched amino acids the functional prefix 'amino' is included in the symbol as the letter A (diamino as D). The location of the amino group in amino acids in positions other than α is shown by the appropriate Greek letter prefix. Examples:

β-Alanine	= β-Ala
α-Aminobutyric acid	= α-Abu
α-Aminoadipinic acid	= α-Aad
α-Aminopimelic acid	= α-Apm
α,γ-Diaminobutyric acid	= α,γ-Dbu
ε-Aminocaproic acid	= ε-Acp

Special symbols are in use for less common amino acids. The N-alkyl amino acids that occur in depsipeptides are signified by the use of Me and Et as prefix: for example, MeVal = N-methylvaline and Etgly = N-ethylglycine.

1.2.1. Representation of Amino Acid Residues and Substituted Amino Acids

The symbols referred to above represent the amino acids complete with all their functional groups intact as in the neutral state. Lack of hydrogen on the α-amino group or a side chain and replacement of the hydroxyl moiety of a carboxyl group must be specially indicated. Lack of hydrogen on the α-amino group is represented by a dash to the left-hand side of the amino acid symbol and lack of hydroxyl on the α-carboxyl by a dash to the right-hand side of the symbol. Replacement of a side chain hydrogen atom or hydroxyl function is represented by a line from above or below the symbol.

Examples:

$-\mathrm{Ala} = -HN-CH(CH_3)-COOH \qquad \mathrm{Ala}- = H_2N-CH(CH_3)-CO-$

$$\overset{|}{\mathrm{Asp}} \text{ or } \underset{|}{\mathrm{Asp}} = \begin{array}{l} H_2N-CH-COOH \\ \qquad\;\; | \\ \qquad CH_2 \\ \qquad\;\; | \\ \qquad CO- \end{array}$$

$$\overset{|}{\mathrm{Lys}} \text{ or } \underset{|}{\mathrm{Lys}} = \begin{array}{l} H_2N-CH-COOH \\ \qquad\;\; | \\ \qquad (CH_2)_3 \\ \qquad\;\; | \\ \qquad CH_2-NH- \end{array}$$

$$\overset{|}{\mathrm{Cys}} \text{ or } \underset{|}{\mathrm{Cys}} = \begin{array}{l} H_2N-CH-COOH \\ \qquad\;\; | \\ \qquad CH_2 \\ \qquad\;\; | \\ \qquad S- \end{array}$$

$$\overset{|}{\mathrm{Ser}} \text{ or } \underset{|}{\mathrm{Ser}} = \begin{array}{l} H_2N-CH-COOH \\ \qquad\;\; | \\ \qquad CH_2-O- \end{array}$$

$$\overset{|}{\mathrm{His}} \text{ or } \underset{|}{\mathrm{His}} = \begin{array}{l} H_2N-CH-COOH \\ \qquad\;\; | \\ \qquad CH_2 \\ \qquad\;\; | \\ \qquad / \!\!= \backslash \\ \quad N \qquad | \\ \quad \llcorner\!\!\_\_ N- \end{array}$$

$$\overset{|}{\mathrm{Arg}} \text{ or } \underset{|}{\mathrm{Arg}} = \begin{array}{l} H_2N-CH-COOH \\ \qquad\;\; | \\ \qquad (CH_2)_2 \\ \qquad\;\; | \qquad\qquad\quad\; NH_2 \\ \qquad CH_2-NH-C \begin{smallmatrix} \diagup \\ \diagdown\!\!\diagdown \end{smallmatrix} \\ \qquad\qquad\qquad\qquad\quad N- \end{array}$$

In amino acid derivatives the various functional groups of the amino acids may bear substituents. These substituents can also be represented by abbreviated symbols and so the derivatives can be described by the various symbols joined by hyphens. The abbreviations allow a simple and comprehensive representation of amino acid sequences and chemical reactions in the field of peptide and protein chemistry.

Examples of derivatives formed by the substitution of several functional groups:

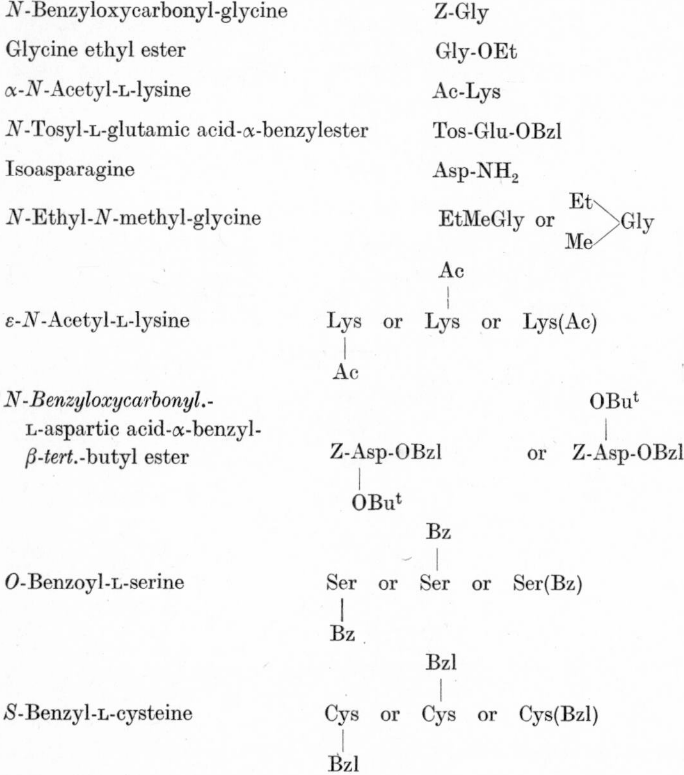

Derivative	Abbreviation
N-Benzyloxycarbonyl-glycine	Z-Gly
Glycine ethyl ester	Gly-OEt
α-*N*-Acetyl-L-lysine	Ac-Lys
N-Tosyl-L-glutamic acid-α-benzylester	Tos-Glu-OBzl
Isoasparagine	$Asp\text{-}NH_2$
N-Ethyl-*N*-methyl-glycine	EtMeGly or Et, Me >Gly
ε-*N*-Acetyl-L-lysine	Lys (Ac below) or Lys (Ac above) or Lys(Ac)
N-Benzyloxycarbonyl.-L-aspartic acid-α-benzyl-β-*tert*.-butyl ester	Z-Asp-OBzl (OBu^t below) or Z-Asp-OBzl (OBu^t above)
O-Benzoyl-L-serine	Ser (Bz below) or Ser (Bz above) or Ser(Bz)
S-Benzyl-L-cysteine	Cys (Bzl below) or Cys (Bzl above) or Cys(Bzl)

All these abbreviations conform to the rules laid down by the IUPAC–IUB Commission on Biochemical Nomenclature[14]. Recently, a series of one-letter symbols, (see table 1) which are particularly appropriate for concise sequence representation and computer use, have been recommended[15,16]. The first system for the abbreviation of amino acids and peptides was published by BRAND and EDSAL[17]. Structural features of the amino acid side chain are demonstrated by the graphic symbols devised by WELLNER and MEISTER[18]. These are illustrated in Figure 2, in which they are used to describe a hypothetical peptide sequence with N-terminal glycine and C-terminal hydroxylysine.

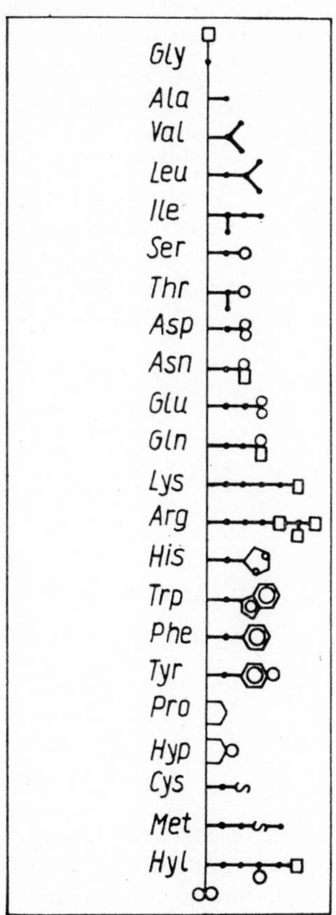

Figure 2. Structural features of the amino acid side chain

1.3. Stereochemistry of Amino Acids[19,20]

In his projection formulae for the natural amino acids, E. FISCHER (1891) postulated spatial structures which were confirmed as a true representation of their absolute configuration 60 years later.

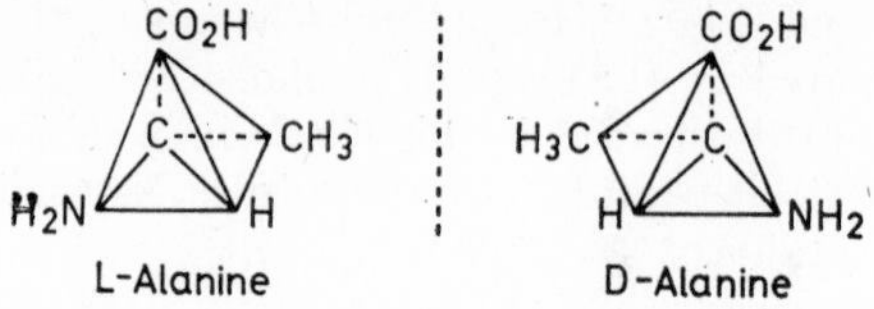

$$\begin{array}{ccc} \text{L-Alanine} & & \text{D-Alanine} \end{array}$$

Three-dimensional projection of the optical antipodes of alanine

Early investigations were largely concerned with proving that all the amino acids found in proteins belonged to the same steric series. Initial arguments for the steric uniformity of the amino acids centred around their similar behaviour towards specific enzymes and the uniform changes of optical rotation with changes in pH that were observed by PASTEUR. Furthermore it proved possible to interconvert numerous amino acids employing reactions that did not involve the asymmetric centre. The first sequence of this type was the conversion of L-serine into L-alanine by E. FISCHER:

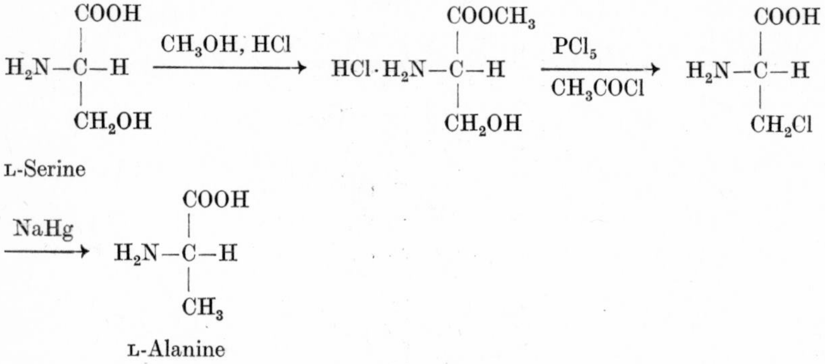

$$\underset{\text{L-Serine}}{H_2N-\overset{\displaystyle COOH}{\underset{\displaystyle CH_2OH}{C}}-H} \xrightarrow{CH_3OH,\ HCl} HCl\cdot H_2N-\overset{\displaystyle COOCH_3}{\underset{\displaystyle CH_2OH}{C}}-H \xrightarrow[CH_3COCl]{PCl_5} H_2N-\overset{\displaystyle COOH}{\underset{\displaystyle CH_2Cl}{C}}-H$$

$$\xrightarrow{NaHg} \underset{\text{L-Alanine}}{H_2N-\overset{\displaystyle COOH}{\underset{\displaystyle CH_3}{C}}-H}$$

The absolute configuration of methyl-ethyl carbinol (and also that of lactic acid) was deduced by rotatory dispersion calculations by W. KUHN (1935), and this allowed the determination of the absolute configuration of the amino acids once they could be unambiguously related to lactic acid. This involved subjecting the asymmetric C-atom to substitution reactions that did not carry the risk of WALDEN inversion. In 1933 W. KUHN and K. FREUDENBERG observed a correlation between the optical rotation and substitution in derivatives of alanine and lactic acid which were similarly substituted. The values for the molecular rotations $[M]_D$ given in Table 4 show the same trend from positive to negative only for (+)alanine and L(+)-lactic acid, suggesting that they are of the same series:

Table 4. Molecular rotation $[M]_D$ of alanine and lactic acid derivatives

R_1	R_2	(+)Alanine-derivative $[M]_D$	L-Lactic acid deriv. $[M]_D$	D-Lactic acid deriv. $[M]_D$
C_6H_5CO-	$-NH_2$	+70	+120	−120
C_6H_5CO-	$-OC_2H_5$	+12	+49	−49
C_6H_5CO-	$-OCH_3$	0	+35	−35
CH_3CO-	$-OC_2H_5$	−74	−76	+76
Configuration of the derivatives		COR_2 / $R_1-HN-C-H$ / CH_3	COR_2 / $R_1-O-C-H$ / CH_3	COR_2 / $H-C-O-R_1$ / CH_3

The final evidence that the protein amino acids do, in fact, belong to the L-series came from K. INGOLD *et al.*, in 1950. They correlated L(+)-lactic acid with L(+)-alanine directly via D-(+)brompropionic and L-azidopropionic acid as intermediates:

$$\underset{\text{L(+)-Lactic acid}}{\begin{array}{c} COOH \\ | \\ HO-C-H \\ | \\ CH_3 \end{array}} \xleftarrow[S_N2\ \text{Reaction}]{OH^-} \begin{array}{c} COOH \\ | \\ H-C-Br \\ | \\ CH_3 \end{array} \xrightarrow[S_N2\ \text{Reaction}]{N_3^-} \begin{array}{c} COOH \\ | \\ N_3-C-H \\ | \\ CH_3 \end{array} \xrightarrow{H_2} \underset{\text{L-Alanine}}{\begin{array}{c} COOH \\ | \\ H_2N-C-H \\ | \\ CH_3 \end{array}}$$

The displacement of bromide by hydroxyl and azide are both S_N2 reactions which are known to be accompanied by optical inversion at the asymmetric C-atom. The catalytic hydrogenation step from L-azidopropionic acid to L-alanine does not involve the asymmetric centre.

Threonine, isoleucine and hydroxyproline have a second centre of asymmetry and can therefore exist in four stereoisomeric forms. The relative configurations of the β-C atoms of threonine and isoleucine and at the γ-C atom of hydroxyproline are known. L-Threonine has been related to L-tartaric acid (the absolute configuration of which was determined by BYVOET in 1951, using an X-ray technique) via D-threose, so that the β-C atom has the D-configuration. The four stereoisomeric forms of threonine are:

L-Threonine	D-Threonine	D-Allo-threonine	L-Allo-threonine
COOH	COOH	COOH	COOH
H_2N-C-H	$H-C-NH_2$	$H-C-NH_2$	H_2N-C-H
$H-C-OH$	$HO-C-H$	$H-C-OH$	$HO-C-H$
CH_3	CH_3	CH_3	CH_3

Isoleucine has an analogous configuration and differs from threonine only in having an ethyl residue in place of the hydroxyl group:

L-Isoleucine	D-Isoleucine	D-Allo-isoleucine	L-Allo-isoleucine
COOH	COOH	COOH	COOH
H_2N-C-H	$H-C-NH_2$	$H-C-NH_2$	H_2N-C-H
H_3C-C-H	$H-C-CH_3$	H_3C-C-H	$H-C-CH_3$
CH_2	CH_2	CH_2	CH_2
CH_3	CH_3	CH_3	CH_3

The γ-C atom of hydroxyproline has the L-configuration relative to the glyceraldehyde standard. Intramolecular lactonisation fails (in contrast to allohydroxyproline) since the hydroxyl group is *trans* relative to the carboxylic group.

L-Hydroxyproline D-Hydroxyproline

L-Allohydroxyproline D-Allohydroxyproline

In recent years the configurations of amino acids have generally been determined by investigation of the optical rotatory dispersions of suitable derivatives—for example, *N*-alkylthiothiocarbonyl amino acids (R—S—CS—NH—CHR—COOH). Derivatives of the L-acids show positive Cotton effects, whereas those of their antipodes show negative Cotton effects.

A correlation between configuration and taste has been observed for several amino acids. L-Tryptophan, L-phenylalanine, L-tyrosine and L-leucine, for instance, have a bitter taste in contrast to the 'sweet' D-compounds. The sweet taste of glycine (Greek *glykys* = sweet) is well known. Monosodium glutamate is used on a large scale as a taste improver in the food industry. Aspartic acid, on the other hand, is tasteless. There was widespread interest in the observation that a dipeptide derivative of aspartic acid and phenylalanine has an intense sweet taste (see p. 210).

Table 5 gives the molecular rotations $[M]_D$ of the proteinogenic amino acids in various solvents. $[M]_D$ values are defined by the formula:

$$[M]_D^t = \frac{[\alpha]_D^t \times M}{100} = \frac{\alpha \times M}{c \times l}$$

where $[\alpha]_D^t$ = specific rotation in degrees for the sodium D-line at t°C;
M = molecular weight of the amino acid;
α = observed rotation;
c = concentration in grams of solute per 100 ml solution;
l = path length in dm.

At concentrations between $c = 1$ and $c = 2$ optical rotation of amino acids is essentially independent of concentration.

Recently there has been a great deal of interest in the conformational analysis of amino acids. The results of various physical measurements, particularly high-resolution NMR[21–25], have shown that the substituents on the α- and β-C atoms prefer to adopt staggered conformations. The preferred conformation may be considered in terms of the relative populations of the three rotamers corresponding to the energy minima about any C—C bond. The NEWMAN projection formulae show the three staggered rotamers about the C_α—C_β bond of an amino acid viewed in the $C_\beta \rightarrow C_\alpha$ direction:

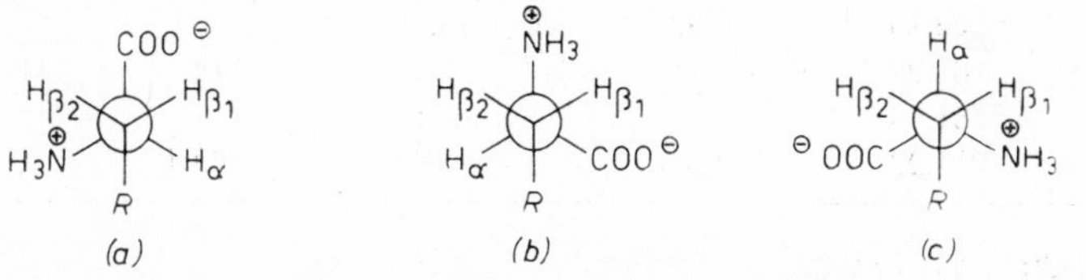

(a) (b) (c)

Consideration of coupling constant values in the NMR spectra of histidine and cysteine indicate that, in these cases, conformation (a) is preferred. In this conformation the bulky (imidazole or SH) groups are adjacent to the ammonium group and placed *trans* to the carboxylic group.

The results provided by the conformational analysis of amino acids are important for the study of the conformations of peptides and proteins. The native conformations of proteins—those required for their biological activity—is determined solely by the amino acid sequence within the polypeptide chains. Molecular orbital calculations on the conformation of protein amino acid residues have recently been published by B. and A. PULLMAN.

Table 5. Molecular rotations $[M]_D$ of the proteinogenic amino acids

Amino acid	Molecular weight (M)	$[M]_D^{25}$			Concentration (c)
		in H_2O	in 5N HCl	in glacial acetic acid	
Ala	89.10	+1.6	+13.0	+29.4	2
Val	117.15	+6.6	+33.1	+72.6	1–2
Leu	131.18	−14.4	+21.0	+29.5	2
Ile	131.18	+16.3	+51.8	+64.2	1
aIle	131.18	+20.8	+51.9	+55.7	1
Ser	105.10	−7.9	+15.9*		2
Thr	119.12	−33.9	−17.9	−35.7	1–2
aThr	119.12	+11.9	+37.8	+45.3†	1–2
Cys Cys (bridged)	240.31	−509.2*	−557.4		1
Cys	121.16	−20.0	+7.9	+15.7	2
Met	149.22	−14.9	+34.6	+29.8	1–2
Pro	115.14	−99.2	−69.5	−92.1	1–2
Hyp	131.14	−99.6	−66.2	−100.9†	2
aHyp	131.14	−78.0	−24.7	−39.3	2
Lys	146.19	+19.7	+37.9		2
Arg	174.21	+21.8	+48.1	+51.3	2
His	155.16	−59.8	+18.3	+11.6	2
Asp	133.11	+6.7	+33.8		2
Asn	132.12	−7.4	+37.8*		2
Glu	147.14	+17.7	+46.8		2
Gln	146.15	+9.2	+46.5*		2
Phe	165.20	−57.0	−7.4	−12.4	1–2
Tyr	181.20		−18.1		2
Trp	204.23	−68.8	−5.7*	−69.4	1–2

* in N HCl; † $c = 0.25$.

1.4. Physical and Chemical Properties of the Amino Acids

1.4.1. Solubility

With a few exceptions, amino acids are readily soluble in water, ammonia and other polar solvents, but they are poorly soluble in non-polar and less polar solvents, such as ethanol, methanol, acetone, etc. The reason for this is that amino acids which are often written as the uncharged molecules (I) really exist almost exclusively in the zwitterion forms (II) in neutral aqueous solution.

$$\underset{\text{(I)}}{H_2N-\overset{\overset{\large R}{|}}{C}H-COOH} \rightleftharpoons \underset{\text{(II)}}{H_3N^{\oplus}-\overset{\overset{\large R}{|}}{C}H-COO^{\ominus}}$$

For glycine and alanine, for instance, the proportion of (II) to (I) in aqueous solution is about 260000. The transition into the uncharged molecule (I) that is necessary for solubility in less polar solvents is energetically highly unfavourable.

The solubility of amino acids also depends on their side chains.

The aromatic amino acids, tyrosine and phenylalanine, and the amino dicarboxylic acids have particularly low solubilities. The imino acids proline and hydroxyproline have relatively high solubilities. Proline is also readily soluble in some non-aqueous solvents — for instance, in alcohol (67 parts will dissolve in 100 ml whereas only 0.06 parts of glycine will dissolve in the same volume). The high solubility of virtually all amino acids in hot acetic acid was suggested, by WEYGAND, to be due to the formation of hydrogen-bonded complexes between the amino acid and the solvent. In this way the equilibrium is shifted in favour of the uncharged form of the amino acid.

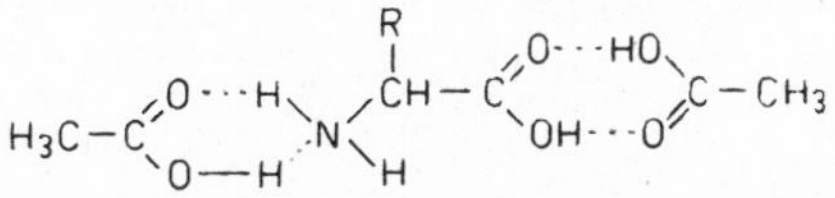

$$H_3C-C\begin{matrix}\nearrow O\cdots H \searrow \\ \searrow O-H \nearrow\end{matrix}N\begin{matrix}\nearrow \overset{\overset{\large R}{|}}{C}H-C\begin{matrix}\nearrow O\cdots HO\searrow \\ \searrow OH\cdots O\nearrow\end{matrix}C-CH_3 \\ \searrow H\end{matrix}$$

In chromatographic solvent systems glacial acetic acid (or its higher homologues) and water facilitate the dissolution of the amino acids which are to be separated.

1.4.2. Acid–Base Behaviour

Because of the bifunctional nature of amino acids their state of dissociation depends strongly on the pH of their environment. Only in neutral solution do amino acids have a zwitterionic structure. In strongly acidic media the dissociation of the COOH group will be suppressed and the amino acids form cations $H_3\overset{\oplus}{N}-CHR-COOH$, and, unlike ordinary carboxylic groups, they will move to the cathode in an electric field. Correspondingly in strongly alkaline solution anions $H_2N-CHR-COO^{\ominus}$ are formed, and their migration to the anode is in marked contrast to the behaviour of simple amines. The acid–base behaviour of amino acids may be summarised in the following equations:

$$\underset{\text{anion}}{\begin{matrix}\text{CHR}-\text{C}\overset{\displaystyle /\!\!/ \text{O}}{\underset{\displaystyle \diagdown \overline{\underline{|O|}}^{\ominus}}{}}\\ |\\ \text{NH}_2\end{matrix}} \underset{\text{OH}^-}{\overset{\text{H}^{\oplus}}{\rightleftharpoons}} \underset{\text{dipolar zwitterionic structure}}{\begin{matrix}\text{CHR}-\text{C}\overset{\displaystyle /\!\!/ \text{O}}{\underset{\displaystyle \diagdown \overline{\underline{|O|}}^{\ominus}}{}}\\ |\\ \text{NH}_3^{\oplus}\end{matrix}} \underset{\text{OH}^{\ominus}}{\overset{\text{H}^{\oplus}}{\rightleftharpoons}} \underset{\text{cation}}{\begin{matrix}\text{CHR}-\text{C}\overset{\displaystyle /\!\!/ \text{O}}{\underset{\displaystyle \diagdown \text{OH}}{}}\\ |\\ \text{NH}_3^{\oplus}\end{matrix}}$$

The dissociation constant of the cation is given by

$$K_1 = \frac{[\text{H}_3\text{N}^{\oplus}-\text{CHR}-\text{COO}^{\ominus}]\,[\text{H}^{\oplus}]}{[\text{H}_3\text{N}^{\oplus}-\text{CHR}-\text{COOH}]}$$

and the dissociation constant of the zwitterion by

$$K_2 = \frac{[\text{H}_2\text{N}-\text{CHR}-\text{COO}^-]\,[H^+]}{[\text{H}_3\text{N}^{\oplus}-\text{CHR}-\text{COO}^-]}$$

For determination of the pK values the titration curves of the amino acid must be constructed. For this an aqueous solution of free amino acid is gradually mixed with acid (or base) and resulting pH values are recorded. The pH at the midpoint of the titration gives the pK. Figure 3 shows (full lines) the titration curves of glycine, lysine and glutamic acid. It is conventional to assign the pK values numbers (pK_1, pK_2, pK_3, etc.) in the order in which the respective dissociations are encountered on ascending the pH scale. Neutral amino acids show only two dissociations and we thus have only pK_1 and pK_2 values to consider for these, but with trifunctional amino acids such as lysine and glutamic there are three dissociations (of pK_1, pK_2, pK_3) to take into account.

Glycine (pK_1 = 2.34) is a stronger acid than acetic acid (pK = 4.7), because of the —I-effect of the ammonium group. As the distance between the amino and carboxyl groups in the molecule increases, this effect becomes weaker. Thus the pK_1 of β-alanine $H_2NCH_2CH_2COOH$ rises to 3.6 and the pK_1 of 6-aminohexanoic acid $H_2NCH_2CH_2CH_2CH_2CH_2COOH$, at 4.43, differs by only 0.27 pH units from that of acetic acid. In the monoamino dicarboxylic acids the α-carboxyl group is the more acidic of the two and is the one which is predominantly involved in the zwitterionic structure.

On the other hand, the basic character of the amino group is weakened by the COO^- group. Thus glycine (pK_2 = 9.72) is less basic than the comparable ethylamine (pK = 10.75). Amino acid esters (glycine ethyl ester, for example, pK 7.7) are even less basic.

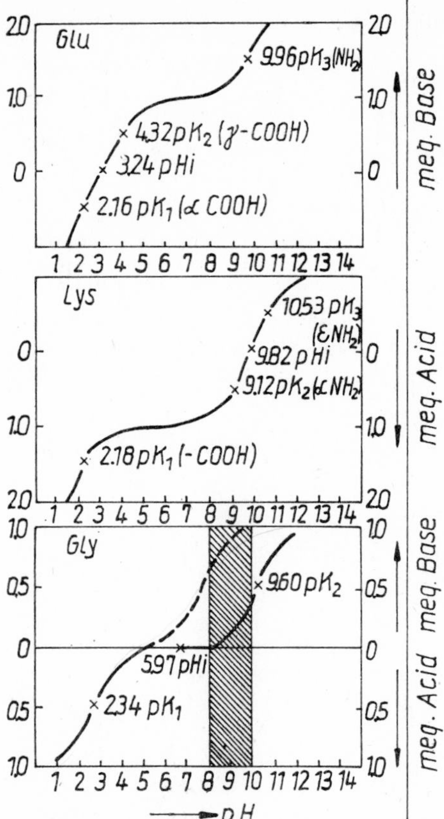

Figure 3. Titration curve of glycine, lysine and glutamic acid.
The dashed curve in the glycine diagram was obtained by SORENSEN's 'Formol titration', in which the amino groups are converted to hydroxymethyl functions which are weaker bases

$$H_3N^{\oplus}{-}CHR{-}COO^{\ominus} + OH^{\ominus} + CH_2O \rightarrow HOH_2C{-}NH{-}CHR{-}COO^{\ominus} + H_2O$$

The ω-amino group of diamino carboxylic acids is more basic than the α-amino group, and therefore takes the major role in the zwitterionic structure—for example:

$$H_3\overset{\oplus}{N}{-}CH_2{-}(CH_2)_x{-}CH(NH_2)(CO_2^{\ominus})$$

Because it is protonated to form the resonance-stabilised guanidino group (here formulated using the guanidinium cation as an example), arginine is the most basic amino acid:

$$\left[\begin{array}{c} H_2N \\ \quad \overset{\oplus}{C{=}NH_2} \\ | \\ NH_2 \end{array}\right]^{\oplus} \longleftrightarrow \left[\begin{array}{c} \overset{\oplus}{H_2N} \\ \quad C{-}NH_2 \\ H_2N \end{array}\right]^{\oplus} \longleftrightarrow \left[\begin{array}{c} H_2N \\ \quad \overset{\oplus}{} C{-}NH_2 \\ H_2N{=} \end{array}\right]^{\oplus}$$

A further characteristic index of the acid–base behaviour of amino acids is the isoelectric (isoionic) point, sometimes abbreviated to I.P. or pH_i. This is the pH at which the maximum number of amino acid molecules are present as zwitterions.

In neutral and acidic amino acids the pH_i value is the arithmetic mean of the pK_1 and the pK_2 values, in basic amino acids that of pK_2 and pK_3 values. Thus for glutamic acid the isoelectric point is given by:

$$pH_i = \frac{2.16 + 4.32}{2} = 3.24 \quad \text{and for lysine} \quad pH_i = \frac{9.12 + 10.53}{2} = 9.82$$

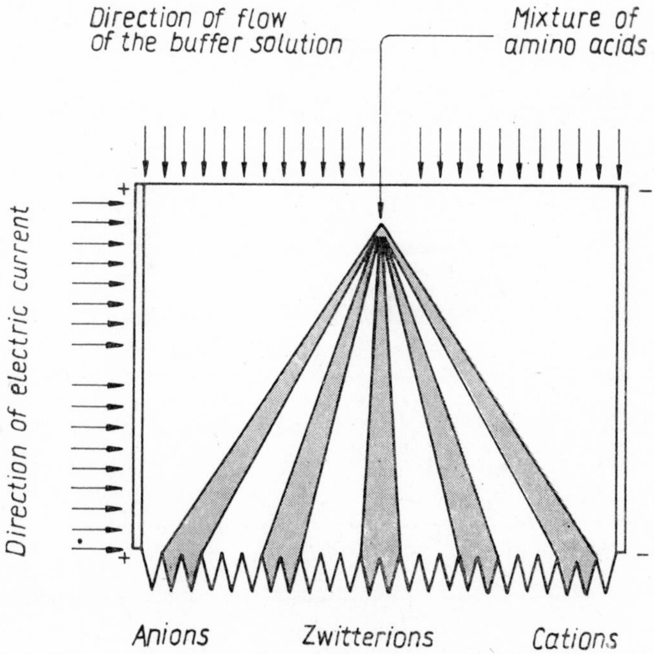

Figure 4. Schematic representation of the action of an apparatus for continuous preparative electrophoresis (diversion electrophoresis)

Table 6. Decomposition points (D.P.), solubilities and dissociation constants of the amino acids found in proteins

Amino acid	D.P., °C	Solubility, g/100 ml 25°C	in water 100°C	pK_1	pK_2	pK_3	pH$_i$
Gly	292	24.99	67.17	2.34	9.60		5.97
Ala	297	16.65	37.3	2.34	9.69		6.01
Val	315	8.85	18.8	2.32	9.62		5.96
Leu	337	2.43	5.64	2.36	9.60		5.98
Ile	284	4.12	8.26	2.36	9.68		6.02
Ser	228	5.0	32.2	2.21	9.15		5.68
Thr	253	20.5		2.71	9.62		6.16
Cys	178*			1.71	8.27 (SH—)	10.78	5.02
Cys Cys (disulfide-linked)	260	0.01	0.11	1.04	2.05 (COOH)	8.0 10.25 (NH_2)	5.03
Met	283	3.5	17.6	2.28	9.21		5.74
Pro	222	16.23	23.9 (70°C)	1.99	10.6		6.30
Hyp	270	36.1	51.6 (65°C)	1.92	9.73		5.83
Lys	224			2.18	9.12 (α-NH_2)	10.53	9.82
Arg	238			2.17	9.04 (α-NH_2)	12.84 (guanidino group)	10.76
His	277	0.43		1.82	6.00 (imidazole)	9.17	7.59
Asp	270	0.5	6.9	1.88	3.65 (β-COOH)	9.60	2.77
Asn	236	2.98	55.1	2.02	8.80		5.41
Glu	249	0.86	14.0	2.16	4.32 (γ-COOH)	9.96	3.24
Gln	185	3.6		2.17	9.13		5.65
Phe	284	2.96	9.9	1.83	9.13		5.48
Tyr	344	0.045	0.56	2.20	9.11	10.07 (OH)	5.66
Trp	282	1.14	4.99	2.38	9.39		5.89

* as hydrochloride.

The knowledge of the pK values and the pH_i values of the amino acids is important for their separation by electrophoresis (ionophoresis)[26] and ion exchange chromatography. The electrophoretic separation of amino acids (and the same applies for peptides and proteins) depends on the fact that compounds with different isoelectric points will migrate at different rates in a solvent at a given pH value. Because of the differences in electrophoretic mobility which arise from structural peculiarities of the individual ions, separation is also possible even when the pH_i values are close. The mode of operation of an apparatus for continuous preparative electrophoresis is shown in Figure 4.

Paper, synthetic polymers, agar gel and especially starch gel have all proved suitable carrier materials.

It is conventional to distinguish between low-voltage (up to 200 V) and high-voltage electrophoresis (up to 10000 V) techniques.

1.4.3. UV[27,28], IR[29,30] and NMR Spectra[20–24,31] of Amino Acids

While the aliphatic amino acids do not absorb in the ultra-violet region above wavelengths of 220 nm, the aromatic amino acids tryptophan, tyrosine, phenylalanine and histidine show characteristic maxima above 250 nm.

The high molar extinctions of tyrosine and tryptophan facilitate rapid quantitative determination of these amino acids in proteins.

In the IR spectra of amino acids the normal NH-stretching frequencies in the region of 3300–3500 cm^{-1} are absent. Instead, absorption at 3070 cm^{-1}

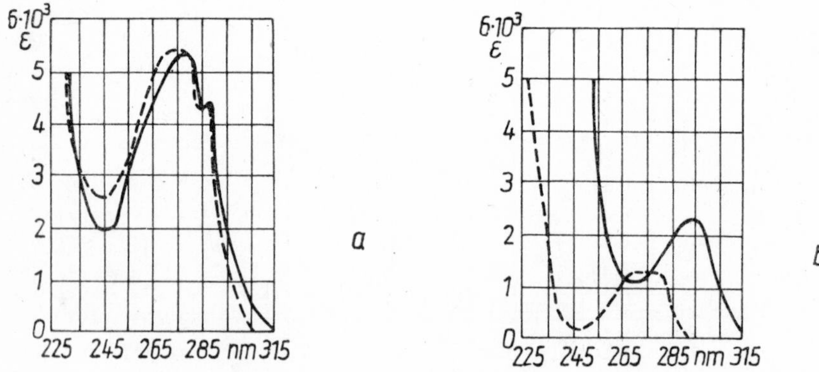

Figure 5. UV absorption spectrum of tryptophan (a) and tyrosine (b): ------- in 0.1 N hydrochloric acid; ——— in 0.1 N sodium hydroxide

is observed which may be attributed to the $NH_3^{\oplus}$ group. This band also appears in amino acid hydrochlorides, but not in N-substituted amino acids or imino acids. Two further characteristic bands of the $NH_3^{\oplus}$ group are observed in the region from 1500 to 1600 cm^{-1}. All amino acids and their carboxylate salts show the absorption typical for the $COO^{\ominus}$ group in the region of 1500–1600 cm^{-1}, but this absorption is absent in the amino acid hydrochlorides. In the hydrochlorides the normal carbonyl absorption of the COOH group (1700–1730 cm^{-1}) is shifted to higher wavenumbers by about 20 cm^{-1} because of the $-NH_3^{\oplus}$ group.

The first high-resolution proton NMR spectra of protein amino acids were obtained in 1957 by TAKEDA and JARDETZKY. Their investigations showed that the chemical shifts and also the proton–proton coupling constants depend on the charge status of the amino acid. The relationship

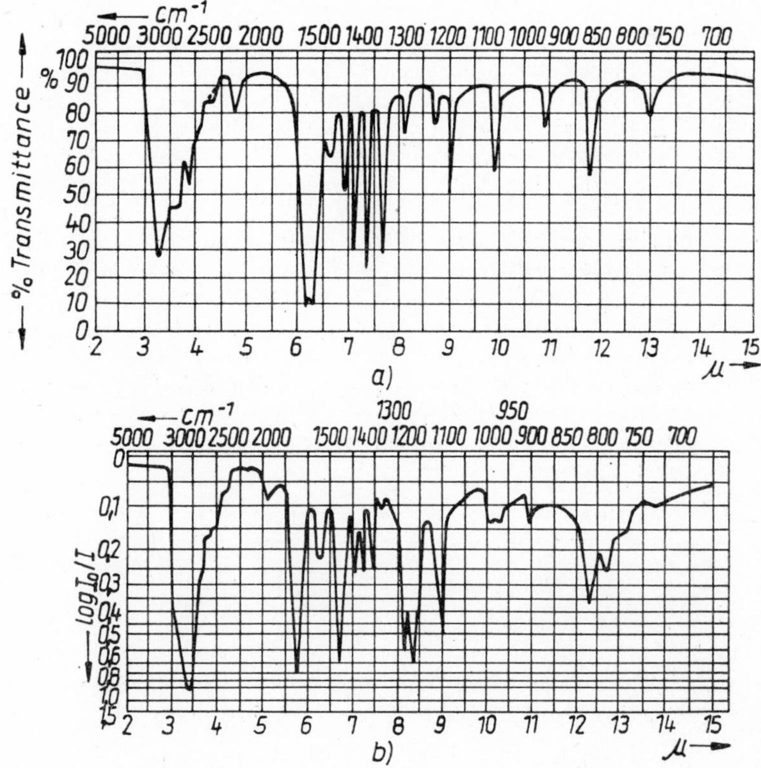

Figure 6. IR spectra of L-alanine (a) and L-alanine hydrochloride (b)

between chemical shift and acidity looks like a typical titration curve. Tetramethylsilane (TMS), hexamethyldisiloxan (HMDS) and 2,2-dimethyl-2-silapentan-5-sulphonate (DDS) can be used as internal standards usually with D_2O as solvent. Trifluoroacetic acid is also a convenient solvent for free amino acids.

The proton resonances of the methyl groups of aliphatic amino acids lie between 0.9 and 1.5 ppm from DDS, the resonances of the other protons of the aliphatic side chains between 1.5 and 3.5 ppm, those of the protons of the α-C atom between 3.5 and 4.5 ppm, those of the aromatic CH protons and those of the CO—NH protons between 6.7 and 9.0 ppm, those of the imidazole-C_2 protons between 8.0 and 9.2 ppm and those of the NH protons of indole (trytophan) between 9.0 and 11.0 ppm.

1.5. Preparation of Amino Acids[1–3,32]

Amino acids may be obtained by isolation from natural materials, by microbiological methods and by chemical synthesis. Whereas the first two methods give L-amino acids, chemical synthesis usually produces DL-compounds, so, in this case, an additional optical resolution step is necessary.

With the increasing demand for (for example) glutamic acid, glycine, alanine and aspartic acid for use as flavouring agents in the food industry, of lysine, methionine, threonine, tryptophan and other essential amino acids as additives for nutritionally inferior foodstuffs and of serine, threonine and cysteine as intermediates in the cosmetic industry, the production of amino acids is now performed on an industrial scale.

In the 1950s glutamic acid was isolated exclusively from corn proteins and sweet turnip molasses by hydrolysis. Today synthesis from petrochemical intermediates plays a dominant role, although there are also fermentation methods. In the DUPONT process acetylene is treated with acrylic acid ester to give the starting aldehyde for the STRECKER synthesis of glutamic acid:

$$\mathrm{HC{\equiv}CH} \xrightarrow{\text{CO, ROH}} \mathrm{CH_2{=}CH{-}COOR} \xrightarrow{\text{CO, }\mathrm{H_2}} \mathrm{OHC{-}CH_2CH_2{-}COOCH_3}$$

$$\xrightarrow[\text{synthesis}]{\text{STRECKER}} \text{glutamic acid}$$

In the analogous AJINOMOTO process, acrylonitrile is converted to β-cyanpropionaldehyde and thence to DL-glutamic acid:

$$\mathrm{CH_2{=}CH{-}C{\equiv}N} \xrightarrow{\text{CO, }\mathrm{H_2}} \mathrm{OHC{-}CH_2{-}CH_2{-}C{\equiv}N} \xrightarrow[\text{synthesis}]{\text{STRECKER}} \text{glutamic acid}$$

The optical resolution is performed by spontaneous crystallisation using optically pure inoculation crystals. The D-glutamic acid obtained is racemised and recycled. In Japan in 1972 120000 tons of glutamic acid were produced in this way largely for use as a food additive in the form of monosodium glutamate (MSG).

L-Lysine is produced on a similar scale by a specific fermentation process (Kyowa Fermentation Company, Japan), again for use as a food additive—the growth rate of pigs and poultry may be increased considerably by the addition of 0.1–0.3 per cent lysine to their food. The industrial preparation of methionine (40000 tons in 1972, major use as an additive for poultry food) is by the STRECKER synthesis from β-methylmercaptopropionaldehyde CH_3–S–CH_2CH_2–CHO, which is in turn obtained by the addition of methyl mercaptan to acrolein. Aspartic acid is obtained by a continuous process from maleic acid and ammonia.

1.5.1. Isolation from Protein Hydrolysates

For the preparation of amino acids from proteins, the proteins are first degraded into their constituents by hydrolysis[33]. The classic procedure for acid hydrolysis is by boiling with freshly distilled 6 N hydrochloric acid b.p. 110 °C) or with 8 N sulphuric acid.

This reaction needs between 12 and 17 hours, because of the stability of peptide bonds involving valine, leucine and isoleucine. Complete destruction of tryptophan and 5–10 per cent decomposition of serine and threonine occurs. Asparagine and glutamine are transformed into the amino

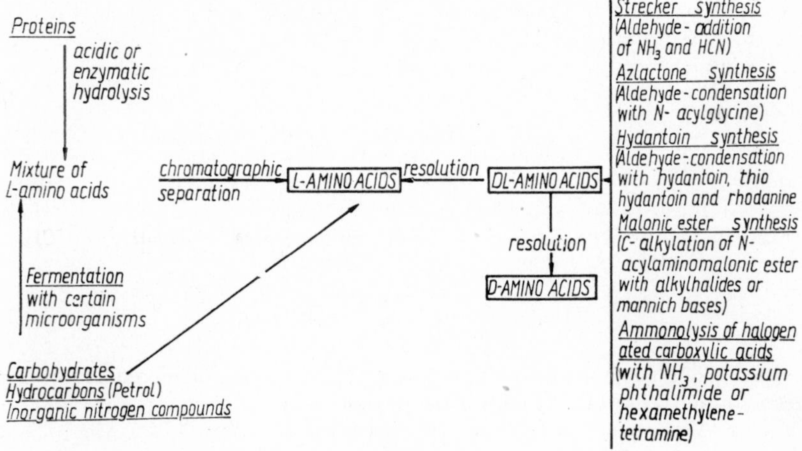

Figure 7. The most important methods for the preparation of amino acids

dicarboxylic acids with loss of ammonia. Further loss of amino acids, particularly of cysteine–cystine, is observed if humin-forming carbohydrates are present. These losses may be minimised by operating *in vacuo* and by using a large excess of acid (proportion protein to hydrochloric acid 1:10000). 4 N methanesulphonic acid has been used for protein hydrolysis (115 °C, 24 h), with excellent results: under these conditions even tryptophan remains unchanged.

In the course of alkaline hydrolysis—normally performed with 6 N $Ba(OH)_2$ solution under 8 atm pressure in an autoclave—serine is destroyed and transformed into glycine and alanine. Arginine is deaminated to give ornithine and citrulline, and finally cysteine is destroyed by decomposition involving loss of H_2S and NH_3. The amino acids that remain structurally intact are more or less completely racemised. In contrast to acid hydrolysis, in the alkaline procedure trytophan survives unchanged.

Enzymic hydrolysis is now more important than classical chemical hydrolysis since it is less wasteful and gives higher yields of amino acids. The specifity of proteolytic enzymes makes it necessary to use several enzymes for the complete hydrolysis of proteins.

Combinations of pepsin, trypsin, erepsin and papain with peptidases such as leucine aminopeptidase and prolidase have proved convenient[34]. Frequently the use of crude enzyme preparations is sufficient: for instance, for the preparation of glutamine and asparagine using pancreatic juice.

The separation of the mixture of amino acids obtained by hydrolysis is usually achieved by chromatographic procedures[35]. Separation into the individual components usually starts with a group separation into acidic, basic and neutral amino acids. In this process electrophoresis and specific anion and cation exchange resins play an important role. Selective isolation of aromatic amino acids is possible by adsorption onto activated charcoal. Glutamic acid can be directly crystallised from a concentrated hydrolysate which has been saturated with hydrogen chloride. The separation methods which were used in the classical era of the subject, such as the fractional distillation of esters (E. FISCHER), the extraction of monoamino carboxylic acids with n-butylalcohol (H. D. DAKIN), the precipitation of lysine, arginine and histidine with phosphotungstic acid or flavianic acid, have lost their significance along with many other more or less specific precipitation reactions.

1.5.2. Microbiological Methods

At the present time, an increasingly large number of microbiological methods for the preparation of optically active amino acids are being published and patented. These methods involve the aerobic cultivation of specific microorganisms in a dilute liquid nutrient which contains simple C- and N-sources such as carbohydrates, hydrocarbons, organic and inorganic nitrogen compounds together with mineral salts and growth substances such as thiamine and biotin.

Particular attention has been paid to the preparation of glutamic acid by fermentation of mixtures containing carbohydrates. This is of special interest because glutamic acid and its precursor α-ketoglutaric acid may be transformed without isolation into other amino acids by transaminases. The starting materials required are easily accessible and it is also possible to use polypeptides and proteins as sources for amino acids. These fermentation processes are likely to become the major method of preparation of amino acids within the very near future.

Table 7. The microbiological production of some amino acids

Microorganism	C,N-Source	Amino acids formed
Corynebacterium	Glucose or starch hydrolysate; urea or NH_3	Glu or α-ketoglutaric acid
Micrococcus *Brevibacterium* *Microbacterium*	Molasses, cane and beet sugar sap, starch hydrolysate, inorg. N-compounds	Glu
Micrococcus glutamicus ATCC 14751	Urea, ammonium sulphate	Glu and Gln
Pseudomonas trifoli	Fumaric acid, NH_3	Asp
Serratia marcescns *Torula utilis* *Pseudomonas* species	Phenyllactic acid α-Hydroxy-β-methyl-n-valeric acid β-Dihydroxybutyric acid Hydroxy-isocaproic acid Hydroxy-β-methylbutyric acid	Phe Ile Thr Leu Val
Corynebacterium oleophilus *Brevibacterium* species *Achromobacter pestifer* *Escherichia coli* *Microbacterium brevicale* *Candida lipolytica* and *tropicalis* *Hansenula anomala*	Petrol with ammonium nitrate, phosphates, magnesium sulphate, manganese-(II)-chloride, calcium chloride and traces of iron-(II)-sulphate	Glu, Ala, Asp, Lys, Arg, Tyr, Ile, Val, Phe, Gly, Trp, Pro, Ser

1.5.3. Synthetic Processes[1–3]

Of the great number of methods available, only a few will be discussed in detail here. These include the aminolysis of halogenated carboxylic acids, the STRECKER synthesis, the azlactone and the hydantoin syntheses and the malonic ester synthesis. Later sections describe methods which can be used for the synthesis of isotopically labelled amino acids and some of the reactions which have been proposed to account for amino acid formation in the course of the development of life on the earth.

Processes such as the catalytic hydrogenation of α-ketoacids in the presence of ammonia, the reduction of α-nitro- or α-azido-carboxylic acids and non-enzymic asymmetric syntheses, which are of less practical importance, will not be mentioned.

1.5.3.1. Aminolysis of Halogenated Carboxylic Acids

The oldest method of synthesising amino acids is the substitution of halogen, in the normally readily available α-halogenated carboxylic acids, by an NH_2 group:

$$R{-}CHCl(Br){-}COOH + NH_3 \rightarrow R{-}CH(NH_2){-}COOH + NH_4Cl(Br)$$

In 1858 glycine was the first amino acid to be prepared in this way by ammonolysis of monochloro- or monobromoacetic acid. Since that time the low yield (10–15 per cent) has been improved by use of alcoholic ammonia and control of temperature and pressure. A tenfold excess of a concentrated solution of ammonia in water with the addition of a five-molar excess of ammonium carbonate and a reaction temperature near 50 °C are the optimum conditions for the reaction. In this way yields between 60 and 70 per cent are attained. The unprotected amino group of the acid formed is converted into the ammonium salt of the corresponding carbamic acid $R{-}CH(NH{-}COONH_4){-}COOH$ by the excess of ammonium carbonate, and further reaction to give secondary and tertiary amino compounds is thus suppressed.

Higher yields and purer products may be obtained by causing halogenated carboxylic acid esters to react with potassium phthalimide (GABRIEL's method). The phthaloyl group of the N-phthaloyl-amino acid can be removed by acid hydrolysis or reaction with hydrazine.

$$\text{(phthalimide-NK)} + Br{-}CHR{-}COOH \longrightarrow \text{(phthalimido)}N{-}CHR{-}COOH + KBr$$

Aminolysis of α-haloacids with hexamethylenetetramine in dioxan followed by treatment of the initial products with hydrogen chloride in absolute ethanol is a useful way of making α-amino ester hydrochlorides (G. HILLMANN, 1948).

1.5.3.2. STRECKER *Synthesis*

This synthesis (A. STRECKER, 1850) is based on the addition of hydrocyanic acid to aldehydes in the presence of ammonia. The α-amino nitriles which are formed as intermediates are hydrolysed without isolation to DL-amino acids:

$$R-CHO + NH_3 + HC{\equiv}N \xrightarrow{-H_2O} R-CH(NH_2)-C{\equiv}N \xrightarrow[H_2O]{H^{\oplus}} R-CH(NH_2)-COOH$$

Iminodinitriles $NH(CHR-CN)_2$, the corresponding trinitriles and carboxylic acids can be formed as side products. The yields from this synthesis are generally about 75 per cent. Instead of hydrocyanic acid sodium cyanide is normally used as it is easier to handle.

In the BUCHERER modification of the STRECKER synthesis the adduct formed by treatment of the aldehyde with the sodium cyanide–ammonium carbonate is converted into a hydantoin which can be isolated easily and subsequently hydrolysed with alkali.

$$R-CH(NH_2)-C{\equiv}N \xrightarrow{+CO_2,\ H_2O} R-CH(NH-CO-OH)-CO-NH_2 \xrightarrow{-H_2O} \text{hydantoin } (R-CH-CO-NH-CO-NH, \text{ ring}) \xrightarrow{OH^{\ominus}} DL\text{-amino acid}$$

1.5.3.3. Aldehyde Condensations

an this type of synthesis the amino acid side chain is introduced by means of an aldehyde which is condensed with a cyclic glycine derivative which has In active methylene group. The two main variants are the azlactone and the hydantoin methods.

1.5.3.3.1. The ERLENMEYER *Azlactone Synthesis*

In this synthesis an aromatic or α-β-unsaturated aliphatic aldehyde is treated with benzoylglycine (hippuric acid) or acetylglycine (aceturic acid) in the presence of acetic anhydride and sodium acetate. The first stage is the formation of an azlactone (oxazolone) whose highly active methylene group then reacts with the carbonyl compound. Alkaline cleavage of the azlactone gives an unsaturated intermediate, which may be converted into the desired benzoyl-amino acid by catalytic hydrogenation (Raney nickel) or by reduction with sodium amalgam. Acid hydrolysis of the benzoyl derivative gives the free amino acid. A simplified procedure for the simultaneous hydrolysis and reduction of the azlactone involves treatment with a mixture of hydrogen iodide and red phosphorus in glacial acetic acid:

$$\underset{\text{Hippuric acid}}{CH_2(COOH)-NH-COC_6H_5} \xrightarrow[CH_3COONa]{(CH_3CO)_2O} \underset{\text{Hippuric acid azlactone}}{CH_2-C(=O)-O-C(C_6H_5)=N \text{ (ring)}}$$

$$\xrightarrow{+\,R-CHO} \underset{\text{Azlactone}}{R-CH=C-C(=O)-O-C(C_6H_5)=N \text{ (ring)}}$$

$$\xrightarrow{OH^{\ominus}} R-CH=C(NH-COC_6H_5)-COOH$$

$$R-CH=C(NH-COC_6H_5)-COOH \xrightarrow{NaHg \text{ or } H_2/\text{Raney-nickel}} R-CH_2-CH(NH-COC_6H_5)-COOH$$

$$R-CH_2-CH(NH-COC_6H_5)-COOH \xrightarrow{H^{\oplus}} R-CH_2-CH(NH_2)-COOH + C_6H_5COOH$$

$$\text{Azlactone} \xrightarrow{P,\ HI;\ CH_3COOH} R-CH_2-CH(NH_2)-COOH + C_6H_5COOH$$

Saturated aliphatic aldehydes give unacceptably low yields of the intermediate azlactones and are not suitable for this synthesis.

The azlactone method is used commercially for the synthesis of *p-N,N*-bis(chlorethyl)-aminophenylalanine (phenylalanine–Lost), which is used as a cytostatic under the names 'Sarcolysin' and 'Melphalan'.

1.5.3.3.2. Hydantoin Synthesis

This synthesis employs hydantoin as the methylene component with a mixture of acetic anhydride and sodium acetate. Aldehydes are allowed to react with this mixture to give condensation products, which can be converted into amino acids by alkaline hydrolysis after sodium amalgam reduction:

$$R-CHO + H_2C-NH-C(=O)-NH-C(=O) \text{ (ring)} \xrightarrow[CH_3COONa]{(CH_3CO)_2O} R-CH=C-NH-C(=O)-NH-C(=O) \text{ (ring)} \xrightarrow{NaHg}$$

$$R-CH_2-CH-NH-C(=O)-NH-C(=O) \text{ (ring)} \xrightarrow{OH^{\ominus}} R-CH_2-CH(NH_2)-COOH$$

Other methylene active compounds, which have been used successfully in this type of amino acid synthesis include thiohydantoin (prepared from acetylglycine and potassium thiocyanate), 2,5-dioxopiperazine (containing two methylene groups) and rhodanine. The last gives high yields in the condensation step in glacial acetic acid. The intermediate condensation products can be reduced with sodium amalgam or by hydrogen iodide/red phosphorus.

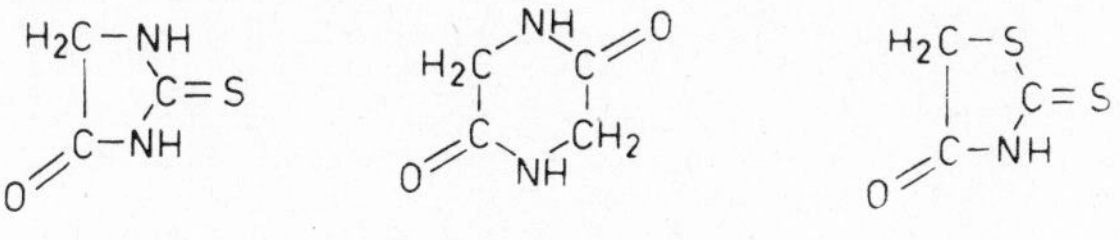

Thiohydantoin 2,5-Dioxopiperazine Rhodanine

1.5.3.4. *Malonic Ester Syntheses*

The malonic ester synthesis, which is widely used for the synthesis of carboxylic acids, has also been applied extensively to the synthesis of amino acids. There are a number of variants of the method which differ in the manner and timing of the introduction of the amino group. In all the procedures, however, the introduction of the amino acid side chain is achieved by C-alkylation of the reactive methylene group of the malonic ester in the presence of strong bases, such as sodium ethoxide. The intermediate, resonance-stabilised, enolate anion is first formed:

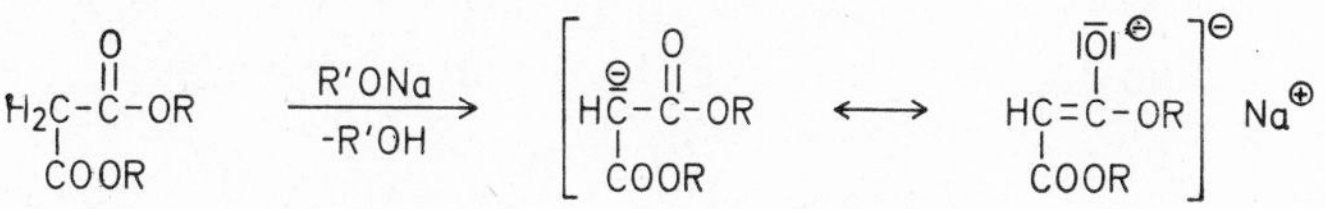

$$H_2C(COOR)-C(=O)-OR \xrightarrow[-R'OH]{R'ONa} \left[H\overset{\ominus}{C}(COOR)-C(=O)-OR \longleftrightarrow HC(COOR)=C(O^{\ominus})-OR \right]^{\ominus} Na^{\oplus}$$

and is then alkylated usually with an alkylhalide:

$$R-Cl + \left[|\overset{\ominus}{C}H(COOR)_2 \right]^{\ominus} Na^{\oplus} \longrightarrow R-CH(COOR)_2 + NaCl$$

In the first synthesis developed by Th. CURTIUS the side chain R was introduced first and then the amino group was introduced by rearrangement of the malonic acid half azide derived from the half ester which is obtained by partial saponification. In a subsequent improvement the partial saponification was avoided by using cyanoacetic ester instead of malonic ester:

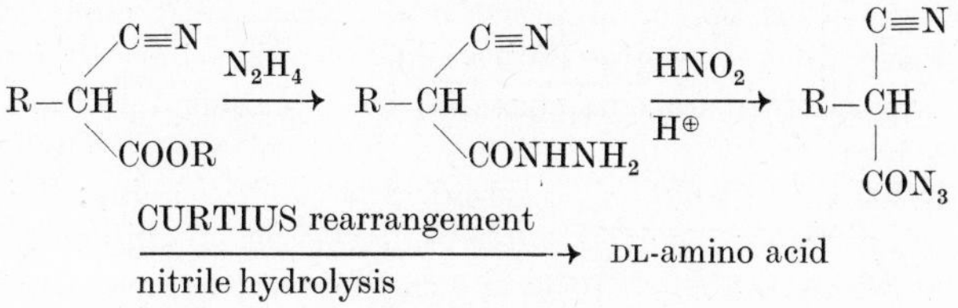

$$R-CH(C\equiv N)(COOR) \xrightarrow{N_2H_4} R-CH(C\equiv N)(CONHNH_2) \xrightarrow[H^{\oplus}]{HNO_2} R-CH(C\equiv N)(CON_3)$$

$$\xrightarrow[\text{nitrile hydrolysis}]{\text{CURTIUS rearrangement}} \text{DL-amino acid}$$

E. FISCHER used the general malonic ester synthesis to obtain halogenated fatty acids which could then be converted into amino acids by ammonolysis.

A valuable simplification of the malonic ester synthesis involves the use of N-protected aminomalonic acid derivatives, in an approach which avoids the lengthy procedure for the introduction of the amino group (SORENSEN, 1903): the sodium salt of phthalimidomalonic ester is treated with an alkylhalide to give the corresponding DL-amino acid (usually in 60–80 per cent yield) after hydrolysis and decarboxylation:

$$C_6H_4(CO)_2NK + CHBr(COOR')_2 \xrightarrow{-KBr} C_6H_4(CO)_2N{-}CH(COOR')_2 \xrightarrow[C_2H_5ONa]{C_6H_5CH_2Cl}$$

$$C_6H_4(CO)_2N{-}C(COOR')_2{-}CH_2{-}C_6H_5 \xrightarrow[-CO_2]{H^{\oplus}} C_6H_5{-}CH_2{-}CH(NH_2){-}COOH$$

DL-phenylalanine

Nowadays *N*-formylamino and *N*-acetylamino malonic esters are more important starting materials. They may be prepared in almost quantitative yield by reduction of oximinomalonic ester with zinc dust and formic acid (formaminomalonic ester) or acetic acid (acetaminomalonic ester):

$$H_2C(COOR)_2 \xrightarrow{HNO_2} HON{=}C(COOR)_2 \xrightarrow[(H_2/CH_3COOH)]{H_2/HCOOH} (CH_3{-}CO{-})NH{-}CH(COOR)_2$$

Both of these esters form readily crystallisable alkylation products, when caused to react with alkylhalides or reactive tertiary bases (MANNICH bases). With MANNICH bases, *C*-alkylation is possible in the presence of NaOH as condensing reagent (for an example see next section).

1.5.3.5. *Synthesis of Labelled Amino Acids*[36,37]

If an amino acid is to be radioactively labelled, the choice of synthetic method is determined by the position of the molecule which is to be labelled.

For such syntheses the malonic ester (or cyanoacetic acid ester) routes are often chosen since they give high yields. Commercially available [2-^{14}C] acetic acid is a suitable starting material for C_2-labelled amino acids while [^{14}C]HCN is used for C_1-labelled amino acids. Using [^{14}C]KCN in the STRECKER synthesis gives carboxyl-labelled amino acids: with $^{15}NH_3$, amino-labelled products result. Double and multiple labelling is also possible with these methods. Labelled products can be used to trace the incorporation of amino acids into natural materials. Non-specifically ^{14}C-labelled amino acids may be produced from microorganisms grown in the presence of $[^{14}C]O_2$ or other simple carbon sources.

$HC^{*}HO$, $(CH_3)_2NH$ → indole-$\overset{*}{C}H_2-N(CH_3)_2$

$CH(COOR')_2$–$NH-COCH_3$, NaOH → indole-$\overset{*}{C}H_2-C(COOR')_2$–$NH-COCH_3$

$OH^{\ominus}$ → indole-$\overset{*}{C}H_2-C(COOH)_2$–$NH_2$

Δ, $-CO_2$ → indole-$\overset{*}{C}H_2-CH(NH_2)-COOH$

DL-Tryptophan $[\beta^{14}C]$

The example shows the synthesis of $\beta[^{14}C]$tryptophan (HEIDELBERGER, 1949), in which $[^{14}C]$formaldehyde was used as the source of the β-carbon (C_3) label.

1.5.3.6. 'Prebiotic Synthesis' of Amino Acids[38]

In recent years scientists have begun to tackle the problem of the origin of life on earth and this has led to an examination of the possible mode of formation of amino acids in the earth's early history. Investigations have shown that the DL-forms of almost all the protein amino acids may be formed from simple carbon and nitrogen compounds under suitable high-energy conditions.

Accurate chromatographic analysis of the Murchison meteorite which fell in Australia in 1969 has provided conclusive evidence for the abiogenetic, extraterrestrial synthesis of amino acids. The meteorite samples contained 6 μg glycine, 3 μg glutamic acid, 3 μg alanine, 2 μg valine and 1 μg proline per gram of sample. Traces of sarcosine, isovaline, pipecolic acid and amino-butyric acid were also found. The optical purity of the amino acids was determined by the gas chromatographic technique (see p. 47) and it was found that D- and L-compounds were present in about equal amounts. The higher proportion of L-amino acids found in earlier tests on meteorites may have been the result of contamination by terrestrial microorganisms before examination.

The moonrock samples brought back by the American Apollo missions were examined for amino acids, and glycine and alanine were detected in aqueous extracts of the samples. Glutamic acid, serine, threonine and aspartic acid were detected, gas chromatographically, in the acid hydrolysate of the aqueous extract.

Recent spectroscopic studies have revealed the presence of ammonia, formaldehyde and HCN in interstellar space, and these molecules may well be the starting materials for the abiogenetic synthesis of amino acids.

Table 8. Abiogenetic formation of amino acids

Starting materials	Energy source	Amino acids formed
CH_4, NH_3, H_2, H_2O	Electric discharge	Gly, Ala, β-Ala, Abu
CH_4, NH_3, H_2, H_2S	Electric discharge	Cys, $\overline{\text{Cys Cys}}$, Met and others
HCHO, NO_3^-, H_2O, $FeCl_3$	UV light	Ser, Asp, Asn, Gly, Ala, Thr, Val, Orn, Arg, Pro, Glu, His, Leu, Ile, Lys
Glucose, NH_3	UV light, $V_2O_5(H_2O_2)$	Gly, Ala, Asp, Val, Lys
CH_4, C_2H_6, NH_3, H_2S, HCN	Electric discharge and UV light	Phe, Tyr
HCHO, KNO_3, H_2O	Sunlight (24–300 h)	Asp, Lys, Ala, Gly, Orn, Arg, Glu, His, Ser, Thr
Tartaric acid, KNO_3, H_2O	Sunlight (500 h)	Asp, Ala
CH_3COONH_4, H_2O	β-rays	Asp, Glu
$(NH_4)_2CO_3$	γ-rays	Gly, Ala
HCHO, NH_2OH	heat	Gly, Ala, Ser, Thr, Asp
$HCONH_2$, N_2	230 °C	Gly, Ala, Asp, Thr, Ser, Glu, Val, Leu
CH_4, NH_3, $H_2O(SiO_2)$	950 °C	Gly, Ala, Asp, Thr, Ser, Glu, Val, Leu, Pro, Ile, aIle, Tyr, Abu
HCN, NH_3, H_2O	90 °C, 18 h	Arg, Ala, Gly, Ser, Asp, Glu, Leu, Ile, Abu, Thr

Once formed, some amino acids may be converted into others. For instance, if an aqueous solution of aspartic acid is exposed to UV light for a short time, serine, β-alanine and glycine may be detected in the solution.

HCN seems to be of particular importance in the abiogenetic synthesis of amino acids. Its presence not only explains the synthesis of a series of amino acids if aldehydes and ammonia are present (STRECKER synthesis) but by itself it can form the C—N structure of certain amino acids (K. HARADA, 1967) by oligomerisation:

$$
\begin{array}{ccccc}
\begin{array}{c} 2H{-}C{\equiv}N \\ \downarrow \\ H{-}C{=}NH \\ | \\ C{\equiv}N \end{array}
& \xrightarrow{HCN} &
\begin{array}{c} C{\equiv}N \\ | \\ H{-}C{-}NH_2 \\ | \\ C{\equiv}N \end{array}
& \xrightarrow{HCN} &
\begin{array}{c} C{\equiv}N \\ | \\ HC{-}NH_2 \\ | \\ C{=}NH \\ | \\ C{\equiv}N \end{array} \\
\text{(dimerisation)} & & \text{(trimerisation)} & & \text{(tetramerisation)} \\
\downarrow & & \downarrow & & \downarrow \\
\text{Gly} & & \text{Ala} & & \text{Asp}
\end{array}
$$

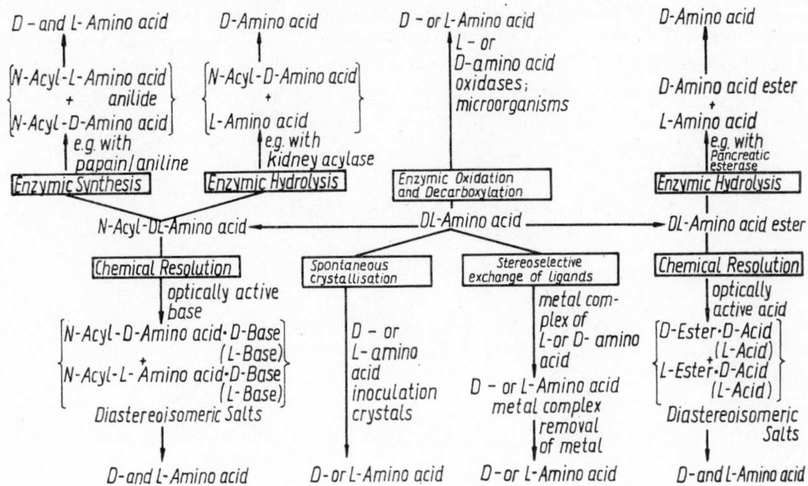

Figure 8. Methods for preparation of optically active amino acids

1.6. The Optical Resolution of Amino Acids[39,40]

In 1851 L. PASTEUR showed that synthetic amino acids and amino acids obtained from natural sources differ in their optical behaviour. The suggestion that synthetic amino acids might be an equimolar mixture of D- and L-compounds was advanced by SCHULZE and BOSSHARDT in 1886 as a result of microbiological experiments. In the same year A. PIUTTI separated racemic DL-aspartic acid into its optical isomers by crystallisation from an aqueous solution. F. EHRLICH developed the method of optical resolution using enzymes and successfully isolated the pure D-forms of a number of important amino acids in good yield.

At about this time E. FISCHER achieved the first optical resolution by chemical methods. He resolved *N*-acylamino acids into their optical components by selective crystallisation of the diastereoisomeric salts they formed with various alkaloids. These methods of resolution are still in current use. For resolution on a commercial scale, the enzymic method and the method of spontaneous crystallisation are the methods of choice, although the latter is not practicable in every case.

The origin of life on earth must have involved a selective preference for one optical antipode over the other. Since the abiogenetic synthesis of amino acids produces DL-mixtures, it is necessary to explain why one isomer was preferred as the basis of natural systems. It has been suggested that circularly polarised light was the agent that performed this selection, possibly by selectively activating one isomer or the other for chemical reactions. Some support for this theory was provided by experiments performed by A. S. GARAY in 1968. He used the β-radiation of ^{90}Sr as a model for circularly polarised light and found that after exposure of DL-tyrosine to this radiation for 18 months, destruction of the D-antipode was favoured over destruction of the L-antipode.

Other possibilities include differential solubility of the optical antipodes and differentiation between isomers at optically active inorganic surfaces such as quartz.

Once optical activity is generated it can be transferred spontaneously into other DL-compounds. This was demonstrated, in 1965, by HARADA in his method of optical resolution by 'stereoselective ligand-exchange'[41]. HARADA has shown that if an aqueous solution of a metal complex of an L-amino acid is treated with a DL-mixture of another amino acid ligand, then exchange takes place selectively to give the metal complex of the D-form of the second amino acid, liberating the L-form of the original amino acid:

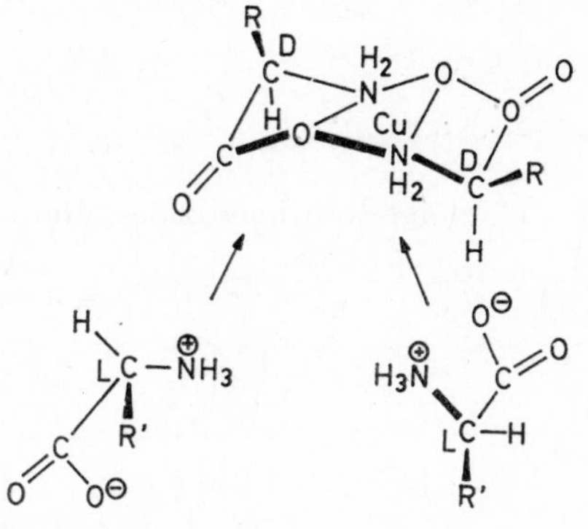

In the following sections the various methods of optical resolution will be discussed in detail.

1.6.1. *Optical Resolution by Spontaneous Crystallisation*[42]

In a few instances the optical antipodes of amino acids do not crystallise out of solution as crystals of the DL-mixture, but rather as a eutectic mixture of D- and L-forms, and in favourable cases manual separation of the crystals is possible. In practice, however, only glutamic acid and asparagine form good enough crystals for this method to be practicable. Nowadays, increasing use is being made of a related method in which a supersaturated solution of the DL-mixture is seeded with crystals of the pure D- or L-form. This induces selective crystallisation onto the seed crystals. The quantity produced is usually two or three times the amount of the seed portion, and this procedure has been successfully applied for resolving histidine, threonine, glutamic acid, aspartic acid, glutamine and asparagine.

1.6.2. Chemical Methods

The chemical methods of resolution are nearly all based on the preparation of diastereoisomers with optically active reagents. These derivatives can be separated by, for example, fractional crystallisation. Alkaloids can be used to prepare diastereoisomeric salts of amino acids, but they are expensive and other bases such as phenylethylamine, fenchylamine, chloramphenicol (and its precursor, L(+)-threo-1-*p*-nitrophenyl-2-amino-propan-1,3-diol) are more frequently used.

The amino acid may be resolved in the form of an N-protected derivative: if the benzyloxycarbonyl derivative is used, this is advantageous since this can be used in peptide synthesis directly after the resolution.

The formation of salts of amino acids with optically active acids was first used by Pyman, in 1911, for the resolution of histidine with D-tartaric acid. Dibenzoyl-D-tartaric acid (Losse and co-workers), D-camphor sulphonic acid and corresponding halogen derivatives have also proved useful as resolving agents. In this method the amino acids are usually used as basic derivatives such as amides and hydrazides, and in particular as the esters.

The diastereoisomer that crystallises directly out of the reaction solution is the one that is obtained in highest purity. Rather than try to obtain the other diastereoisomer from the mother-liquors it is better to try using the antipode of the resolving agent because this often produces the effect of crystallising out the other diastereoisomer.

1.6.3. Biochemical Processes

The biochemical methods of optical resolution rely for their efficacy on the strict chemical and steric specificity of enzymes. There are three different biochemical methods of resolution. The first method employs the observation first made by Pasteur, that certain microorganisms (yeast, *Penicillium glaucum*, *Aspergillus niger*, *Escherichia coli*, etc.) as a result of the specificity of their oxidases and decarboxylases only metabolise one optical antipode, and leave the other, usually the D-form, unchanged. The second method has only limited preparative application. It uses the enzyme papain, which catalyses the specific formation of amides, in an interesting case of enzymic asymmetric synthesis of amino acids.

The most useful enzymic method of optical resolution uses an enzyme to regenerate the free amino acid from a suitable derivative. Because of the high optical specificity of the enzyme, only one optical antipode (usually the L-form) can be a substrate so the process yields a mixture of free amino acid and the unchanged derivative of the D-form. Suitable substrates include esters (especially isopropyl, methyl and ethyl esters) and -acyl derivatives such as *N*-acetyl and *N*-chloracetyl amino acids. The most frequently used enzyme system is the acylase obtained from kidney, which was first employed by J. P. GREENSTEIN.

1.6.4. Chromatographic Separation of Diastereoisomers[43]

Amino acids may be separated into their optical antipodes by paper and thin-layer chromatography. The separation can be achieved either by the use of optically active eluants or by interaction with an asymmetric support such as cellulose fibre.

ROGOZHIN and DAVANKOV[44] reported the chromatographic resolution of DL-proline using a chlormethylated polystyrene resin incorporating L-proline. Optically pure L-proline was eluted by an ammoniacal solution of copper sulphate: D-proline followed after treatment with aqueous ammonia. Since this work, this procedure has developed into the general technique known as 'ligand chromatography' which can be used successfully to separate the optical antipodes of all the monoamino monocarboxylic acids. The analytical optical resolution of all the amino acids found in proteins has also been achieved by gas chromatography. For example, the optical isomers of the amino acid constituents of a meteorite sample have been separated as *N*-Tfa amino acid isopropyl esters using a capillary column containing *N*-Tfa-L-valyl-valine cyclohexyl ester as the optically active component of the stationary phase[45].

1.7. Analysis of Amino Acids

The ability to separate and detect individual amino acids and to estimate them quantitatively is an essential prerequisite for any investigations of polypeptide and protein structure and of protein and amino acid metabolism.

Many procedures have been described for the qualitative and quantitative analysis of amino acids, but only the chromatographic, mass spectroscopic, microbiological, enzymic and isotope dilution methods will be discussed here.

1.7.1. Chromatographic Procedures[46,47]

All chromatographic methods are based on the different distribution between two phases of the components of a mixture that is to be separated. One of the phases is firmly bound to a carrier, and is called the stationary phase: the other phase moves over the carrier at a uniform rate and is called the mobile phase.

According to the properties of the solid carrier of the stationary phase, we can distinguish between adsorption and ion exchange chromatography: The terms 'column' and 'thin-layer chromatography' obviously derive from the arrangement of the carrier material. In the case of paper chromatography, water, bound to the cellulose, constitutes the stationary phase, and a variety of other solvents can act as liquid phase. In the case of gas chromatography, a stream of nitrogen, helium or hydrogen is used as the mobile phase. In electrochromatographic procedures, chromatography and electrophoresis are combined[48]. Once separated, the individual amino acids can be identified and characterised by specific colour reactions. The most important colour reactions are those which can be used quantitatively. Ninhydrin and 2,4,6-trinitrobenzene sulphonic acid[49] are the most familiar reagents. Fluorescamine, 4-phenyl-[furan-2H(3H),1′-phthalan]-3,3′-dione, is not fluorescent by itself, but reacts with amino acids (with the exception of Pro and Hyp) to form intensely fluorescing derivatives (WEIGERLE and UDENFRIEND, 1972) which allow the detection of as little as 100 nmol/ml.

Gel filtration is an especially important chromatographic technique. This procedure (largely developed by the Swedish company Pharmacia) uses a three-dimensional network of dextrans as adsorbent and performs the fractionation of substances with different molecular weights by a molecular sieve effect. Using the commercially available 'Sephadex', dextrans, proteins, peptides and amino acids may be separated from one another. The fractionation of proteins from each other according to molecular weight is possible if the differences between the molecular weights are great enough and if Sephadex types with suitable pore sizes are used (cp. p. 259). An important preparative application of Sephadex types G10 and G25 is the desalting of solutions of amino acids, peptides and proteins, which, until recently, could only be achieved by dialysis.

1.7.1.1. Paper Chromatography[50]

This technique was first described in 1944 by CONSDEN, GORDON and MARTIN, and, in its two-dimensional version, is now widely used for the qualitative analysis of amino acids. Besides the original solvents phenol/water (8 : 2) and collidine/lutidine (1 : 1, saturated with water) numerous other combinations have been found suitable for the resolution of the constituents of protein hydrolysates.

A very good separation of the amino acids in proteins is obtained by a one-dimensional technique known as the 'passage process', which involves multiple elution with the same acidic–alcoholic solvent mixture[51]. In this way even leucine and isoleucine can be separated from each other. The determination of R_f values is impossible under these conditions, so measurements are made relative to the amino acids with the greatest migration speed. Table 9 shows the $R_{leucine}$ values of a series of amino acids, using n-butanol/isobutyric acid/glacial acetic acid/water (5 : 0,5 : 0,7 : 5) and SCHLEICHER-SCHÜLL-2043 b paper with threefold elution. The R_f values of the same amino acids in butanol/glacial acetic acid/water (4 : 1 : 1) are given for comparison.

Table 9. $R_{leucine}$ and R_f values of some amino acids

Amino acid	$R_{leucine}$	R_f	Amino acid	$R_{leucine}$	R_f
Cys/Cys Cys	0.04	0.07/0.08	Ala	0.44	0.38
Lys	0.09	0.14	Pro	0.50	0.43
His	0.11	0.20	Tyr	0.56	0.45
Arg	0.14	0.20	Trp	0.63	0.50
Asp	0.21	0.19	Met	0.75	0.55
Ser	0.23	0.27	Val	0.77	0.60
Gly	0.27	0.26	Phe	0.91	0.68
Glu	0.32	0.30	Ile	0.98	0.72
Thr	0.35	0.35	Leu	1.00	0.73

The scope of paper chromatographic separation has been extended by the introduction of ion exchange papers. Ion-exchangers such as Amberlite-SA-2 or Zeokarb 225 have been fixed on cellulose papers and instead of common eluant systems buffer solutions of different pH values are used to effect the resolution. There appear to be no special advantages associated with the complementary approach which uses liquid ion-exchangers (for example, *N*-lauryl-*N*-trialkylmethylamine) as mobile phase.

The colour reaction with ninhydrin (cp. p. 56) is usually used for the development of amino acid chromatograms. The composition of the spray mixture has been modified in many ways: ninhydrin (0.2 g) and collidine (0.2 ml) in 100 ml chloroform gives a colour that will last for 14 days. The red complexes formed by subsequent spraying with a dilute solution of a copper(II) salt are also relatively stable. Characteristic coloured spots are produced if a chromatogram which has been run in butanol/glacial acetic acid/water is treated with diethylamine before being developed.

There is a correlation between structure and chromatographic mobility. Amino acids with hydrophilic side chains (amino-dicarboxylic acids, diaminocarboxylic acids and hydroxyacids) favour the stationary 'aqueous' phase, so that they have lower R_f values than amino acids with similar carbon skeletons but unsubstituted side chains. Amino acids with lipophilic side chains prefer the mobile organic phase with the result that there is an increase in R_f value with increasing lipophilic character (i.e. Gly < Ala < Val < Leu).

1.7.1.2. Thin-layer Chromatography[52–54]

Thin-layer chromatography has now become a serious rival to paper chromatography for work with amino acids. Silica gel G is the most commonly used support, but alumina, DEAE cellulose, cellulose MN 300 and a variety of starch and dextran gels are also widely employed.

In general, the eluants are the same as in paper chromatography. The main advantages of the thin-layer procedures, which can be used in one-dimensional or multi-dimensional separations, are: (a) increased sensitivity; (b) lowering of the time required (for example, for analysis of a protein hydrolysate from 2–3 days down to 4 h); (c) the spots obtained generally have smaller areas than are observed when paper is used; (d) the availability of a wide selection of adsorbents.

Thin-layer chromatography is not only useful in the qualitative analysis of peptides and protein hydrolysates in sequence analysis and in protein end-group determination but also in the analytical monitoring of peptide syntheses. Figure 9 shows the two-dimensional separation of the hydro-

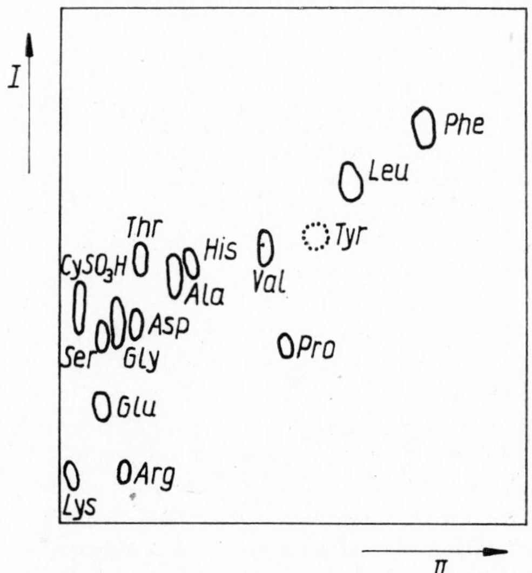

Figure 9. Thin-layer chromatographic separation of an insulin B-chain hydrolysate. Eluant I: $CHCl_3$/MeOH/17% NH_4OH solution (2 : 1 : 1); 75 min. Eluant II: PhOH/H_2O (3 : 1); 180 min

lysate of an insulin B-chain on silica gel G. Cysteine/cystine appear as cysteine acid [Cys(SO_3H)] after preliminary oxidation with performic acid[55].

1.7.1.3. *Ion exchange Chromatography*

After systematic investigation aimed at defining optimum conditions for ninhydrin determinations, Moore and Stein (1948) succeeded in separating and analysing the constituents of 2.5 mg of a beef serum albumin hydrolysate by partition chromatography on a starch column. The eluate was

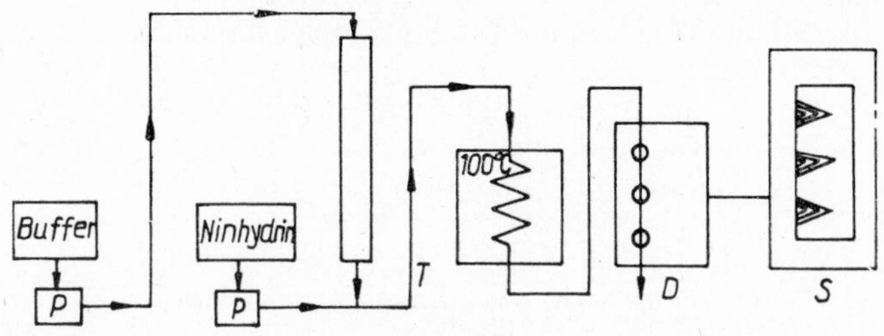

Figure 10. Scheme of the first automatic amino acid analyser of Spackman, Moore and Stein

divided, using a fraction collector, into about 400 portions. The amino acid content of each fraction was determined manually by the ninhydrin reaction and the molar proportions of each amino acid were calculated from the integrated curve of intensity versus fraction number. This was a considerable achievement. The accuracy (= 3 per cent) was the highest achieved up to that time and very little material was used. However, the analysis of each sample took a week. Moore and Stein were able to perform the analysis in a shorter time by changing to ion exchange chromatography. The most suitable ion-exchangers for amino acid analysis are the Dowex resins (Dowex 50 × 8 and Dowex 50 × 4) which are crosslinked with 8 and 4 per cent divinylbenzene, respectively, and sulphonated polystyrene resins of small bead size (Amberlite resins), which were used in the first automatic amino acid analysers[56]. Figure 10 illustrates the principle of the first automatic analysers developed by Spackmann, Moore and Stein. The separation of the components of a mixture of nanomole amounts of amino acids is shown in Figure 11.

In the early automatic analysers, the separation of amino acids was performed in a 9.9 × 1.50 cm column by means of buffer solution pumped through continuously (flow rate: 30 ml/h; sodium citrate buffer solutions pH 3.25 and 4.25). The eluate was made to react with ninhydrin (15 min at 100 °C); the intensity of the ninhydrin colour at 570 nm and at 440 nm (for proline) was determined in a colorimeter (D) and automatically recorded (S). The analysis time for a sample was about 24 hours. The MOORE–STEIN analysis has been considerably improved in recent years[58] until, today, only 20 μg of protein hydrolysate is necessary for a quantitative amino acid analysis and shorter times are required. Recent developments include the demonstration that with a capillary column under high pressure a complete analysis can be performed within 7 min.

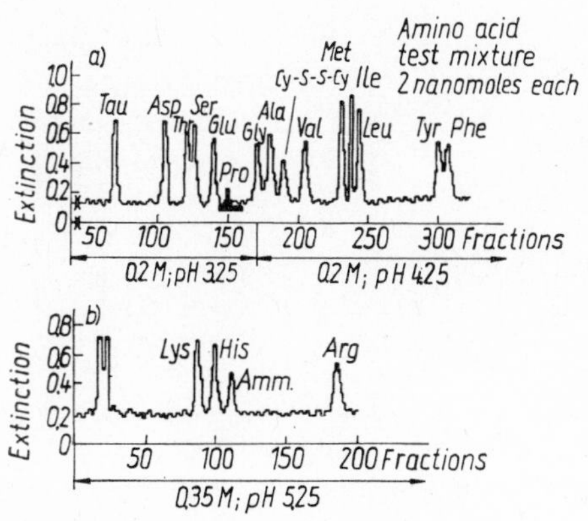

Figure 11. Chromatogram of an amino acid test mixture: (a) acidic and neutral, (b) basic amino acids

1.7.1.4. Gas Chromatographic Analysis[59–61]

Because of their low volatility, direct gas chromatographic analysis of amino acids is impossible. However, mixtures may be separated after converting the amino acids into suitable volatile derivatives. Table 10 lists the most important derivatives suitable for gas chromatography.

The separation of the ester derivatives is generally achieved using polyesters, polyethyleneglycols or silicone polymers as liquid phases.

In the gas chromatographic analysis of Dnp-amino acid methyl esters, amounts as small as 10^{-14} mol of amino acid may be detected. This extreme sensitivity is achieved with an electron capture detector, making use of the high electron affinity of the Dnp group.

The separation of diastereoisomeric amino acid derivatives by gas chromatography has already been described (on p. 42).

Table 10. Amino acid derivatives which may be separated by gas chromatographic analysis

N-Trifluoracetyl methyl esters	N-Trimethylsilyl trimethylsilyl esters
N-Trifluoracetyl isopropylesters	N-2,4-Dinitrophenyl methyl esters
N-Trifluoracetyl amyl esters	N-Acetyl n-propyl esters
N-Trifluoracetyl 2-octyl esters	N-Acetyl n-amyl esters
N-Isobutyl methyl esters	N-Formyl methyl esters

Degradation products suitable for chromatography	Formed by
Aldehydes	Ninhydrin reaction
Amines	Decarboxylation
Aminoalcohols	Reduction with lithium aluminium hydride
Nitriles	Reaction with iodosobenzene; hypobromite oxidation

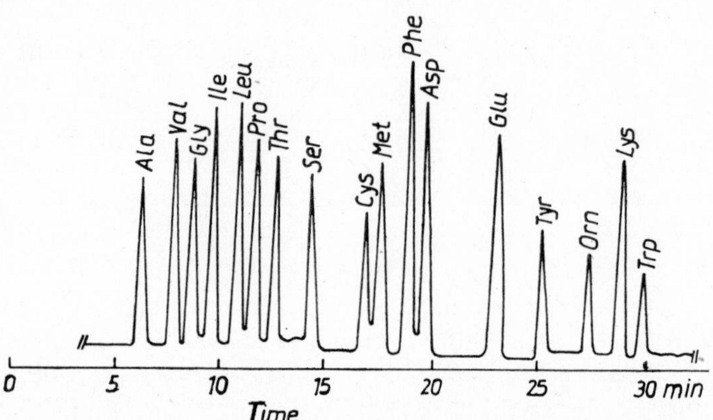

Figure 12. Separation of amino acids as their trifluoro-acetyl-n-butyl esters by gas chromatography (1 m/2 mm column with 0.65% EAG on Chromosorb W, 80–100 mesh)

1.7.2. Mass Spectral Analysis of Amino Acids[62]

Improved instrumentation has enabled mass spectroscopy to become increasingly valuable in amino acid and peptide chemistry in recent years.

The detection and identification of fragments generated by the decay of the molecular ion formed in the electron beam is of prime interest. The characteristic decay of α-amino acid molecular ions is into resonance-stabilised iminium ions, with loss of the relatively stable COOH and side chain radicals R, as shown in the following reaction scheme:

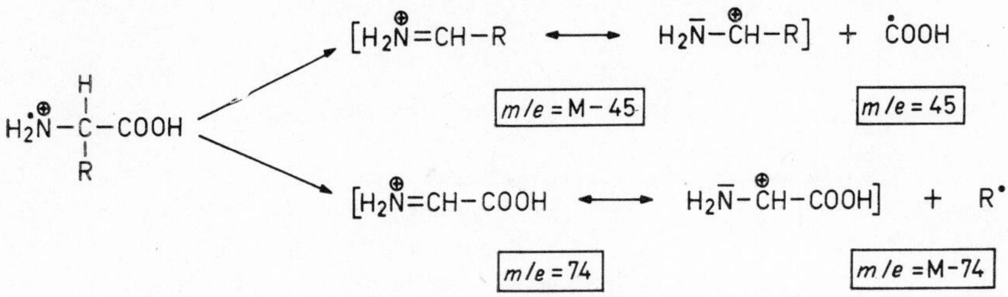

$$H_2\overset{\bullet\oplus}{N}-\underset{R}{\overset{H}{C}}-COOH \rightarrow [H_2\overset{\oplus}{N}=CH-R \longleftrightarrow H_2\bar{N}-\overset{\oplus}{C}H-R] + \dot{C}OOH$$

m/e = M − 45 m/e = 45

$$H_2\overset{\bullet\oplus}{N}-\underset{R}{\overset{H}{C}}-COOH \rightarrow [H_2\overset{\oplus}{N}=CH-COOH \longleftrightarrow H_2\bar{N}-\overset{\oplus}{C}H-COOH] + R^{\bullet}$$

m/e = 74 m/e = M − 74

Other fragmentations are possible, depending on the structure of the side chain. In Figure 13 the mass spectrum of threonine, measured at 70 eV, shows in addition to the normal fragments (*m/e* 74 and 75) two further peaks of high relative intensity, which may be explained by the following fragmentation:

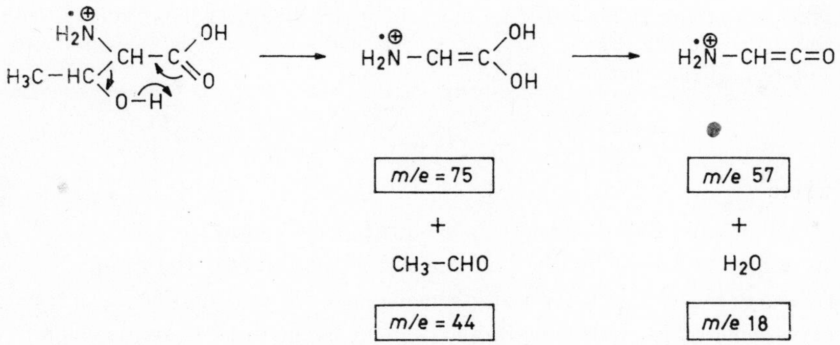

$$H_3C-HC(OH)-CH(\overset{\bullet\oplus}{N}H_2)-C(=O)OH \longrightarrow H_2\overset{\bullet\oplus}{N}-CH=C(OH)_2 \longrightarrow H_2\overset{\bullet\oplus}{N}-CH=C=O$$

m/e = 75 + CH_3-CHO m/e = 44

m/e 57 + H_2O m/e 18

Because of the low volatility of amino acids, it is necessary to introduce samples into the mass spectrometer at 10^{-6}–10^{-7} Torr and 100–150 °C. Better results are obtained by using volatile amino acid derivatives, such as those used for gas chromatography. The combination of gas chromatographic separation and mass spectroscopic identification of Tfa-methyl esters has been of enormous use in the determination of rare naturally occurring amino acid structures. Another important application of mass spectrometric analysis is in the sequence determination of peptides (cp. p. 281).

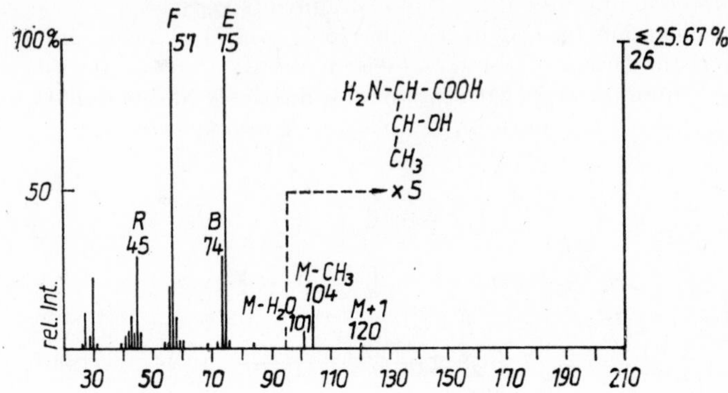

Figure 13. Mass spectrum of threonine (70 eV)

1.7.3. Methods of Analysis Using Isotopes[63–65]

After ion exchange chromatography, isotope methods, with an accuracy of 1–2 per cent, are the most important standard methods of quantitative analysis of amino acids.

In the isotope dilution method the sample for analysis is added to a precise amount of the same amino acid which has been labelled and then the degree of dilution of the radioactive isotope content is measured after isolation of the pure amino acid:

$$B = \left(\frac{C_0}{C} - 1\right) A$$

where

A = amount of labelled amino acid added to the sample;
B = amount of the 'normal' amino acid which contains the sample;
C_0 = concentration of the isotopic element in the added amino acid;
C = concentration of the isotopic element in the isolated amino acid.

In the labelled derivative method the amino acid to be estimated is first labelled by a quantitative reaction with reagents containing radioactive or stable isotopes. Then a large excess of the normal amino acid derivative is added to the mixture. Then the sample is purified until the molar isotope proportion (C_c) stays constant. The amount w of the amino acid being estimated is given by:

$$w = W \frac{C_c}{C_r}$$

W is the amount of the unlabelled derivative added and C_r the molar concentration of the isotope in the derivative resulting from the reaction between pure amino acid and reagent.

[^{131}I] and [^{35}S]*p*-iodophenylsulphonyl chlorides (pipsyl chlorides) have been widely used as radioactive reagents.

1.7.4. Enzymic Processes

The enzymic methods are of limited applicability because of their low sensitivity. They are based on the estimation of the products formed on treatment of amino acids with specific enzymes. For example, a specific decarboxylase (or microorganism producing this decarboxylase) is added to the mixture of amino acids to be analysed: if the CO_2 formed in the decarboxylation is measured in a WARBURG apparatus, 0.5–1.5 mg amino acid can be detected.

Table 11 shows the most important enzymatic processes used in amino acid determination.

Table 11. Enzymatic procedures for the determination of amino acids

Amino acid	Enzyme	Reaction products
Arg	Arginase	Urea (colorimetric or as CO_2 after treatment with urease)
His	Histidase	NH_3 (titration)
Lys	Lysine decarboxylase	CO_2 (manometric or volumetric)
Orn	Ornithine decarboxylase	CO_2 (manometric or volumetric)
Tyr	Tyrosine decarboxylase	CO_2 (manometric or volumetric)
Glu	L-Glutamic acid decarboxylase	CO_2 (manometric or volumetric)
Asp	Aspartase	NH_3 (titration)
Gln	Glutaminase	NH_3 (titration)

1.7.5. Microbiological Processes[66]

Several microorganisms—especially lactic acid bacteria—may be used for qualitative and quantitative analyses of amino acid which depend on their different amino acid requirements.

The amino acid to be estimated is incorporated into a growth medium alone with all the other requirements for normal growth of the micro-organism. The growth produced by this nutrient mixture is estimated either by photoelectric nephelometric analysis or by titration of the lactic acid formed: comparison with an independently determined growth curve gives an indication of the amount of the amino acid added. For complete analysis of a protein hydrolysate, about 100 mg of sample is needed.

1.8. Characteristic Reactions of Amino Acids

The chemical reactions of the amino acids are, in the main, those characteristic of their functional groups. These are numerous and so only the most important reactions of the amino and carboxyl groups will be discussed here. Those reactions which lead to derivatives that are important to peptide synthesis, such as the preparation of acid chlorides, acid hydrazides, azides and certain esters, are discussed in the peptide section.

1.8.1. The Formation of Metal Complexes[67]

Amino acids form chelate complexes with a wide range of heavy metal ions. The best-known are the deep-blue-coloured, generally well-crystalline complexes with bivalent copper. A method of separating amino acids based on the different solubilities of these complexes was at one time suggested (M. B. Brazier, 1930).

X-ray crystal structure analyses aimed at throwing light on the bonding in protein metal complexes have revealed the octrahedral structure of the metal amino acid complexes. Two amino acid molecules are bound to the central metal ion through the carboxyl and amino groups, and the two remaining ligand positions are taken by water molecules. Amino acids having side chains with functional groups often form particularly stable complexes; for instance, histidine is additionally bound to the central atom through the imidazole nitrogen.

The structure of amino acid–metal ion complexes in solution has been explored by high-resolution NMR spectroscopy, UV spectroscopy, ORD, etc.

The copper complexes of the diamino monocarboxylic acids in which the α-amino and the α-carboxylic groups participate in the formation of the complex and the ω-amino group stays free have been used in peptide syntheses for the preparation of ω-modified derivatives. The amino acid–Cu–EDTA complex, formed in the presence of copper sulphate and EDTA, has been employed in the polarographic estimation of amino acids.

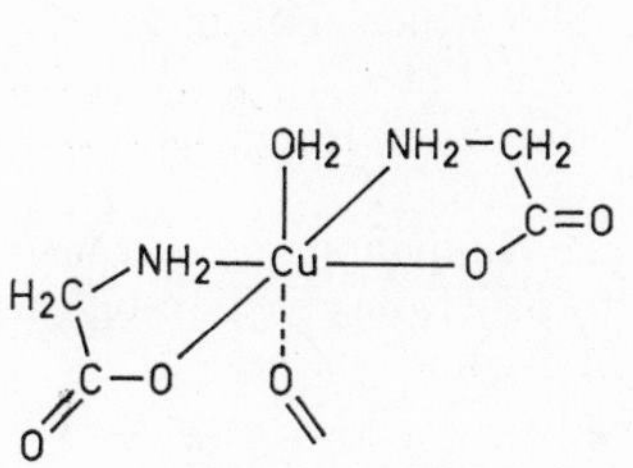

Bis-glycinato-copper(II)-hydrate $Cu(Gly)_2 \cdot H_2O$

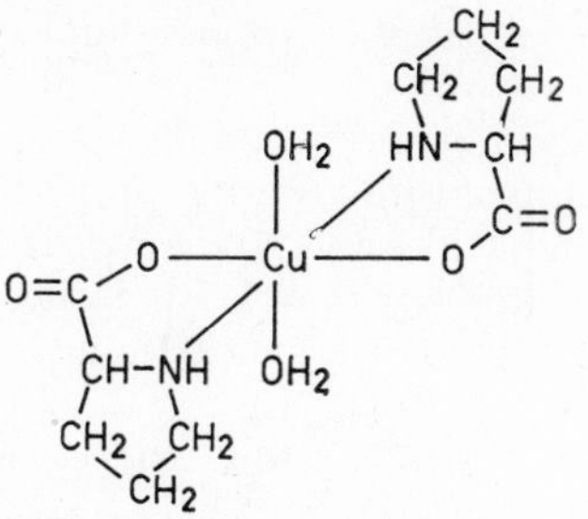

Bis-*DL*-prolinato-copper(II)-dihydrate $Cu(Pro)_2 \cdot 2\,H_2O$

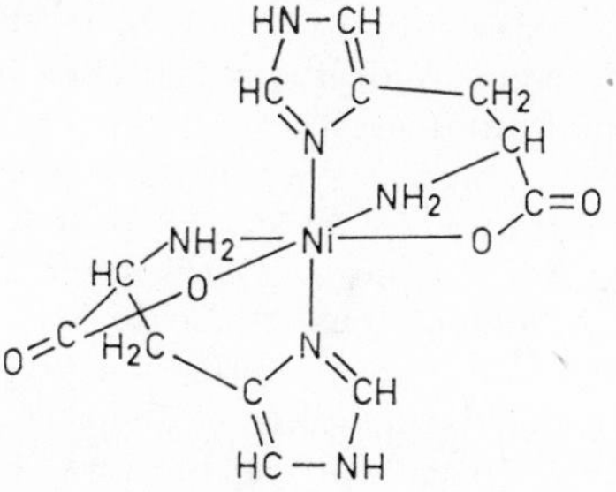

Bis-L-histidinato-nickel(II)-hydrate $Ni(His)_2 \cdot H_2O$

1.8.2. Reactions with Nitrous Acid

Free amino acids react with nitrous acid like primary amines and evolve nitrogen, the amino group being converted to a hydroxyl group. The reaction proceeds with retention configuration at the asymmetric carbon atom. Measurement of the volume of nitrogen evolved in this reaction has been used for the quantitative estimation of amino groups (D. D. van Slyke, 1910):

$$R{-}CH(NH_2){-}COOH + HONO \longrightarrow R{-}CH(OH){-}COOH + H_2O + N_2$$

Sydnones can be made by the action of nitrous acid on *N*-alkyl(aryl)-amino acids followed by treatment of the resulting nitrosamino acid with acetic anhydride:

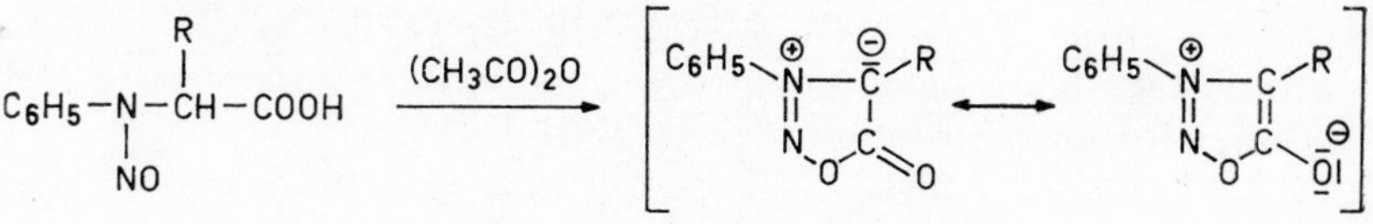

$$C_6H_5{-}N(NO){-}CH(R){-}COOH \xrightarrow{(CH_3CO)_2O} \left[C_6H_5{-}\overset{\oplus}{N}{-}\overset{\ominus}{C}(R){-}C(=O){-}O{-}N \longleftrightarrow C_6H_5{-}\overset{\oplus}{N}{-}C(R){=}C(O^{\ominus}){-}O{-}N \right]$$

Phenylsubstituted sydnones and sydnonimines[68] which contain a $-C{=}NH$ group instead of a carbonyl group have been the subject of particular interest in recent years, because of their tumour-retarding, bacteriostatic and temperature-lowering properties.

Amino acid esters are converted into diazo compounds by nitrous acid: the best-known example is diazoacetic ester (formed from glycine ester), which is widely used as a synthetic intermediate in organic synthesis:

$$HNO_2 + H_2N{-}CH_2{-}COOR' \mid N{\equiv}\overset{\oplus}{N}{-}\overset{\ominus}{\bar{C}}H{-}COOR + 2H_2O$$

1.8.3. *Decarboxylation*

On gradual heating to a temperature of about 200 °C amino acids lose CO_2 and are converted into primary amines:

$$H_2N{-}CHR{-}COOH \xrightarrow{\Delta} H_2N{-}CH_2R + CO_2$$

The reaction is catalysed by metal ions. Aspartic acid, for example, is decarboxylated readily in the presence of cupric ions, giving alanine.

Decarboxylation of amino acids in living systems is catalysed by decarboxylase enzymes: it takes place via the SCHIFF's base formed from pyridoxal phosphate:

HC=N−CHR−COO⊖ | Ⓟ−O−CH₂ | OH | N⊕H | CH₃ —(−CO₂)→ [HC−N=CHR | Ⓟ−O−CH₂ | OH | NH | CH₃ ←→

HC̄⊖−N=CHR | Ⓟ−O−CH₂ | OH | N⊕H | CH₃ ←→ HC=N−C̄⊖HR | Ⓟ−O−CH₂ | OH | N⊕H | CH₃] —(H⊕)→

HC=N−CH₂R | Ⓟ−O−CH₂ | OH | N⊕H | CH₃ —(H₂O)→ HC=O | Ⓟ−O−CH₂ | OH | N | CH₃ + H₂N−CH₂R

Ⓟ = PO_3H_2

The aliphatic amino acids yield on decarboxylation simple aliphatic amines which are of little interest, but the amines derived from other amino acids, especially the aromatic and heterocyclic amino acids, display a wide range of biological properties including pharmacological, coenzyme, hormone and vitamin activity.

Table 12. The most important biogenic amines

Amino acid	Decarboxylation products	Effect and occurrence
His	Histamine	Tissue hormone, active upon blood pressure
Lys	Cadaverine }	Ribosome component, bacterial metabolic product
Orn	Putrescine }	
Arg	Agmatine	Metabolic product of intestinal bacteria
Asp	β-Alanine	Coenzyme A component
Glu	γ-Amino butyric acid	Metabolic product of brain
Ser	Ethanolamine	Component of phosphatides and choline
Thr	Propanolamine	Vitamin B_{12} component
Cys	Cysteamine	Coenzyme A component
Tyr	Tyramine	Tissue hormone, causes uterine contraction
Trp	Tryptamine	Tissue hormone

Tryptamine derivatives include well-known narcotic agents: e.g. the 'snuff' used by some South American Indian tribes contains *N*,*N*-dimethyl and *N*,*N*-diethyl tryptamine, and Psilocin and Psilocybin (the active constituents of the Mexican halucinogenic fungus *Psilocybe mexicana*) are 4-hydroxy-*N*,*N*-dimethyl-tryptamine and its *O*-phosphoryl derivative, respectively. The properties of these so-called 'biogenic amines' are summarised in Table 12.

The stereospecificity of decarboxylase enzymes is exploited in their use for preparation of D-amino acids from DL-mixtures.

1.8.4. Deamination

Chemical oxidation of amino acids leads to loss of the amino group and the formation of carbonyl compounds. This was first observed by STRECKER in 1862 in the reaction between alanine and alloxan; suitable oxidants include di- and triketones, *N*-bromosuccinimide, hypochlorite and silver oxide.

The reaction proceeds with decarboxylation and leads to the aldehyde having one carbon atom less than the starting amino acid:

$$R{-}CH(NH_2){-}COOH \xrightarrow{-2H;\ -CO_2} R{-}CH{=}NH \xrightarrow{H_2O} R{-}CHO + NH_3$$

The biological deamination of amino acids is mediated by amino acid oxidases (which are generally flavoproteins) and follows a course which is quite different from the chemical oxidation. In the first step of the enzymic reaction the amino group is dehydrogenated by hydrogen transfer to the flavin coenzyme (FAD or FMN) and the resulting imino acid is then hydrolysed to a keto acid:

$$R{-}CH(NH_2){-}COOH \xrightarrow[-H_2]{\text{flavin}} R{-}C({=}NH){-}COOH \xrightarrow{H_2O} R{-}C({=}O){-}COOH + NH_3$$

The oxidative deamination of amino acids by ninhydrin[69] is important since it forms the basis of the standard detection reaction for amino acids. In this reaction ninhydrin is first reduced to the corresponding diketoalcohol one molecule of which condenses with one of ninhydrin and one of ammonia giving a characteristic blue-violet dye which has an indigo-like structure:

$$R{-}CH(NH_2){-}COOH + \text{Ninhydrin} \longrightarrow R{-}CH{=}O + NH_3 + CO_2 + \text{diketoalcohol}$$

Ninhydrin

$$\text{Ninhydrin} + NH_3 + \text{diketoalcohol} \xrightarrow{\Delta} \text{dye} + 3H_2O$$

1.8.5. *Transamination*

Transamination plays a fundamental role in amino acid biochemistry. It is the reversible exchange of an amino group between an amino acid and a keto acid, producing a new amino acid and a new keto acid:

$$R{-}CH(NH_2){-}COOH + R_1{-}C({=}O){-}COOH \rightleftharpoons R_1{-}CH(NH_2){-}COOH + R{-}C({=}O){-}COOH$$

The enzymes which catalyse this process are found in all cells: they are called transaminases or aminotransferases. The coenzyme pyridoxal phosphate plays a central role in the reaction, which proceeds through the SCHIFF's base as shown below.

The metabolic pathways of most of the amino acids and their related α-keto acids are interconnected by transaminations—glutamic acid and aspartic acid are widely involved and the corresponding keto acids (α-ketoglutaric and oxalacetic acids) are a link between carbohydrate metabolism and the citric acid cycle on the one hand, and amino acid and protein metabolism on the other:

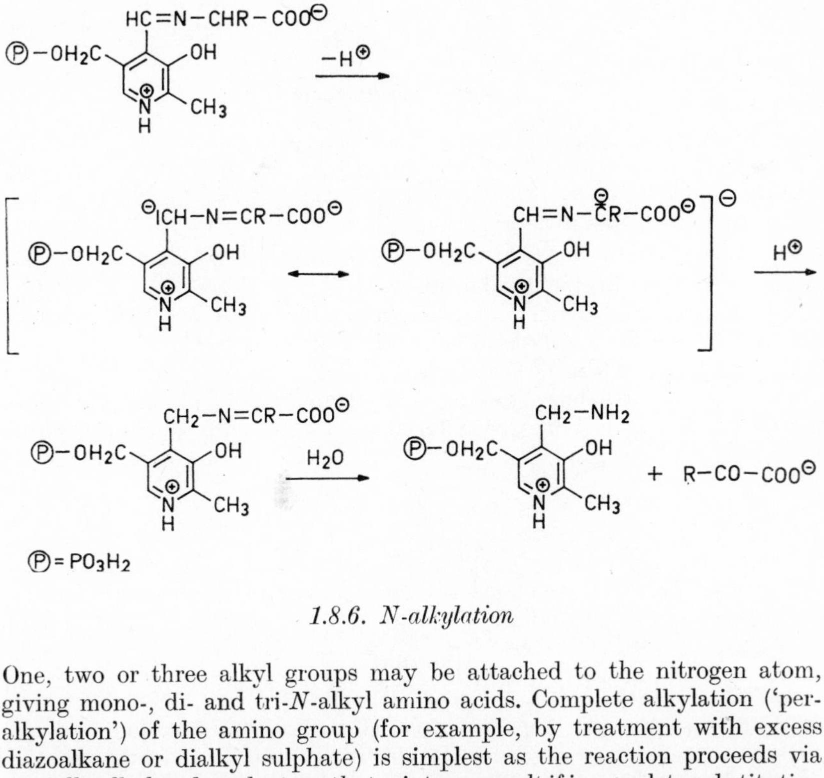

HC=N−CHR−COO⊖
Ⓟ−OH$_2$C, OH, N⊕ H, CH$_3$ $\xrightarrow{-H^{\oplus}}$

[⊖|CH−N=CR−COO⊖ Ⓟ−OH$_2$C, OH, N⊕ H, CH$_3$ ⟷ CH=N−C⊖R−COO⊖ Ⓟ−OH$_2$C, OH, N⊕ H, CH$_3$]⊖ $\xrightarrow{H^{\oplus}}$

CH$_2$−N=CR−COO⊖ Ⓟ−OH$_2$C, OH, N⊕ H, CH$_3$ $\xrightarrow{H_2O}$ CH$_2$−NH$_2$ Ⓟ−OH$_2$C, OH, N⊕ H, CH$_3$ + R−CO−COO⊖

Ⓟ = PO$_3$H$_2$

1.8.6. N-alkylation

One, two or three alkyl groups may be attached to the nitrogen atom, giving mono-, di- and tri-*N*-alkyl amino acids. Complete alkylation ('peralkylation') of the amino group (for example, by treatment with excess diazoalkane or dialkyl sulphate) is simplest as the reaction proceeds via partially alkylated products so that mixtures result if incomplete substitution is attempted. Monoalkylamino acids can be synthesised unambiguously from N-protected amino acids by treatment with alkyl halides followed by removal of the amino protecting group. Di- and trialkyl amino acids are obtained by the action of di- and trialkyl amines, respectively, on the corresponding halogenated fatty acids. Mono- and dimethylamino acids occur naturally as components of bacterial peptides. Their biosynthesis involves *N*-methylation of the amino group by *S*-adenosylmethionine, which, for example, also converts ethanolamine into choline, $(CH_3)_3N^{\oplus}-CH_2-CH_2OH$, being itself converted into adenosylhomocysteine in the process.

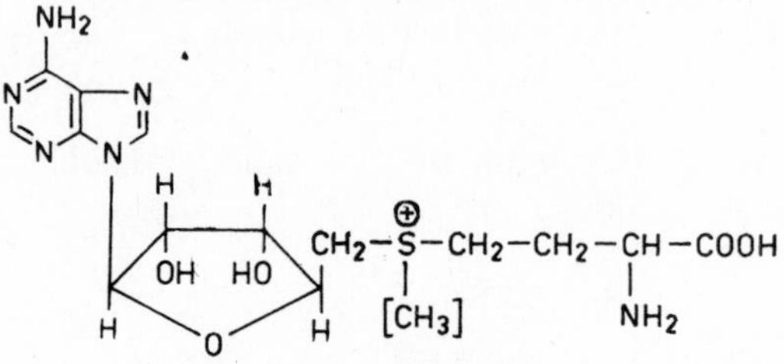

Adenosylmethionine

A range of permethylated amino acids have been isolated from natural sources. The most common of these is $(CH_3)_3N^{\oplus}-CH_2-CO_2^{\ominus}$, which is found in sugar beet (*Beta vulgaris*)—hence the term 'betaine'. Herzynine (the betaine of histidine) is found in mushrooms, hypaphorine (the betaine of tryptophan) in the seeds of *Erythrina hypaphorus* and stachydrine (the betaine of proline) in stachys and other plants.

N-benzyl-, *N*-benzhydryl- and *N*-tritylamino acids are useful intermediates in peptide synthesis because the alkyl residues may be cleaved easily by hydrogenolysis or acidolysis.

The *N*-(2,4-dinitrophenyl)-amino acids (Dnp-amino acids) are the basis of a method for identifying the N-terminal amino acid residue in peptides (see p. 274).

1.8.7. Acylation

Treatment of activated carboxylic acid derivatives (acid chlorides, acid anhydrides, etc.) with amino acids gives *N*-acylamino acids through a SCHOTTEN–BAUMANN reaction:

$$R_1COCl + H_2N-CHR-COOH \xrightarrow{NaOH} R_1CO-NHR-COOH + NaCl$$

Usually the *N*-acyl compounds formed are highly crystalline and are therefore suitable derivatives for the characterisation and identification of amino acids. The *N*-acyl residues may often be removed easily and specifically so that they may be used for the protection of the amino group. Such *N*-acyl-protecting groups will be discussed in Section 2.2.2.1. Trifluoroacylamino acids are used for the preparation of volatile esters for gas chromatography. Acetyl- and chloracetylamino acids are important starting materials for the optical resolution of DL-amino acids by enzymic methods.

The *N*-acyl derivatives of amino acids with fatty acids—for example, *N*-lauroyl- and *N*-stearoyl-glutamic acids—are important detergents.

1.8.8. Esterification

The earliest method of esterifying amino acids was devised by E. FISCHER; he passed hydrogen chloride into ethanolic suspensions of the amino acids. The amino acids dissolve when so treated and the reaction mixture becomes warm. From the resulting ester hydrochlorides formed in this process the esters themselves can be isolated by addition of base, extraction into ether and distillation *in vacuo*. Fractional distillation of an ester mixture formed in this way was the earliest method of obtaining single compounds from amino acid mixtures:

$$H_2N-CHR-COOH + R_1OH \xrightarrow[-H_2O]{HCl} \left[H_3\overset{\oplus}{N}-CHR-COOR_1\right]^{\oplus} Cl^{\ominus} \xrightarrow{-HCl} H_2N-CHR-COOR_1$$

Free amino acid esters are usually colourless liquids with a characteristic amine smell and pronounced basic character. They are partially soluble in water and may be detected by the ninhydrin reaction like free amino acids. On storage and especially on heating they form 2,5-dioxopiperazines with the elimination of alcohol:

$$2\ H_2N-CHR-COOR_1 \xrightarrow{-2\,R_1OH} \text{2,5-Dioxopiperazine (cyclo[-NH-CHR-CO-NH-CHR-CO-])}$$

2,5-Dioxopiperazine

The losses which accompany any attempt to distil amino acids are due to this reaction. The isopropyl esters are relatively stable to heat: it has been suggested that they are the most suitable for the separation of amino acid mixtures by distillation of ester mixtures (V. HOLEYSOVSKY and F. ŠORM, 1955).

Reducing agents convert amino acid esters to amino aldehydes or amino alcohols. Lithium aluminium hydride gives complete reduction and high yields of optically active amino alcohols are obtained.

Various ester functions such as the benzyl, tertiary butyl ester and other groups are used as carboxyl-protecting groups in peptide syntheses. The use of 'activated' esters in peptide synthesis will be discussed in the chapter on peptides. Amino acid amides and hydrazides are readily obtained from amino acid esters by ammonolysis and hydrazinolysis, respectively.

1.9. Cyclic Amino Acid Derivatives

Of the various cyclic derivatives of amino acids the most important ones are the five-membered hydantoins, azlactones (oxazolones) and oxazolin-2,5-diones and also the six-membered dioxopiperazines.

Hydantoins have already been mentioned in connection with amino acid synthesis. They are formed from ureido carboxylic acids by acid-catalysed cyclisation, or from alk(aryl)-oxycarbonyl amino acid amides by base-catalysed cyclisation:

```
R-CH-COOH                        R-CH—C=O                    R-CH ——— C=O
  |            H⊕                  |    |         OH⊖          |       |
H-N   NH-R'  ———→                H-N   N-H      ←———         H-N  OR"  NH2
   \ /        -R'OH                 \ /        -R"OH            \ /
    C                                C                           C
    ||                               ||                          ||
    O                                O                           O
```

Creatinine (2-imino-3-methyl-hydantoin) is formed in muscular tissue and excreted in the urine. The phenylthiohydantoin derivatives of amino acids are the final products of the EDMAN degradation of polypeptides (see p. 276):

```
R-CH—C=O                 R-CH—C=O
  |   |                    |   |
 HN   N-C6H5               N   O
   \ /                      \\ /
    C                        C
    ||                       |
    S                        R'

Phenylthiohydantoin      Azlactone
```

Azlactones are formed by dehydration of *N*-acylamino acids by acetic anhydride. They are intermediates in an important amino acid synthesis (see p. 33).

The *N*-carboxylic acid anhydrides, also called LEUCHS anhydrides (H. LEUCHS, 1906), may be synthesised by eliminating benzylchloride from *N*-benzyloxycarbonyl amino acid chlorides:

```
R-CH—C=O                                   R-CH—C=O
  |   |                                      |   |
  |   Cl           -C6H5CH2Cl               HN   O
  |               ——————————→                 \ /
 HN   O-CH2C6H5                                C
   \ /                                         ||
    C                                          O
    ||
    O
```

As they are carboxylic acid anhydrides, they are very reactive towards nucleophilic reagents. They are used in the preparation of polyamino acids and in peptide synthesis (see p. 130).

2,5-Dioxopiperazines (see p. 59) may be formed by intermolecular ester condensation and are sometimes obtained as by-products in peptide synthesis. Naturally occurring examples of this group have been discovered among the metabolic products of several microorganisms (cp. p. 221).

1.10. Phosphoryl- and Phosphatidylamino Acids

These are components of the biochemically important phosphopeptides and phosphoproteins. The naturally occurring ones are mostly derived from the hydroxyamino acids (serine, threonine, hydroxyproline), phosphorus being linked by a covalent P—O bond.

O-Phosphorylamino acids may be prepared by treating amino acids with H_3PO_4/P_2O_5, but better yields and purity are obtained by reaction of N- and C-protected hydroxyamino acids with $POCl_3$ in pyridine or with an alkyl- or arylphosphorylchloride. Reaction of diphenylphosphoryl chloride with a benzyloxycarbonyl amino acid benzyl ester followed by deprotection is the most satisfactory method:

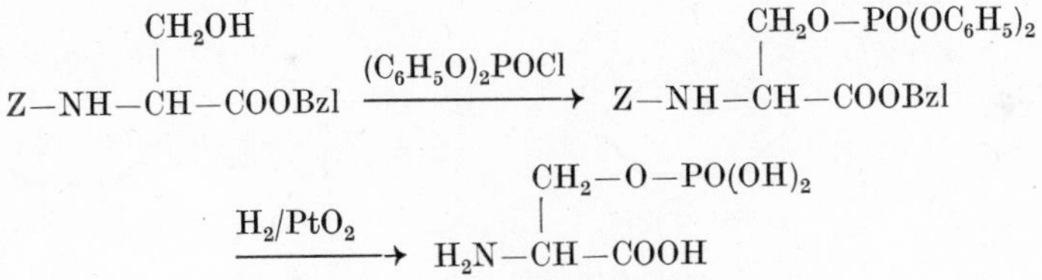

$$\text{Z-NH-}\underset{}{\overset{\displaystyle CH_2OH}{\overset{|}{\text{CH}}}}\text{-COOBzl} \xrightarrow{(C_6H_5O)_2POCl} \text{Z-NH-}\overset{\displaystyle CH_2O-PO(OC_6H_5)_2}{\overset{|}{\text{CH}}}\text{-COOBzl}$$

$$\xrightarrow{H_2/PtO_2} H_2N\text{-}\overset{\displaystyle CH_2-O-PO(OH)_2}{\overset{|}{\text{CH}}}\text{-COOH}$$

phosphorylserine

O-(Palmitoyl-oleyl-glyceryl-phosphoryl)-L-threonine is a naturally occurring phosphatidylamino acid which was isolated from a species of tunny (*Thynnus orinitalis*):

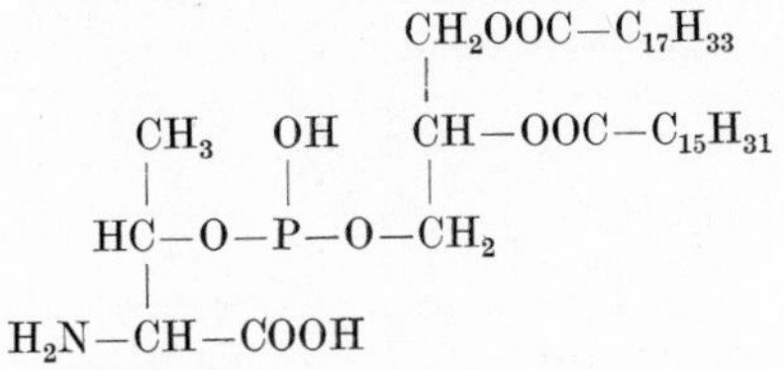

$$\begin{array}{l} \qquad\qquad\qquad\qquad CH_2OOC-C_{17}H_{33} \\ \qquad\qquad\qquad\qquad | \\ \quad CH_3 \quad\; OH \quad\; CH-OOC-C_{15}H_{31} \\ \quad | \qquad\quad | \qquad\;\; | \\ HC-O-P-O-CH_2 \\ \quad | \\ H_2N-CH-COOH \end{array}$$

The base-stable *N*-phosphorylamino acids which are formed by reaction of dialkyl- or diarylphosphorylchlorides with amino acid derivatives with a free amino group are of relatively little importance.

1.11. Amino Acids Linked to Nucleosides and Nucleotides

There are several modes of linking amino acid residues to nucleosides and nucleotides. In the most important type the amino acid is linked via an ester bond to the 2′ or 3′ hydroxyl group of ribose (I). In a second type the carboxyl group of the amino acid forms an amide bond to an amino group of the heterocyclic base (II). In nucleotides the phosphate residue offers additional points of attachment to the amino and carboxyl groups of amino acids (IIIa and IIIb).

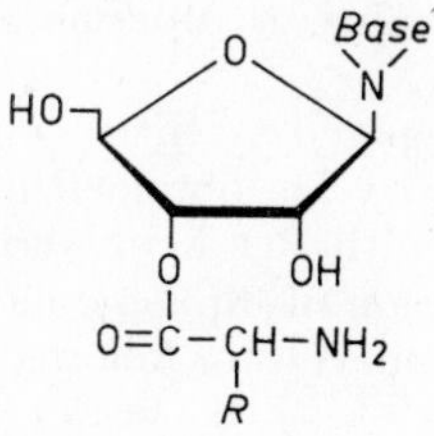

I 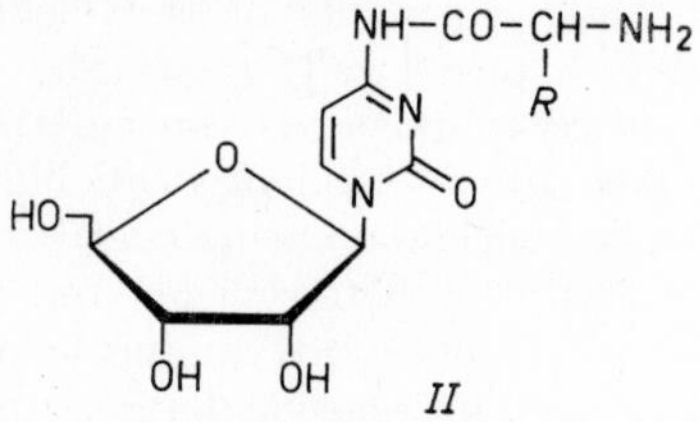 *II*

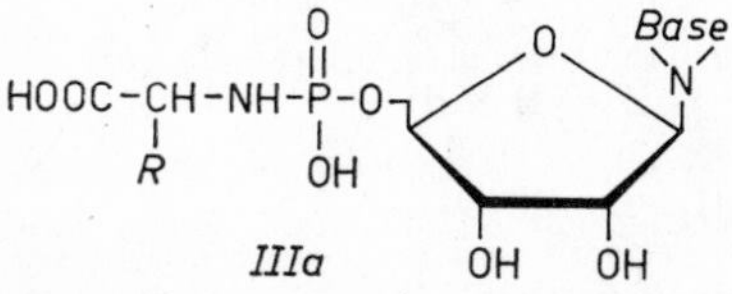

IIIa 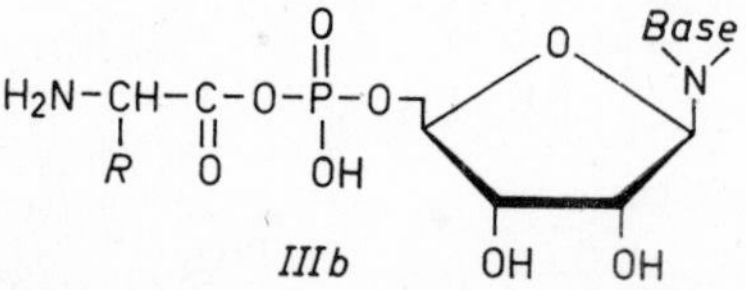 *IIIb*

The type of linkage shown in type I is that present in tRNA—the critical intermediate of protein biosynthesis. There is a structural similarity between nucleoside amino acids and the antibiotic puromycin which inhibits protein synthesis:

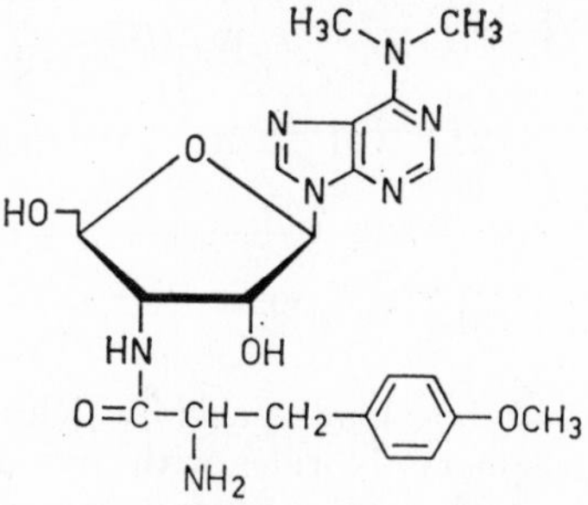

References

1 WIELAND, TH. *et al.* (1958). Methoden zur Herstellung und Umwandlung von Aminosäuren und Derivaten, *Houben-Weyl*, Vol. **11**/2, 279—509

2 GREENSTEIN, J. P. and WINITZ, M. (1960). *Chemistry of the Amino Acids*, Wiley, New York

3 TURBA, F. (1955). Aminosäuren und Peptide, Hoppe/Seyler/Thierfelder, *Handbuch der Physiologisch- und Pathologisch-Chemischen Analyse*, Vol. III/2, 1648, Springer-Verlag, Berlin-Göttingen-Heidelberg
4 BLOCK, R. J. (ed.) (1956). *Amino Acid Handbook*, C. C. Thomas, Springfield, Illinois
5 MEISTER, A. (1957). *Biochemistry of the Amino Acids*, Academic Press, New York
6 MUSSO, H. (1956). *Angew. Chem.*, **68**, 313
7 FOWDEN, I. (1962). *Endeavour*, **21**, 35
8 TSCHIERSCH, B. (1962). *Pharmazie*, **17**, 721
9 KÜHNAU, J. (1949). *Angew. Chem.*, **61**, 357
10 SCHUPHAN, W. and SCHWERDTFEGER, E. (1965). *Die Nahrung*, **9**, 755
11 SAFONOVA, E. N. and BELIKOW, V. M. (1967). *Uspekhi Khim.*, **36**, 913
12 ROSE, W. C. *et al.* (1948). *J. Biol. Chem.*, **176**, 753; (1951). **188**, 49; (1954). **206**, 421; (1955). **216**, 763 and **217**, 987
13 CHAMPAGNAT *et al.* (1963). *Nature*, **197**, 13
14 IUPAC–IUB Commission on Biochemical Nomenclature (1967). *Hoppe-Seyler's Z. Physiol. Chem.*, **348**, 245; (1972). *J. Biol. Chem.*, **247**, 977
15 IUPAC–IUB Commission on Biochemical Nomenclature (1968). *Biochem. Biophys. Acta*, **168**, 6
16 REVESZ, G. S. (1968). *Nature*, **219**, 1113
17 BRAND, E. and EDSAL, J. T. (1947). *Ann. Rev. Biochem.*, **16**, 224
18 WELLNER, D. and MEISTER, A. (1966). *Science*, **151**, 77
19 NEUBERGER, A. (1948). *Advan. Protein Chem.*, **4**, 297
20 GREENSTEIN, J. P. and WINITZ, M. (1960). *Chemistry of the Amino Acids*, Vol. 1, p. 46, Wiley, New York
21 TADDEL, F. and PRATT, L. (1964). *J. Chem. Soc. (London)*, 1553
22 ABRAHAM, R. J. and THOMAS, U. A. (1964). *J. Chem. Soc. (London)*, 3739
23 MARTIN, R. B. and MATHUR, R. (1965). *J. Amer. Chem. Soc.*, **87**, 1065
24 CAVANAUGH, J. R. (1967). *J. Amer. Chem. Soc.*, **89**, 1558
25 ROBERTS, G. C. K. and JARDETZKY, O. (1970). *Advan. Protein Chem.*, **24**, **460**
26 DITTNER, A. (1961). *Papierelektrophorese*, 2nd edn, VEB Gustav Fischer Verlag, Jena
27 BEAVEN, G. H. and HOLIDAY, E. R. (1952). *Advan. Protein Chem.*, **7**, 319
28 WETLAUFER, D. B. (1962). *Advan. Protein Chem.*, **17**, 303
29 SUTHERLAND, G. B. B. M. (1952). *Advan. Protein Chem.*, **7**, 291
30 reference 3, pp. 1685–1698
31 TAKDEA, M. and JARDETZKY, O. (1957). *J. Chem. Phys.*, **26**, 1346
32 WIELAND, TH. (1949). *Fortschr. Chem. Forschung*, **1**, 211
33 HILL, R. L. (1965). *Advan. Protein Chem.*, **20**, 37
34 HILL, R. L. and SCHMIDT, W. R. (1962). *J. Biol. Chem.*, **237**, 389
35 PARTRIDGE, S. M. and BRIMLEY, R. C. (1952). *Biochem. J.*, **51**, 628
36 SCHÜTTE, H. R. (1966). *Radioaktive Isotope in der organischen Chemie und Biochemie*, Deutscher Verlag der Wissenschaften and Verlag Chemie GmbH, Weinheim/Bergstraße

37 Murray, A. and Williams, D. L. (1958). *Organic Synthesis with Isotopes*, Interscience, New York
38 Böhm, R. and Losse, G. (1967). *Z. Chem.*, **7**, 409
39 Greenstein, J. P. (1954). *Advan. Protein Chem.*, **9**, 121
40 Losse, G. and Jeschkeit, H. (1960). *Pharmazie*, **15**, 164
41 Harada, K. (1965). *Nature*, **205**, 590
42 Harada, K. (1965). *Nature*, **206**, 1354
43 Losse, G. and Kuntze, K. (1970). *Z. Chem.*, **10**, 22
44 Rogozhin, S. V. and Davankov, V. A. (1970). *Dokl. Akad. Nauk SSSR*, **192**, 1288; (1972). *Ideen des exakten Wissens*, p. 319, DVA Stuttgart
45 Oro, J., Nakaparksin, S. and Lichtenstein, H. (1971). *Nature* **230**, 105
46 Hrapia, H. (1965). *Einführung in die Chromatographie*, Akademie-Verlag, Berlin
47 Matthias, W. and Wagner, J. (1965). *Z. Chem.*, **5**, 171; (1967). *J. Chromatogr.*, **26**, 41
48 Ansorge, S. (1968). *Pharmazie*, **23**, 16
49 Harmeyer, J. and Sallmann, H. P. (1967). *Hoppe-Seyler's Z. Physiol. Chem.*, **348**, 1215
50 Hais, I. M. and Macek, K. (1958). *Handbuch der Papierchromatographie*, Vol. 1—4, VEB Gustav Fischer Verlag, Jena
51 Grütte, F. K. and Kohnke, B. (1967). *J. Chromatogr.*, **26**, 325
52 Stahl, E. (1962). *Dünnschichtchromatographie*, Springer-Verlag
53 Randerath, K. (1962). *Dünnschichtchromatographie*, Verlag Chemie GmbH, Weinheim/Bergstraße
54 Pataki, G. (1969). *Techniques of Thin-Layer Chromatography in Amino Acid and Peptide Chemistry*, Humphrey Science Publishers, Ann Arbor, Michigan
55 Fahmy, A. R., Niederwieser, A., Pataki, G. and Brenner, M. (1961). *Helv. Chim. Acta*, **44**, 2022
56 Spackmann, D. H., Stein, W. H. and Moore, S. (1958). *Anal. Chem.*, **30**, 1190
57 Kirsten, E. and Kirsten, R. (1964). *Biochem. Z.*, **339**, 287
58 Braunitzer, G. (1960). *Angew. Chem.*, **72**, 485
59 Landowne, R. A. and Lipsky, S. R. (1963). *Nature*, **199**, 141
60 Lamkin, W. M. and Gehrke, C. W. (1965). *Anal. Chem.*, **37**, 383
61 Pollok, G. E. and Oyama, V. C. (1966). *J. Gas Chromatogr.*, **4**, 126
62 Heyns, K. and Grützmacher, H. F. (1966). Massenspektrometrische Analyse von Aminosäuren und Peptiden, *Fortschr. Chem. Forschung*, **6**, 536
63 Foster, G. L. (1945). *J. Biol. Chem.*, **159**, 431
64 Shemin, D. (1945). *J. Biol. Chem.*, **159**, 439
65 Keston, A. S., Udenfriend, S. and Cannan, R. K. (1949). *J. Amer. Chem. Soc.*, **71**, 249
66 Snell, E. S. (1946). *Advan. Protein Chem.*, **2**, 85
67 Freeman, H. C. (1967). *Advan. Protein Chem.*, **22**, 258
68 Ackermann, E. (1967). *Pharmazie*, **22**, 537
69 Coldin, D. J. M. (1960). *Chem. Rev.* (*Baltimore*), **60**, 39

2. Peptides

2.1. General Properties of Peptides

Peptides are compounds consisting of two or more amino acids which are linked together by amide functions. The term 'peptide' was proposed by EMIL FISCHER. On the one hand, the term is reminiscent of the nomenclature of carbohydrates (mono-, di-, trisaccharides, etc.) and, on the other hand, it calls to mind the relation to 'peptones', which are products of protein degradation by pepsin. Peptides fall between the high-molecular-weight proteins and the amino acids both in size and with respect to their

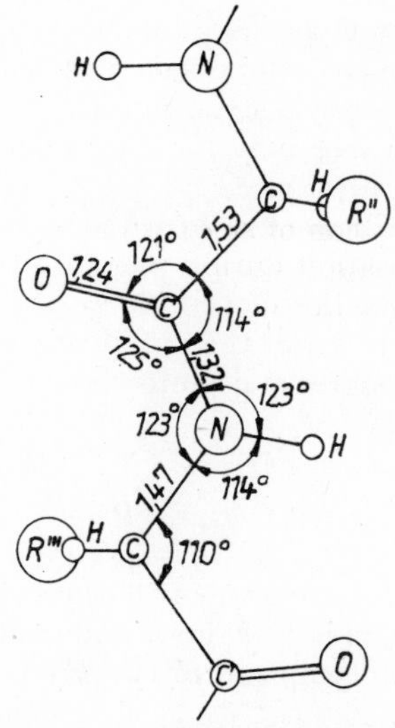

Figure 14. Atomic distances (Å) of an extended peptide chain, determined by X-ray structure analysis[1]

physicochemical properties. Naturally, the general behaviour of peptides depends on the nature of their constituent amino acids. The solubility behaviour depends, among other things, on the number of amino acids with hydrophobic side chains. The acid–base behaviour is dependent upon the total number of acidic and basic groups present in the molecule. Modern techniques have revealed that peptides are widely distributed in nature.

Particular interest has been excited by the isolation, structure elucidation and synthesis of numerous polypeptide hormones as well as by the discovery and use of polypeptide antibiotics produced by microorganisms.

The separation and isolation of peptides, which often differ only slightly in molecular weight and solubility, is accomplished by one of the many chromatographic methods now available, by countercurrent distribution (CRAIG), or by preparative electrophoresis, and increasingly by gel filtration methods.

Characterisation of larger peptides is achieved by amino acid analysis of chromatographically and electrophoretically uniform material. The amino acid sequence is investigated by enzymatic degradation.

More than half a century of intensive research has passed from the first synthesis of a peptide by E. FISCHER and E. FOURNEAU (glycylglycine, 1901) to the automated polypeptide synthesis of today. Many methods for the rational synthesis of peptides have been developed. The most important coupling procedures, as well as the common methods for the protection of amino, carboxyl and side chain functions, will be dealt with in detail. Likewise, the position of peptide synthesis on polymeric supports, a methodology which started a new era (not only for the formation of peptides and proteins), will be assessed critically.

After considering special matters such as the synthesis of cyclic peptides and questions of racemisation during peptide formation, a section is devoted to 'Strategy and tactics of peptide synthesis', in which it is shown that it is often difficult to select the optimum combination from the great variety of coupling procedures and protective methods available.

2.1.1. Structural Principles

As already stated, peptides are compounds consisting of amino acids which are linked together by amide functions. A dipeptide is built up from two amino acids. The peptide composed of two glycine units is represented by the formula

$$H_2N-CH_2 - \boxed{CO-NH} - CH_2-COOH$$

The acid amide grouping —CO—NH— is termed a *peptide bond*. Crystallographic studies show that the C—N bond length in the peptide bond is considerably shorter than a common C—N single bond. This is due to the mesomeric nature of the peptide bond. This mesomerism has the effect of abolishing the free rotation about the C—N bond and results in the existence of two stable conformations, *cis* and *trans*, in which there is a planar distribution of the atoms:

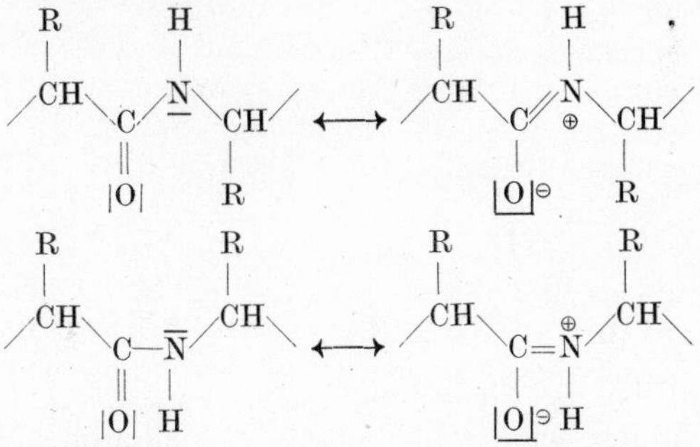

```
  R      H                   R      H
  |      |                   |      |
 /CH\   _N\                 /CH\    N\
     C/ −  \CH/   ←→             C//  ⊕ \CH/      trans-form
     ||     |                    |       |
    |O|     R                   |O|⊖     R

  R          R               R          R
  |          |               |          |
 /CH\       /CH\            /CH\   ⊕   /CH\
     C—N̄/          ←→           C=N/          cis-form
     ||  |                       |  |
    |O|  H                      |O|⊖ H
```

In 2,5-dioxopiperazines, which are the simplest cyclic peptides, being composed of two amino acids, *cis*-peptide bonds are necessarily present. Generally, in native peptides and proteins the *trans*-form occurs.

A second important covalent link is the *disulphide bond,* which is formed by oxidation of the functional SH groups of cysteine residues:

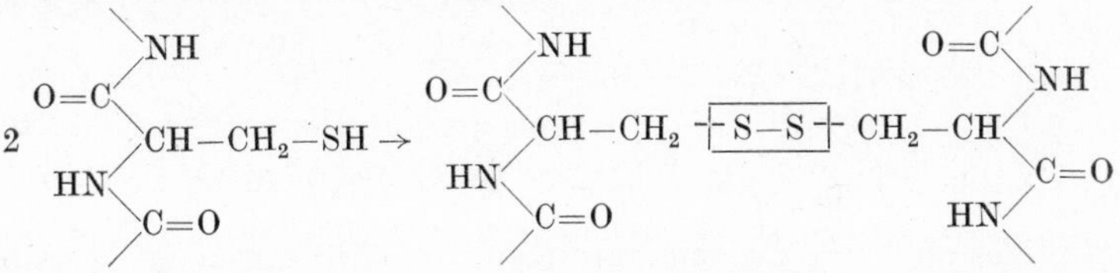

```
      \                      \                                     /
       NH                     NH                              O=C
  O=C/                   O=C/                                     \NH
2      \CH—CH₂—SH →          \CH—CH₂ ┤S—S├ CH₂—CH/
  HN/                    HN/                                      \C=O
     \C=O                   \C=O                              HN/
     /                      /                                    \
```

A distinction is made between *intrachain disulphide bonds,* which occur within a peptide chain, and *interchain disulphide bonds,* which occur between different peptide chains.

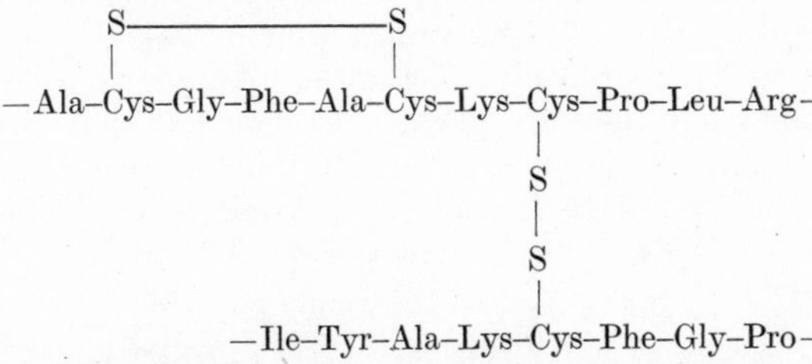

```
          S——————————————S
          |              |
—Ala–Cys–Gly–Phe–Ala–Cys–Lys–Cys–Pro–Leu–Arg—
                              |
                              S
                              |
                              S
                              |
              —Ile–Tyr–Ala–Lys–Cys–Phe–Gly–Pro—
```

In oxytocin, vasopressin, in the A-chain of insulin (see p. 200) or in ribonuclease, etc., we find examples of intrachain S—S linkages. These disulphide bonds can bring parts of the molecule which are apparently remote from each other in terms of sequence into close proximity, as in ribonuclease (see p. 299). In the oxidised form of glutathione (I), two identical peptide chains are joined by covalent disulphide bonds. In insulin two different chains are linked in the same way.

```
HOOC–CH–CH2–CH2–CO–NH–CH–CO–NH–CH2–COOH
     |                 |
     NH2               CH2
                       |
                       S
                       |
                       S                          (I)
                       |
     NH2               CH2
     |                 |
HOOC–CH–CH2–CH2–CO–NH–CH–CO–NH–CH2–COOH
```

Besides these covalent bonds, the acidic, basic, aromatic and aliphatic side chains of other amino acids contribute to the adoption of definite conformations by interacting with one another or with the *backbone* of the peptide chain. These additional forms of stabilisation usually become fully effective only at the protein level, but they are also significant for maintaining the effective conformations of several biologically active polypeptides. Moreover, short peptide chains and even cyclic peptides in certain solvents form definite structures. The nature of the solvent has a great influence on the strength of the non-covalent interactions.

Hydrogen bonds (H-*bonds*) are interactions between molecules which involve, on the one hand, O- and N-atoms and, on the other hand, NH and OH functions. The term 'hydrogen bonding' was suggested by LATIMER and RODEBUSCH in 1920.

In peptides and proteins there are many structural elements capable of forming hydrogen bonds and in both aqueous solution and colloid systems these interactions abound and are of both structural and biochemical significance.

According to recent studies in aqueous solution the hydrogen bonds from hydrogen atoms on nitrogen to oxygen atoms, —N—H···O=C—, have a maximum binding energy of only 1.5 kcal/mol. In apolar solvents they can have higher values but they are nevertheless weaker than previously supposed (6–8 kcal/mol). A distance of *ca.* 2.8 Å between the atoms involved is presumed necessary for the formation of H-bonds, which is optimal only when the atoms —N—H···O=C— are in a straight line:

```
 |O|      H      R
 ||       |      |
 C        N      CH
/  \    /   \   /   \   N/
    CH        C        |
    |         ||       H
    R        |O|       :
              :       |O|
    R         H        ||
    |         |        C
\   CH        N       /  \
 N/    \    /   \   CH
 |        C          |
 H        ||         R
         |O|
```

Besides the H-bonds involving the structural elements of the peptide backbone, there are further hydrogen bonds to which the hydroxyl groups of tyrosine, the carboxyl groups and the imidazole-nitrogen of histidine, etc., can contribute.

For the stabilisation of conformation the *hydrophobic* (*apolar*) *bond* is important. When two apolar groups approach, there is a small attractive force between them arising from distortion of the electron cloud. This effect is termed VAN DER WAALS or dispersion forces. The attractions between the methyl groups of alanine (a), aliphatic branched side chains (b), aromatic residues (c), as well as hydrophobic side chains and the backbone of peptide chain (d) are based on this phenomenon. In aqueous solution the hydrophobic bond arises less from VAN DER WAALS forces than from a disruption of the ordered state of the water molecules in the environment

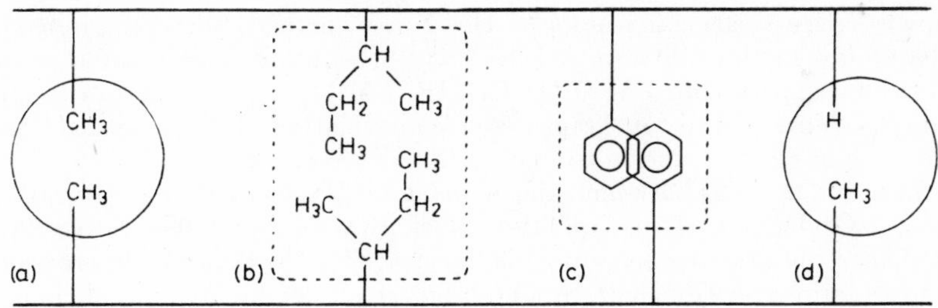

of apolar groups. In the neighbourhood of a hydrophobic group, clusters of water molecules bridged by hydrogen bonds are formed. This enhances the degree of order in the water and therefore increases the free energy in the environment of those groups. On the other hand, when two hydrophobic residues conglomerate, the hydrophobic area exposed to the water is minimised. This minimises the structuring of the water and, hence, the free energy is less than for the two separate groups. This is the principle of hydrophobic bonding.

Finally, the *ionic bond* has to be mentioned. This represents an additional stabilisation factor for peptide and protein conformations, and it arises from electrostatic interactions between positive and negative charges on terminal and side chain amino and carboxyl groups and the guanidine function of arginine.

In 1951, PAULING and COREY[1] deduced the basic structural principles of the spatial configuration of peptide chains (Figure 14) from crystallographic studies on amino acids and small peptides (see Section 3.6.2.1). In spite of these results it is very difficult to predict the stable conformation of a peptide under physiological conditions[2]. But because the native conformations of biologically active peptides determine the specific activity, the elucidation of the three-dimensional structure is an essential prelude to understanding the molecular mechanisms of activity. Such an investigation is, however, very laborious and extremely complicated. Therefore the state of knowledge about the mechanism of the interaction between biologically active peptides (hormones, antibiotics, etc.) and the related receptors is poor.

Although by X-ray analysis of single crystals of peptides and proteins (see also Sections 3.6.2 and 3.6.3) some sensational structural information has been gained, there are some limits to this approach. The growing of proper single crystals is frequently very difficult and has been successful only in a few cases so far. Above all, however, it has been doubted whether the spatial structure in single crystals corresponds to the preferred conformation under physiological conditions. For these reasons studies on peptide and protein conformations in solution should be an essential complement to X-ray investigations. Although the conformational analysis of peptide and protein systems in physiological environments has not yet shown such sensational results as the X-ray method, there are inherent advantages in the universal applicability of solution methods and possibilities for experiment. UV, ORD, CD, IR, NMR and ESR spectra can give valuable hints of the topography and conformation of peptides (and also proteins) in solution and in contact with receptors.

Since interactions between side chains doubtless contribute substantially to the stabilisation of conformation, R. Schwyzer and collaborators have established interesting spectroscopic methods for the study of interactions between aromatic side chain residues and an artificially introduced charge-transfer-[3,4] or a proton-nuclear-resonance-labelling group[5,6], respectively. Better results were obtained by fluorescence labelling[7]. By measuring the energy of intramolecular transfer between fluorescent side chain groups, separations in the range 10–100 Å can be measured, whereas in the case of the labelling method first mentioned, the signal limit is reached at 3–6 Å. Using the fluorescent dansyl group, Schwyzer and Schiller[8] have reported conformational studies of N-dansyl-lysine[21]-adrenocorticotropin-(1–24)-tetracosipeptide (see p. 192) on a potential receptor by means of the intramolecular resonance–energy transfer and fluorescence–depolarisation methods.

2.1.2. Nomenclature and Classification

According to the number of amino acid residues forming a peptide, di-, tri-, tetra-, penta-, ..., octa-, nona-, decapeptides, etc., are distinguished. For systematic chemical designation peptides are regarded as acylamino acids, and the ending *-yl* is given to that amino acid whose carboxyl group has reacted. Only the amino acid residue at the end of a peptide chain, bearing a free carboxyl group, keeps the original name—for example,

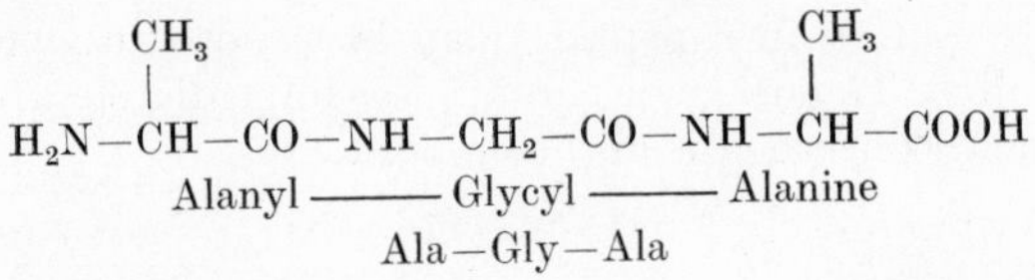

```
       CH3                              CH3
        |                                |
H2N—CH—CO—NH—CH2—CO—NH—CH—COOH
     Alanyl ———— Glycyl ———— Alanine
              Ala—Gly—Ala
```

According to the abbreviations of amino acids already illustrated the name of this tripeptide is reduced to Ala–Gly–Ala.

In writing formulae of linear peptides the amino acid bearing a free α-amino group is commonly written on the left. K. Bailey suggested designating such an amino acid *N-terminal*, whereas the amino acid at the other end of the peptide chain written horizontally is to be denoted as *C-terminal*. On the other hand, C. Fromageot recommended designating the residue bearing the free amino group as the initial residue and the corresponding residue with the free carboxyl group as the terminal residue. Though this proposal is the simpler one, Bailey's suggestion has prevailed.

The indication of the direction of linkage by an arrow (→), whose tail points to the nitrogen of the peptide bond, is not usually necessary for the representation of linear peptides.

The abbreviation Ala–Gly–Ala, for example, represents the tripeptide alanylglycylalanine, no matter what the state of ionisation. But if one wants to emphasise that a peptide is unsubstituted, according to the suggestions by J. P. Greenstein and M. Winitz the amino and the carboxyl grouping may be amplified with H and OH, respectively (I). For the ionised forms the related additionally modified symbols result (II or III):

H–Ala–Gly–Ala–OH (I) $H_2^{\oplus}$–Ala–Gly–Ala–OH (II) H–Ala–Gly–Ala–$O^{\ominus}$ (III)

If not indicated otherwise, polyfunctional amino acids such as aspartic acid, glutamic acid, lysine, ornithine, etc., are assumed to be linked by α-peptide bonds. The possibilities for symbolising α- and ω-peptide bonds are shown in the following examples:

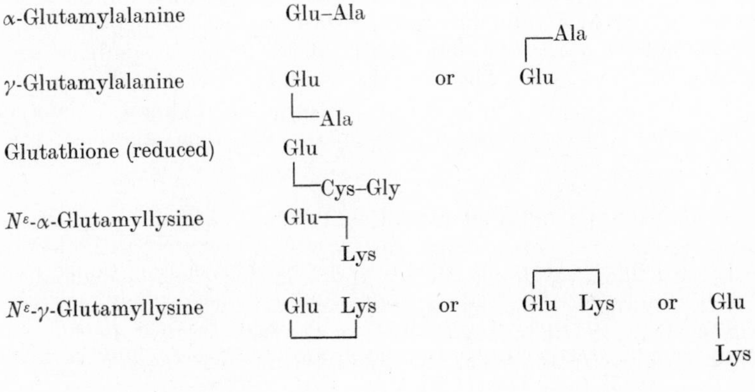

```
α-Glutamylalanine        Glu–Ala
                                          ┌—Ala
γ-Glutamylalanine        Glu        or    Glu
                         └—Ala
Glutathione (reduced)    Glu
                         └—Cys–Gly
Nε-α-Glutamyllysine      Glu—┐
                             Lys
                                          ┌————┐
Nε-γ-Glutamyllysine      Glu  Lys   or    Glu  Lys   or   Glu
                         └————┘                            |
                                                           Lys
```

Side chain substitution in a peptide may be denoted above or below the amino acid symbol. The usage of parentheses immediately after the symbol of the amino acid concerned is also common:

```
                                      Bzl
                                       |
Glycyl-S-benzyl-cysteinylalanine  Gly–Cys–Ala        or

                                  Gly–Cys(Bzl)–Ala
```

The following possibilities of abbreviation result for the substituted peptide sequence alanylaspartyl-(*β-tert*-butylester)-glycyl-*O-tert*-butyl-tyrosyl-phenylalanyl-N^{ε}-*tert*-butyloxycarbonyl-lysylalanine

```
                                        CH3                               O     CH3
                                        |                                 ||    |
           CH3                     CH3-C-CH3                       H-N-C-O-C-CH3
           |                            |                            |       |
      CH3-C-CH3                         O                            CH2     CH3
           |                            |                            |
           O                      (benzene ring)    (benzene ring)   CH2
           |                            |                 |          |
           C=O                          |                 |          CH2
           |                            |                 |          |
  CH3      CH2                          CH2               CH2        CH2        CH3
  |        |                            |                 |          |          |
H2N-CH-CO-NH-CH-CO-NH-CH2-CO-NH-CH-CO-NH-CH-CO-NH-CH-CO-NH-CH
                                                                                |
                                                                                COOH

      OBut    But     Boc
      |       |       |
Ala-Asp-Gly-Tyr-Phe-Lys-Ala          Ala-Asp-Gly-Tyr-Phe-Lys-Ala
                                         |       |       |
                                         OBut    But     Boc

              Ala-Asp-(OBut)-Gly-Tyr(But)-Phe-Lys(Boc)-Ala
```

The number and sequence of amino acids joined in a peptide is called its *primary structure*. If the sequence of a peptide is completely known, the symbols of the residues are written one after another and connected to each other by hyphens. Finally one distinguishes between the peptide itself—for example Ala–Gly–Ala (without dashes at the ends of symbols) and the sequence—for example, –Ala–Gly–Ala– (with additional dashes at the ends of symbols). If partial sequences of a peptide are unknown, the symbols of the amino acids concerned are separated by commas and placed in parentheses:

Gly–Glu–Ala–Ser–Phe–(Tyr, Phe, Pro, Arg)–Lys–Pro–Gly–Ala

For cyclic peptides there are several modes of abbreviation. Concerning the constitution, *homodetic cyclic peptides* and *heterodetic cyclic peptides* are distinguished. The former consist of amino acid residues joined only by peptide bonds, whereas in the second group other bonds may be involved.

The sequence of a homodetic cyclic peptide may be formulated as follows. On the one hand, the sequence is placed in parentheses and indicated by a preceding italic cyclo:

cyclo-(–Val–Orn–Leu–D-Phe–Pro–Val–Orn–Leu–D-Phe–Pro–)

On the other hand, the sequence may be written on one line and joined by a long stroke, as shown on the example of gramicidin S:

└Val–Orn–Leu–D-Phe–Pro–Val–Orn–Leu–D-Phe–Pro┘ (joined by a stroke above)

or

┌Val–Orn–Leu–D-Phe–Pro–Val–Orn–Leu–D-Phe–Pro┐ (joined by a stroke below)

If the residues are written on more than one line, the CO—NH direction must be indicated by arrows:

$$\begin{array}{l} \rightarrow \text{Val} \rightarrow \text{Orn} \rightarrow \text{Leu} \rightarrow \text{D-Phe} \rightarrow \text{Pro} \\ \text{Pro} \leftarrow \text{D-Phe} \leftarrow \text{Leu} \leftarrow \text{Orn} \leftarrow \text{Val} \leftarrow \end{array}$$

Heterodetic cyclic peptides are denoted by analogy with the substituted amino acids. For example, in oxytocin:

$$\overline{\text{Cys–Tyr–Ile–Gln–Asn–Cys}}\text{–Pro–Leu–Gly–NH}_2$$

The number of ways of combining a given number of amino acids to form oligo- and polypeptides and proteins is immense. Two amino acids—for example, proline and alanine—can give only two dipeptide structures:

Pro–Ala and Ala–Pro

Adding a third to these two amino acids—for example, glycine—gives six combinations:

Pro–Gly–Ala Ala–Pro–Gly Gly–Pro–Ala
Pro–Ala–Gly Ala–Gly–Pro Gly–Ala–Pro

The number of structural combinations (P_n) is expressed by the formula:

$$P_n = n! = 1 \times 2 \times 3 \times 4 \cdots \times n$$

Thus 720 different linear peptide structures can result from 6 different amino acids. The biologically active peptide [8-arginine]-vasopressin is composed of nine amino acids, of which two are identical:

$$\overline{\text{Cys–Tyr–Phe–Gln–Asn–Cys}}\text{–Pro–Arg–Gly–NH}_2$$

Thus the general formula for the possible number of sequences with this composition is:

$$P = \frac{n!}{x!\,y!\,z!}$$

where n indicates the total number of amino acids, and x, y, z represent the numbers of the amino acid residues that occur more than once. According to this calculation 181 440 different sequences are possible for [8-arginine]-vasopressin:

$$P = \frac{9!}{2!} = \frac{362\,880}{2} = 181\,440$$

In principle, all these analogues could be synthesised. However, the preparation of all these compounds for elucidation of the relationship between constitution and biological activity would involve an enormous preparative effort. However, in recent decades a great number of biologically important peptide analogues have been synthesised with systematic modifications of the amino acid sequence. Some of these exhibit a higher biological activity than the related natural products. Therefore it is necessary to explain the rules for the designation of synthetic analogues, suggested by the IUPAC–IUB Commission on Biochemical Nomenclature[9].

When a residue in a natural polypeptide is replaced, the name of the replacing amino acid and the position of exchange are placed in brackets before the trivial name of the polypeptide concerned—for example, [8-arginine]-vasopressin. The abbreviated designation to be used only for tables is [Arg^8]vasopressin. In the case of multiple replacement the nomenclature is analogous—for example, [5-isoleucine, 7-alanine]angiotensin II = [Ile^5, Ala^7]angiotensin II.

Elongation of a polypeptide can be accomplished from either the N-terminal end or the C-terminal end; in this case the general rules of nomenclature already described are used. The elongation of the amino end of bradykinin by valine results in the semi-trivial name valyl-bradykinin (Val-bradykinin), whereas the elongation on the carboxyl end by the same amino acid leads to bradykinyl valine, abbr. bradykinyl Val.

Insertion of an additional amino acid residue may be most conveniently demonstrated by an example. The incorporation of alanine into the position between the third and fourth amino acid residue of bradykinin gives the term *endo*-3a-alanine bradykinin or *endo*-Ala^{3a}-bradykinin, respectively. Multiple insertions are indicated by extending this rule appropriately.

A *lack of amino acid residues* is marked by the prefix *des* with indication of the position. The formal removal of proline from the oxytocin in the position 7, for example, leads to the designation *des*-7-proline oxytocin or *des*-Pro^7-oxytocin.

Side chain substitutions on amino and carboxyl groups in a polypeptide yield analogues whose nomenclature can be related to the trivial name using the common nomenclature rules. If, for example, eledoisin is substituted on the amino group of the side chain by valine, those symbols which indicate the position of substitution (number of residue and atom) are placed before the name valyl eledoisin: $N^{\varepsilon 4}$-valyl eledoisin. In a similar way one denotes compounds with side chain substitution on the carboxyl group. $C^{\gamma 5}$-adrenocorticotropinyl valine means that a valine residue is acylated by the γ-carboxyl group of the glutamic acid in position 5 of the adrenocorticotropin.

Partial sequences derived from peptides with well-known trivial names can be named by writing the positions in the named peptide of the first and last amino acid that make up the partial sequence together with the size (that is, di-, tri-, tetra-, etc.) of the peptide being described. Thus

$$\lfloor\!\!\!\!\square\text{Glu–Pro–Ser–Lys–Asp–Ala–Phe–Ile–Gly–Leu–Met–NH}_2$$

1 2 3 4 5 6 7 8 9 10 11

the name: eledoisin-(6–11)-hexapeptide amide is given to the partial sequence Ala–Phe–Ile–Gly–Leu–Met–NH_2 of eledoisin. The symbol 1 means the pyroglutamic acid residue (usually also indicated by $<$ Glu– or Pyr–).

The above rules refer only to analogues of natural peptides, involving normal α-peptide bonds.

Peptides containing less than about 10 amino acids are termed *oligopeptides*, whereas those with more amino acids are *polypeptides*. From the chemical point of view differentiation between *peptides* and *proteins* is purely arbitrary, since the methods that have been developed over recent years in the field of peptide chemistry are directly applicable to the synthesis of proteins. Only didactic considerations justify the old distinction between peptides and proteins as compounds with molecular weights of below or above 10 000, respectively (about 100 amino acids), based on their ability to be separated by dialysis on natural membranes. On the other hand, the characteristic physicochemical properties which distinguish the proteins cannot be ignored.

Peptides may be classified on a structural basis as well as according to the number of amino acid residues they contain. Thus peptides consisting exclusively of amino acid residues are named homomeric, whereas those which contain additional units not found in proteins, such as hydroxy acids, are termed heteromeric. According to the kinds of bonds between the units one distinguishes between homodetic peptides, which contain only peptide bonds, and heterodetic peptides, in which additional bonds—for example, ester, disulphide or thioether bonds—occur.

In Figure 15, which gives a classification of homomeric peptides, the amino acid units are represented by the symbol AA in a circle. Only at points of conjunction of sections of the peptide molecule are individual residue symbols used.

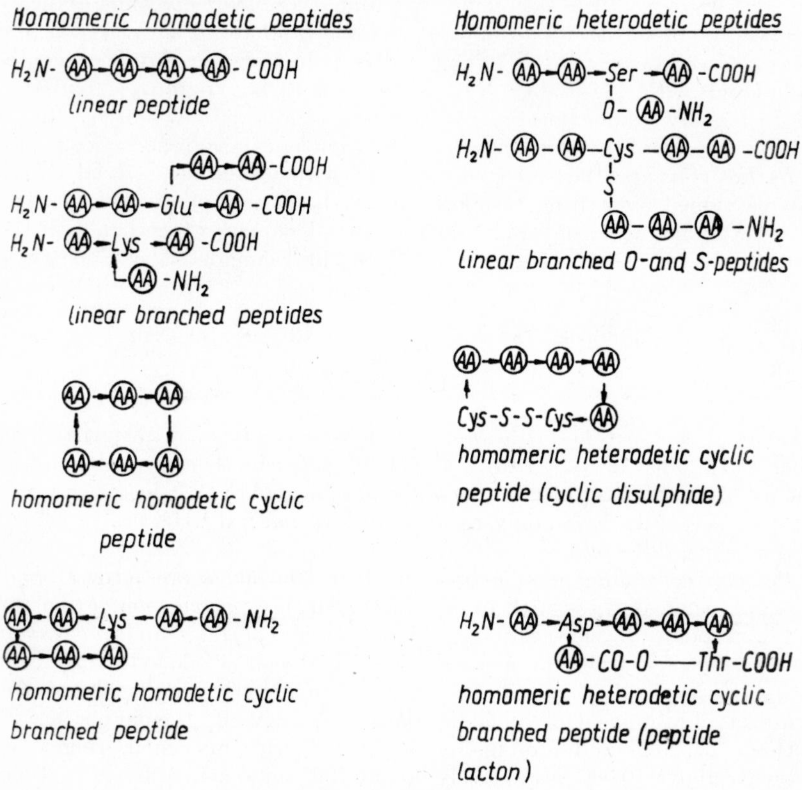

Homomeric homodetic peptides

H_2N-(AA)-(AA)-(AA)-(AA)-COOH

linear peptide

(AA)-(AA)-COOH
H_2N-(AA)-(AA)-Glu-(AA)-COOH

H_2N-(AA)-Lys-(AA)-COOH
(AA)-NH_2

linear branched peptides

(AA)-(AA)-(AA)
(AA)-(AA)-(AA)

homomeric homodetic cyclic peptide

(AA)-(AA)-Lys-(AA)-(AA)-NH_2
(AA)-(AA)-(AA)

homomeric homodetic cyclic branched peptide

Homomeric heterodetic peptides

H_2N-(AA)-(AA)-Ser-(AA)-COOH
O-(AA)-NH_2

H_2N-(AA)-(AA)-Cys-(AA)-(AA)-COOH
S
(AA)-(AA)-(AA)-NH_2

linear branched O-and S-peptides

(AA)-(AA)-(AA)-(AA)
Cys-S-S-Cys-(AA)

homomeric heterodetic cyclic peptide (cyclic disulphide)

H_2N-(AA)-Asp-(AA)-(AA)-(AA)
(AA)-CO-O——Thr-COOH

homomeric heterodetic cyclic branched peptide (peptide lacton)

Figure 15. Structure of homomeric peptides (products of hydrolysis are amino acids exclusively)

According to SHEMYAKIN, *depsipeptides* are peptides which contain ester bonds in addition to peptide bonds. Therefore they can belong to the homomeric as well as the heteromeric series. Cyclic depsipeptides—*peptolides*—are of particular importance.

The term *peptoides* was suggested by SCHROEDER and LUEBKE. Important members of this class are the lipo-, glyko-, nucleo-, phospho- and chromopeptides, in which a hetero component is covalently linked to a peptide chain through terminal amino or carboxyl groups or side chain functions, respectively.

References

1 Pauling, L. and Corey, R. B. (1951). *Proc. Nat. Acad. Sci., U.S.A.*, **37**, 729
2 Ramachandran, G. N. and Sasikharan, V. (1968). *Advan. Protein Chem.*, **23**, 283
3 Schwyzer, R. *et al.* (1967). *Proc. 8th Europ. Peptide Symp.*, Noordwijk, p. 177, North-Holland, Amsterdam
4 Schwyzer, R. *et al.* (1968). *Helv. Chim. Acta*, **51**, 459
5 Schwyzer, R. and Ludescher, U. (1968). *Biochemistry*, **7**, 2514
6 Schwyzer, R. and Ludescher, U. (1968). *Biochemistry*, **7**, 2519
7 Stryer, L. (1968). *Science*, **162**, 526
8 Schwyzer, R. and Schiller, P. W. (1971). *Helv. Chim. Acta*, **54**, 897
9 IUPAC–IUB Commission on Biochemical Nomenclature: Regeln für die Benennung synthetischer Analoga natürlicher Peptide (1967). *Z. Physiol. Chem.*, **348**, 262; Definitive rules for naming synthetic modifications of natural peptides (1972). *Pure Appl. Chem.*, **31**, 647

2.2. Peptide Synthesis

2.2.1. General Fundamentals of Peptide Synthesis

The formation of the peptide bond may be formally represented as the acylation of the amino group of an amino acid by a second amino acid:

$$H_2N-\underset{\displaystyle R}{\overset{}{CH}}-COOH + H_2N-\underset{}{CH}(R)-COOH \rightarrow H_2N-CH(R)-CO-NH-CH(R)-COOH + H_2O$$

In principle, the peptide bond is formed by nucleophilic attack of the N-atom of the second component on the carboxyl C-atom of the first. However, the ability of amino acids to react together is much restricted for reasons that will be explained below.

Only by use of elevated temperatures can amino acids themselves be joined to form polypeptides. Such drastic conditions, of course, are not to be recommended and do not allow rational peptide synthesis. For this reason it is appropriate to start from correspondingly 'activated' amino acid derivatives.

Since any reacting amino acid includes two functional groups either of which can participate in peptide formation, it is not enough merely to activate the amino acid acting as acylating agent (*carboxyl component*), in order to obtain a definite dipeptide. The amino acid to be acylated (*amino component*) has to be released from the zwitterion state by a proper carboxyl-protecting group in order to recover the nucleophilic power necessary for the reaction.

But in the reaction

$$H_2N{-}\underset{\displaystyle R}{\overset{}{CH}}{-}COX + H_2N{-}\underset{\displaystyle R}{CH}{-}COOR' \xrightarrow{-HX} H_2N{-}\underset{\displaystyle R}{CH}{-}CO{-}NH{-}\underset{\displaystyle R}{CH}{-}COOR'$$

it is possible that, in addition to the carboxyl-protected dipeptide, the cyclic dipeptide (dioxopiperazine) derived by twofold condensation of the activated carboxyl component will be produced:

$$R\,CH\begin{cases}COX\\NH_2\end{cases} + \begin{matrix}H_2N\\XOC\end{matrix}\Big\rangle CH{-}R \xrightarrow{-2HX} R{-}HC\Big\langle\begin{matrix}CO{-}NH\\NH{-}CO\end{matrix}\Big\rangle CH{-}R$$

In addition, the formation of numerous linear and cyclic peptides is theoretically possible because of the unprotected amino function of the carboxyl component. Therefore all the amino and carboxyl groups which are not to react in the chosen coupling should be blocked selectively. The situation becomes even more complicated in syntheses involving trifunctional amino acids. For this reason the problem is dealt with separately in the next section.

The methods for formation of peptide bonds are formally divided into three groups. Besides methods based on the activation of the carboxyl or amino group, synthesis effected by so-called condensing agents holds a commanding position. However, there is little point in such a differentiation since all the condensing agents operate by carboxyl activation.

The isocyanate method recommended for peptide synthesis by S. GOLDSCHMIDT is often cited as prototype of 'amino activation'. However, according to mechanistic studies this reaction proceeds via the mixed anhydride of a carbamic acid. Thus, even in this case, a carboxyl activation takes place (see Section 2.2.3.3):

$$Y{-}NH{-}\overset{R}{\overset{|}{CH}}{-}COOH + O{=}C{=}N{-}\overset{R}{\overset{|}{CH}}{-}COOR' \rightarrow \begin{matrix}Y{-}NH{-}\overset{R}{\overset{|}{CH}}{-}C{\begin{matrix}{\nearrow}O\\{\searrow}O\end{matrix}}\\ R'OOC{-}\underset{R}{CH}{-}NH{-}C{\begin{matrix}{\nearrow}\\{\searrow}O\end{matrix}}\end{matrix}$$

$$\xrightarrow{-CO_2} Y{-}NH{-}\overset{R}{\overset{|}{CH}}{-}CO{-}NH{-}\overset{R}{\overset{|}{CH}}{-}COOR'$$

When a mixture of the amino and carboxyl components is treated with the condensing agent, ethoxyacetylene, an intermediate *O*-acylketene hemiketal is formed. This can either couple with the amino component to give the peptide derivative (A) or react with the second molecule of the carboxyl component to give a symmetrical anhydride (B). The latter is then cleaved by the amino component (C):

$$\text{Y-NH-}\underset{}{\overset{\text{R}}{\overset{|}{\text{C}}}}\text{H-COOH} + \text{HC}\equiv\text{C-OC}_2\text{H}_5$$

$$\downarrow$$

$$\text{Y-NH-}\overset{\text{R}}{\overset{|}{\text{C}}}\text{H-}\overset{\text{O}}{\overset{\|}{\text{C}}}\text{-O-}\underset{\text{CH}_2}{\underset{\|}{\text{C}}}\text{-OC}_2\text{H}_5$$

(A) $+ \text{H}_2\text{N-}\overset{\text{R}}{\overset{|}{\text{C}}}\text{H-COOR}'$ $\downarrow$

$$\text{Y-NH-}\overset{\text{R}}{\overset{|}{\text{C}}}\text{H-}\overset{\text{O}}{\overset{\|}{\text{C}}}\text{-NH-}\overset{\text{R}}{\overset{|}{\text{C}}}\text{H-COOR}' + \text{CH}_3\text{-COOC}_2\text{H}_5$$

(B) $+ \text{Y-NH-}\overset{\text{R}}{\overset{|}{\text{C}}}\text{H-COOH}$ $\downarrow$

$$\begin{matrix}\text{Y-NH-}\overset{\text{R}}{\overset{|}{\text{C}}}\text{H-}\overset{\text{O}}{\overset{\|}{\text{C}}} \\ \text{Y-NH-}\underset{\text{R}}{\underset{|}{\text{C}}}\text{H-}\underset{\text{O}}{\underset{\|}{\text{C}}}\end{matrix}\!\!>\text{O} + \text{CH}_3\text{-COOC}_2\text{H}_5$$

(C) $\downarrow + \text{H}_2\text{N-}\overset{\text{R}}{\overset{|}{\text{C}}}\text{H-COOR}'$

$$\text{Y-NH-}\overset{\text{R}}{\overset{|}{\text{C}}}\text{H-COOH} + \text{Y-NH-}\overset{\text{R}}{\overset{|}{\text{C}}}\text{H-CO-NH-}\overset{\text{R}}{\overset{|}{\text{C}}}\text{H-COOR}'$$

It is clear that the activation of the carboxyl component is dependent on the fact that an electron-attracting substituent (X) will diminish the electron density on the carboxyl C-atom and consequently enhance its susceptibility to nucleophilic attack:

$$\text{R}_1\text{-}\underset{\text{H}}{\underset{|}{\overset{\text{H}}{\overset{|}{\text{N}}}}}| + \underset{\text{R}_2}{\underset{|}{\overset{|\text{O}|}{\overset{\|}{\text{C}}}}}\text{-X} \rightleftharpoons \text{R}_1\text{-}\underset{\text{H}}{\underset{|}{\overset{\text{H}}{\overset{|\oplus}{\text{N}}}}}\text{-}\underset{\text{R}_2}{\underset{|}{\overset{\overline{|\text{O}|}^{\ominus}}{\overset{|}{\text{C}}}}}\text{-X} \rightleftharpoons \text{R}_1\text{-NH-CO-R}_2 + \text{HX}$$

(R_1 = residue of the amino component; R_2 = residue of the carboxyl component)

Naturally the rate of aminolysis also depends on the nucleophilic power of the attacking amino component. A real amino activation would be provided only by factors which increased the nucleophilicity of the attacking agent. Because of the +I-effect *tert*-butyl esters of amino acids might be expected to possess a slightly greater nucleophilic activity than the methyl, ethyl or benzyl esters, but steric influences also play an important role.

If bulky groups are present in the environment of the nucleophilic N-atom, the aminolysis rate decreases, and an analogous situation arises when the carboxyl carbon atom is not easily accessible to the attacking reagent. Secondary amines and amino acid derivatives with very bulky α-substituents therefore react more slowly.

Of great importance for each coupling reaction is the problem of racemisation. Almost all reactions of a functional group joined to an asymmetric centre carry a risk of racemisation. The formation of peptide bonds without accompanying racemisation is essential, especially for the synthesis of large peptide sequences, since the resultant mixtures of diastereoisomers cannot be separated by crystallisation. For this reason the problem of racemisation during peptide bond formation is discussed in detail on p. 163.

The first attempts to synthesise polypeptides were made by Th. CURTIUS in Heidelberg in 1881. His investigations on the self-condensation of free amino acid esters naturally led to products of indeterminate chain length, but his studies on hippuric acid yielded results of great methodological importance. With the benzoyl residue an N-protecting group was used in peptide synthesis for the first time: even though this acyl group cannot be removed without destroying the peptide bond, it was an important pointer to the rational synthesis of peptides. Of even more significance was CURTIUS' use of the azide method of coupling for peptide synthesis. Since the introduction of selectively removable N-protecting groups it has been applied with great success up to the present day because it allows coupling without racemisation.

Despite his ambitious plans CURTIUS failed to synthesise free peptides, but a little later E. FISCHER and his laboratory in Berlin were successful and the first synthesis of a dipeptide was described by him in 1901. By opening dioxopiperazine with strong mineral acid, glycylglycine was prepared. Later he found α-halo carboxylic acid chlorides to be suitable starting materials for peptide synthesis. They react smoothly with amino acid esters and yield dipeptides after saponification of the esters and subsequent ammonolysis. Omitting the ammonolysis step, α-halo acyl amino acids can be converted into their acid chlorides and then treated with amino acid esters and amino acids or peptides, respectively, in aqueous alkaline solution. The substitution of halogen by the amino group by means of ammonia is made only at the desired stage. As a further development the use of the acid chlorides of amino acids in the form of their hydrochlorides was established by FISCHER in 1905. These, as well as the acid chlorides of oligopeptides, can be treated with amino acid esters to give peptide esters.

Even the carbethoxy function recommended by FISCHER as an N-protecting group was not removable under mild conditions. In 1926 with the *p*-toluene sulphonyl residue a grouping was found which can be removed by hydrogen iodide in presence of phosphonium iodide. But it was the introduction of the benzyloxycarbonyl group (carbobenzoxy group) as a reversible N-protecting function by FISCHER's pupil M. BERGMANN with L. ZERVAS in 1932 which opened a new era in synthetic peptide chemistry, which reached its first climax with the total synthesis of the neurohypophyseal hormone oxytocin by V. DU VIGNEAUD. This result was recognised by the award of the Nobel prize to DU VIGNEAUD in 1955. In the course of the last two decades there has been rapid development, resulting in the continuing discovery of new protecting groups and efficient coupling methods and culminating in total syntheses of insulin and the S-protein of ribonuclease.

The synthesis of peptides on polymeric supports introduced by R. B. MERRIFIELD in 1962 set new standards with regard to simplicity of preparative operations, time of synthesis and possible automation. Although experience with *solid-phase peptide synthesis* over recent years has considerably diminished the optimism of several investigators, a notable advance in synthetic peptide chemistry has clearly been made, and the syntheses of many biologically active peptides and simple proteins have been achieved in this way.

Summaries of the most important synthetic procedures in the form of reviews, monographs[1–18] and the reports on the *European Peptide Symposia*[19–31] are to be found in the references below. In the following sections, therefore, only particularly new or significant references will be quoted.

References

1 FRUTON, J. S. (1949). *Advan. Protein Chem.*, **5**, **1**

2 WIELAND, TH. (1951). *Angew. Chem.*, **63**, 7

3 WIELAND, TH. (1954). *Angew. Chem.*, **66**, 507

4 GRASSMANN, W. and WUENSCH, E. (1956). *Fortschr. Chem. Org. Naturstoffe (Wien)*, **13**, 444

5 WIELAND, TH. and HEINKE, B. (1957). *Angew. Chem.*, **69**, 362

6 GOODMAN, M. and KENNER, G. W. (1957). *Advan. Protein Chem.*, **12**, 465

7 SCHWYZER, R. (1958). *Chimia*, **12**, 53

8 WIELAND, TH. (1959). *Angew. Chem.*, **71**, 417

9 GREENSTEIN, J. P. and WINITZ, M. (1961). *Chemistry of the Amino Acids*, Vol. 2, Wiley, New York

10 MEIENHOFER, J. (1962). *Chimia*, **16**, 385

11 RYDON, H. N. (1962). *Peptide Synthesis,* Lecture Series, No. 5, Royal Institute of Chemistry, London
12 RUDINGER, J. (1963). *Pure Appl. Chem.,* **7**, 335
13 BODANSZKY, M. and ONDETTI, M. A. (1966). *Peptide Synthesis,* Wiley, New York
14 LUEBKE, K. and SCHROEDER, E. (1966). *The Peptides,* Academic Press, New York
15 KOPPLE, K. D. (1966). *Peptides and Amino Acids,* Benjamin, New York
16 JONES, J. H. (1969—1972). *Peptide Synthesis,* in *Specialist Periodical Reports of Amino Acids, Peptides and Proteins,* Vol. 1–4, The Chemical Society, London
17 LAW, H. D. (1970). *The Organic Chemistry of Peptides,* Wiley, New York
18 WÜNSCH, E. (1974). Synthese von Peptiden, in *Houben-Weyl,* Vol. 15, *Methoden der organischen Chemie,* Müller, E. (ed.), Georg Thieme Verlag, Stuttgart
19 Prague (C.S.R.) 1958 (1959). *Coll. Czech. Chem. Comm.,* Special Issue, **24**, 1–160
20 Munich (F.G.R.) 1959 (1959). *Angew. Chem.,* **71**, 741–743
21 Basle (Switzerland) 1960 (1960). *Chimia,* **14**, 366–418
22 Moscow (U.S.S.R.) 1961 (1962). *Zh. Mendeleyevskovo Obshch.,* **7**, 353–486; (1962). *Coll. Czech. Chem. Comm.,* **27**, 2229–2262
23 Oxford (England) 1962: YOUNG, G. T. (ed.) (1963). *Peptides,* Pergamon Press, Oxford
24 Athens (Greece) 1963: ZERVAS, L. (ed.) (1966). *Peptides,* Pergamon Press, Oxford
25 Budapest (Hungary) 1964: BRUCKNER, V. and MEDZIHRADSKY, K. (editors) (1965). *Acta Chim. Acad. Hung.,* **44**, 1–239
26 Noordwijk (Netherland) 1966: BEYERMAN, H. C., VAN DEN LINDE, A. and MAASEN VAN DEN BRINK, W. (eds.) (1967). *Peptides,* North-Holland, Amsterdam
27 Orsay (France) 1968: BRICAS, E. (ed.) (1968). *Peptides,* North-Holland, Amsterdam
28 Padua (Italy) 1969: SCOFFONE, E. (ed.) (1971). *Peptides,* North-Holland, Amsterdam
29 Vienna (Austria) 1971: NESVADBA, H. (ed.) (1972). *Peptides,* North-Holland, Amsterdam
30 Reinhardsbrunn (G.D.R.) 1972: HANSON, H. and JAKUBKE, H. D. (eds.) (1973). *Peptides,* North-Holland, Amsterdam
31 Kiryat Anavim (Israel) 1974. WOLMAN, Y. (ed.) (1975). *Peptides,* Wiley, New York; Israel Universities Press, Jerusalem

2.2.2. Reversible Protecting Groups

2.2.2.1. Amino protecting Groups[1]

Owing to the lack of reversible amino-protecting groups the methods for peptide formation developed by E. FISCHER and Th. CURTIUS in the early days of synthetic peptide chemistry were inapplicable: hydrochlorides or other salts of amino acids and amino acid derivatives do not provide adequate protection of the amino function.

2.2.2.1.1. Benzyloxycarbonyl (Carbobenzoxy) Group[2]

The benzyloxycarbonyl group, given the abbreviation Z- in honour of L. ZERVAS, is the most used N-protecting group, but the term 'carbobenzoxy

(Cbo- or Cbz-) group' is still often found in the literature. Its introduction into amino acids and peptides is easily effected by treatment with carbobenzoxy chloride under SCHOTTEN–BAUMANN conditions:

$$C_6H_5-CH_2-O-\overset{O}{\overset{\|}{C}}-Cl + NH_2-\overset{R}{\overset{|}{C}H}-COOH \xrightarrow{-HCl} C_6H_5-CH_2-O-\overset{O}{\overset{\|}{C}}-NH-\underset{COOH}{\underset{|}{\overset{R}{\overset{|}{C}H}}}$$

Other methods for the preparation of benzyloxycarbonyl amino acid derivatives are the reaction of amino acids with benzyl-*p*-nitrophenyl carbonate and the reaction of a corresponding α-isocyanato-acid ester (*N*-carbonyl amino acid ester) with benzyl alcohol.

The particular usefulness of this acyl group lies in its easy removability (a) by catalytic hydrogenation, (b) by sodium in liquid ammonia or (c) by acidolysis with 6N hydrobromic acid in glacial acetic acid[2]. There are a number of additional acidolytic de-protection procedures. Using sodium in liquid ammonia, several side reactions have been described—for example, in methionine the methyl group may be removed and the Lys–Pro bond can be cleaved.

The ester bond of the benzyloxycarbonyl group is remarkably stable to nucleophilic attack. The saponification of benzyloxycarbonyl dipeptide esters with C-terminal glycine results in the formation of urea derivatives.

$$C_6H_5-CH_2-O-\overset{O}{\overset{\|}{C}}-NH-R$$

(c) HBr/CH_3COOH → $C_6H_5-CH_2-Br$, CO_2, $[\overset{\oplus}{N}H_3-R]\overset{\oplus}{} Br^{\ominus}$

(b) Na / l.NH_3 → $C_6H_5-CH_2-CH_2-C_6H_5$, Na_2CO_3, NH_2-R

(a) H_2/Pd → $C_6H_5-CH_3$, CO_2, NH_2-R

The most important methods for the de-protection of the benzyloxycarbonyl group (R = residue of amino acid or pepide)

[For the preparation of benzyloxycarbonyl-amino acid chlorides the use of phosphorus pentachloride at low temperatures is necessary, because otherwise *N*-carboxyanhydrides (LEUCHS anhydrides; see p. 60) are formed with elimination of benzyl chloride.

It is particularly interesting that the urethane structure of the benzyloxycarbonyl group protects the correspondingly substituted amino acids from racemisation. In the section on racemisation this protecting effect is discussed in more detail. The polymeric benzyl chloroformate (see p. 151) should also be noted.

2.2.2.1.2. tert-*Butyloxycarbonyl Group*[3]

In recent years the *tert*-butyloxycarbonyl group (Boc-) has attained increasing significance and is the most important protective group in MERRIFIELD synthesis (see p. 140). For preparing Boc-amino acids the reactions between α-isocyanato-acid esters and *tert*-butanol or between amino acids and various activated *tert*-butyloxycarbonyl derivatives (*tert*-butyl *p*-nitrophenyl carbonate, *tert*-butyloxycarbonylimidazole, *tert*-butyloxycarbonylcyanide and especially *tert*-butyloxycarbonylazide) are convenient.

However, the most commonly used route is acylation with *tert*-butyloxycarbonylazide[4]. The best yields are obtained by performing the acylation according to SCHNABEL[5] with a slight excess of *tert*-butyloxycarbonylazide at a fixed pH, which varies according to the individual amino acid, in an autotitrator.

$$(CH_3)_3C{-}O{-}CO{-}N_3 + H_2N{-}CH(R){-}COOH \xrightarrow{-HN_3} (CH_3)_3C{-}O{-}CO{-}NH{-}CH(R){-}COOH$$

$$\xrightarrow{H^{\oplus}} CH_3{-}C(CH_3){=}CH_2 + CO_2 + H_3\overset{\oplus}{N}{-}CH(R){-}COOH$$

The same author has reported[6] the use of *tert*-butyloxycarbonylfluoride (prepared from carbonylchloride fluoride and *tert*-butanol), which—unlike the *tert*-butyloxycarbonylchloride—is stable at 0°. It is more reactive than the azide and gives high yields of Boc-amino acids when used with automatic pH control.

Recently 4-dimethylamino-1-*tert*-butyloxycarbonylpyridinium chloride (I) has been found to react rapidly with the sodium salts of amino acids[7].

$$\left[(CH_3)_3C{-}O{-}CO{-}\overset{\oplus}{N}C_5H_4{-}N(CH_3)_2 \longleftrightarrow (CH_3)_3C{-}O{-}CO{-}NC_5H_4{=}\overset{\oplus}{N}(CH_3)_2\right]^{\oplus} Cl^{\ominus}$$

(I)

The *tert*-butyloxycarbonyl group is distinguished by the high selectivity of its removal by trifluoroacetic acid in the cold. This liberates only volatile co-products (isobutylene and CO_2). Under these conditions the benzyloxycarbonyl group is not cleaved. Moreover, the group is resistant to acetic acid, so that a trityl group may be removed selectively (see p. 89). The group is also stable to hydrolysis. Cleavage of the Boc-group is also possible with 98 per cent formic acid at room temperatures—conditions that do not attack benzyloxycarbonyl, formyl, *O*-benzyl ether groups, alkyl ester or benzyl ester groups[8]. However, the de-protection does not always proceed smoothly[9,10]. SCHNABEL, KLOSTERMEYER and BERNDT[11] observed partial formylation prior to the complete cleavage of the Boc-group. Satisfactory cleavage has been accomplished with boron trifluoride etherate in acetic acid or acetic acid–chloroform mixtures[12,13]. Under these cleavage conditions *S*-trityl and *S*-tetrahydropyranyl groups are not attacked, although aliphatic hydroxyl functions are partially acetylated[11]. Especially for the construction of long peptide chains with a variety of protecting groups the ability to remove protecting groups selectively at different stages of the synthesis is essential.

2.2.2.1.3. Further N-protecting Groups with Urethane Structures

Substituted benzyloxycarbonyl groups have been investigated in order to find more readily crystallisable derivatives as well as graduated reactivity towards cleavage reagents.

The *p-methoxybenzyloxycarbonyl group*, Z(OMe)— can be easily introduced via the crystalline azide, and can be removed selectively by trifluoroacetic acid at 0° in presence of the Z-group (F. WEYGAND, 1962).

The cleavage of *p-nitrobenzyloxycarbonyl group*, Z(*p*-NO_2)— by catalytic hydrogenation is easier than that of the unsubstituted compound, but troubles often arise over the separation of *p*-toluidine formed as a by-product.

The *2-(p-biphenylyl)-isopropyloxycarbonyl (Bpoc-) group* offers interesting possibilities, since it can be easily removed by dilute acetic acid and can therefore be used in combination with other protecting groups deriving from *tert*-butanol. For the introduction of the Bpoc-group, Bpoc-fluoride and (2-*p*-biphenylylisopropyl)-phenyl carbonates substituted in the phenyl group have been suggested[14].

In addition to the photosensitive *3,5-dimethoxybenzylcarbonyl group*[15,16] the *6-nitroveratryloxycarbonyl group* and *o-nitrobenzyloxycarbonyl group*[17] as well as the *α,α-dimethyl-3,5-dimethoxybenzyloxycarbonyl group*[18] are examples of protecting groups removable by photolysis. The last of these may be cleaved by 5 per cent trifluoroacetic acid in methylene chloride. Because of its lability the abbreviation *Lab*- has been suggested for it. Other interesting new groups include the *isobornyloxycarbonyl* group[19] (removable by solvolysis) and the *9-fluorenylmethoxycarbonyl* group[20], which is cleaved under extremely mild basic conditions.

2.2.2.1.4. p-Toluenesulphonyl Group

For the introduction of the *p*-toluenesulphonyl group (Tos-), usually abbreviated to 'tosyl', *p*-toluenesulphochloride is suitable. Sodium in liquid ammonia (V. DU VIGNEAUD, 1937), phosphonium iodide, hydrogen iodide

Table 13. N-Protecting groups of the urethane type $X-\overset{O}{\overset{\|}{C}}-NH-R$

Group	Abbreviation	Formula of X	Cleavage
Cyano-tert.-butoxycarbonyl		$N{\equiv}C-CH_2-C(CH_3)_2-O-$	Weakly basic reagents (aqueous K_2CO_3 or triethylamine at pH 10) via β-elimination
p-Methoxybenzyloxycarbonyl	Z(OMe)–	$CH_3O-C_6H_4-CH_2-O-$	CF_3COOH
p-Bromobenzyloxycarbonyl	Z(p-Br)–	$Br-C_6H_4-CH_2-O-$	Analogous to Z
p-Chlorobenzyloxycarbonyl	Z(*p*-Cl)–	$Cl-C_6H_4-CH_2-O-$	Analogous to Z
p-Nitrobenzyloxycarbonyl	Z(*p*-NO_2)–	$O_2N-C_6H_4-CH_2-O-$	HBr/CH_3COOH with difficulty; easily by H_2/Pd
o-Nitrobenzyloxycarbonyl	Z(*o*-NO_2)–	$o\text{-}NO_2-C_6H_4-CH_2-O-$	By photolysis
3,5-Dimethoxybenzyloxycarbonyl	$Z(OMe)_2$–	$3,5\text{-}(CH_3O)_2C_6H_3-CH_2-O-$	By photolysis
α,α-Dimethyl-3,5-dimethoxybenzyloxycarbonyl	Ddz-	$3,5\text{-}(CH_3O)_2C_6H_3-C(CH_3)_2-O-$	By photolysis; 5% CF_3COOH in CH_2Cl_2 at 20°

Table 13 (cont.)

6-Nitroveratryl-oxycarbonyl	Nvoc-	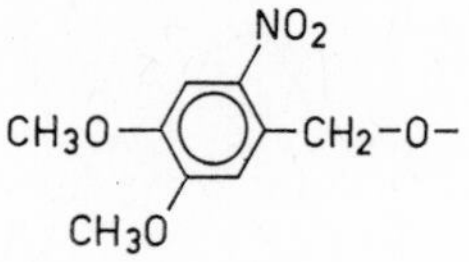 NO_2; CH_3O–; $-CH_2-O-$; CH_3O	By photolysis (350 nm) in presence of aldehydes; H_2/Pd; HBr/CH_3COOH
α,α-Dimethylbenzyl-oxycarbonyl		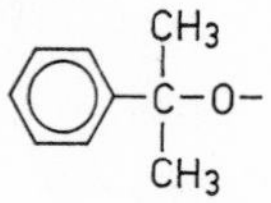 CH_3; $-C-O-$; CH_3	By thermolysis (130°, 10 min)
[2-Biphenyl-(4)-propyl-(2)]-oxycarbonyl	Bpoc-	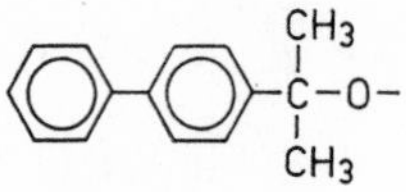 CH_3; $-C-O-$; CH_3	Dilute CH_3COOH and other weakly acidic reagents
tert-Amyloxy-carbonyl	Aoc-	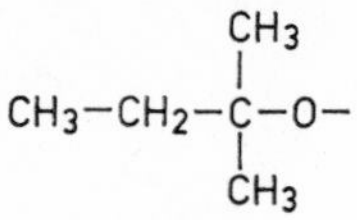 $CH_3-CH_2-C(CH_3)_2-O-$	CF_3COOH; HCl in organic solvents
Adamantyl-oxycarbonyl	Adoc-	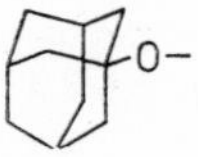 O–	CF_3COOH
Isobornyl-oxycarbonyl	Iboc-	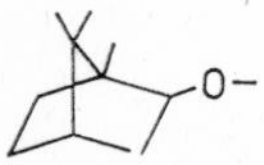 O–	CF_3COOH; HCl/CH_3COOH; resistant to H_2/Pd and alkaline reagents
9-Fluorenyl-methoxy-carbonyl	Fmoc-	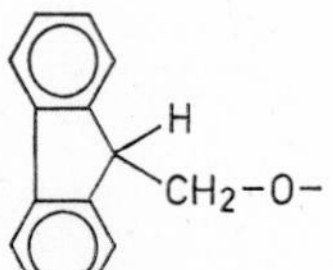 H; CH_2-O-	liq. NH_3; ethanolamine; morpholine
Furfuryloxy-carbonyl	Foc-	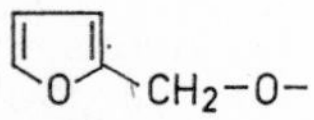 O; CH_2-O-	HCl/CH_3COOH; CF_3COOH
2-Iodo-ethoxy-carbonyl		$I-CH_2-CH_2-O-$	Zinc/methanol

Table 13 (cont.)

2-(*p*-Tosyl)-ethoxycarbonyl		$CH_3-C_6H_4-SO_2-CH_2-CH_2-O-$	Sodium ethoxide in ethanol
Piperidinooxycarbonyl	Pipoc-	$C_5H_{10}N-O-$	Electrolytic reduction or H_2/Pd
p-Phenylazobenzyloxycarbonyl	Pz-	$C_6H_5-N{=}N-C_6H_4-CH_2-O-$	HBr/CH_3COOH; H_2/Pd

or hydrogen bromide and phenol in acetic acid (J. Rudinger, 1955) are recommended for cleavage. A reductive cleavage method by means of tetramethylammonium (TMA) generated by electrolysis was developed by L. Horner in 1965. This method also allows removal of the *N*-benzoyl group under very mild conditions[21]. Okamura and co-workers have shown that the cleavage of *N*-tosyl groups by cathodic reduction is selective in presence of *N*-benzyloxycarbonyl, *tert*-amyloxycarbonyl, *tert*-butyloxycarbonyl and *S*-benzyl groups[22].

In peptide synthesis with Tos-α-amino acids, not all coupling procedures can be employed. The best results are obtained with the DCC method or the anhydride procedure with pivaloyl chloride and pyridine. The acid chloride method in dilute sodium hydroxide generally results in ionisation of the imide hydrogen followed by elimination of chloride ion and carbon monoxide to give a Schiff base, which is decomposed by hydrolysis to give tosylamide and the aldehyde containing one carbon less than the starting amino acid:

$$CH_3-C_6H_4-SO_2-NH-CHR-COCl \xrightarrow{OH^-} CH_3-C_6H_4-SO_2-\overset{\ominus}{N}-CHR-COCl$$

$$\xrightarrow{-Cl^{\ominus},\,-CO} CH_3-C_6H_4-SO_2-N{=}CHR \xrightarrow{+H_2O} CH_3-C_6H_4-SO_2-NH_2 + R-CHO$$

Such decomposition does not take place with Tos-pyroglutamic acid chloride, Tos-proline chloride and Tos-sarcosine chloride. However, this and other restrictions have served to reduce the importance of the tosyl group as a protecting group for the α-amino function.

2.2.2.1.5. Phthaloyl Group

Because of its sensitivity to alkali the phthaloyl group (Pht-) has been used less and less in peptide synthesis. This fact is not influenced by the fast and mild preparations of Pht-amino acids using ethoxyacetylene and phthalic acid[23] or *N*-ethoxycarbonylphthalimide[24]:

$$C_6H_4(CO)_2N{-}CO{-}O{-}C_2H_5 + H_2N{-}CH(R){-}COO^{\ominus} \longrightarrow C_6H_4(CO)_2N{-}CH(R){-}COO^{\ominus} + H_2N{-}COOC_2H_5$$

Recently *o*-methoxycarbonylbenzoyl chloride has been shown to react with amino acid esters in anhydrous media: excellent yields of optically pure phthaloyl derivatives were obtained[25].

The phthaloyl group is resistant to HBr in glacial acetic acid and may be removed most conveniently by hydrazinolysis in weakly acidic solution.

2.2.2.1.6. Trifluoroacetyl Group[26]

The trifluoroacetyl (Tfa-) group, $CF_3{-}CO{-}$, is a second important amino-protecting group removable by alkaline reagents. 0.01–0.02 N sodium hydroxide at room temperature, barium hydroxide and even dilute ammonium hydroxide solution are suitable de-protecting reagents.

Complications in the removal of the *N*-Tfa residue from sterically hindered amino acid or peptide derivatives can be avoided if the cleavage is carried out by reduction with sodium borohydride in ethanol[27]. In such a case the carboxyl group must be protected by a *tert*-butyl ester, since, unlike other esters, it is attacked by the reducing agent only to a minute extent. The reductive deacylation method is not yet in general use for derivatives of trifunctional amino acids.

The introduction of the Tfa residue is effected without racemisation by trifluoroacetic anhydride in anhydrous trifluoroacetic acid, trifluoroacetic acid thioethyl ester, trifluoroacetic acid phenyl ester, or by trichlorotrifluoroacetone in dimethylsulphoxide under neutral conditions[28].

The value of this protecting group is somewhat diminished by the difficulties that often arise in the preparation of carboxyl-activated derivatives.

2.2.2.1.7. Triphenylmethyl Group[29]

The triphenylmethyl residue (Trt-), usually abbreviated to 'trityl', has proved convenient because of its removability under mild conditions by catalytic hydrogenation or dilute acids. For acidolytic cleavage hydrogen chloride in organic solvents, trifluoroacetic acid or even dilute acetic acid is used.

Using diethylamine as HCl acceptor, the tritylation of amino acids produces satisfactory yields. TAMAKI *et al.*[30] treated the amino acids with two equivalents of trityl chloride in an aprotic solvent in the presence of triethylamine:

$$(C_6H_5)_3C{-}Cl + NH_2{-}\underset{|}{\overset{R}{C}}H{-}COOH \xrightarrow{-HCl} (C_6H_5)_3C{-}NH{-}\underset{|}{\overset{R}{C}}H{-}COOH$$

This direct preparation proceeds in good yield. Good results are also achieved if the corresponding esters are used, but the subsequent saponification usually gives rise to difficulties. Similarly in Trt-α-amino acids the ability to activate the carboxyl group is diminished by the steric hindrance of the three bulky phenyl residues. For coupling the carbodiimide method is most satisfactory (see p. 108).

*2.2.2.1.8. o-Nitrophenylthio Group**

The *o*-nitrophenylthio group (Nps-) announced by ZERVAS and co-workers[31]—often still called *o*-nitrophenylsulphenyl group—is a valuable extension of the possibilities of combining various protecting groups. The Nps-group can be removed from amino acids and simple peptide derivatives by hydrogen chloride in apolar organic solvents, by Raney nickel (J. MEIENHOFER, 1965), by various nucleophilic thiol reagents (W. KESSLER and B. ISELIN, 1966, and E. SCOFFONE, 1966), or by thiosulphate and sulphonic acid amide (K. PODUŠKA, J. RUDINGER *et al.*, 1967). It should be noted that the different labilities of the Nps- and Boc-groups towards acids is too small to allow selective cleavage to be carried out. Recent investigations have shown that several nucleophilic agents (thioacetamide, thiourea, hydrogen cyanide, etc.) are suitable for removal of the Nps-group. If *o*-nitrophenylthio peptides are to be de-protected in presence of tryptophan by HCl in organic solvents, the addition of 10–20 equivalents of an indole scavenger is necessary, since otherwise modified tryptophan residues may be formed (E. WUENSCH, 1967). Moreover, the cleavage methods mentioned can cause N$\rightarrow$S-migration of the *N*-*o*-nitrophenylthio group in cysteine peptides (L. ZERVAS, 1967). E. WUENSCH obtained very good results when

* Name recommended by IUPAC–IUB Committee on Nomenclature (1972). *J. Biol. Chem.*, **247**, 977.

he employed $SCN^{\ominus}$ in presence of an equivalent of methylindole to remove the Nps-group.

For the preparation of Nps-amino acids the sodium salts or esters of amino acids are treated with *o*-nitrophenylsulphenyl chloride:

$$o\text{-}NO_2\text{-}C_6H_4\text{-}S\text{-}Cl + H_2N\text{-}CH(R)\text{-}COOR' \xrightarrow{-HCl} o\text{-}NO_2\text{-}C_6H_4\text{-}S\text{-}NH\text{-}CH(R)\text{-}COOR'$$

(R′ = Na or alkyl residue)

o-Nitrophenylsulphenyl thiocyanate is much more stable than *o*-nitrophenylsulphenyl chloride and, in the presence of silver nitrate, gives Nps-amino acids in good yields.

The *2,4-dinitrophenylthio group,* which can be introduced and cleaved in the same way as the Nps-group, increases the ease of crystallising correspondingly substituted derivatives. The application of tritylthio, 2,4,5-trichloro- and pentachloro-phenylthio residues has also been described.

2.2.2.1.9. Further N-protecting Groups

The N-protecting groups listed in Table 14 have been used in special cases, but so far they have attained no great practical significance.

Table 14. Further *N*-protecting groups

Protecting group	Formula (R = amino acid residue)	Cleavage
Formyl group (For-)	$HC(=O)\text{-}NH\text{-}R$	Solvolysis or oxidation; NaH, hydrazine and hydrazine derivatives in presence of weak acids
Benzylsulphonyl group	$C_6H_5\text{-}CH_2\text{-}SO_2\text{-}NH\text{-}R$	Na/liq.NH_3, hydrogenolysis with RANEY nickel
o-Hydroxyarylidene group	$Cl\text{-}C_6H_3(O\text{-}H\cdots)\text{-}CH{=}N\text{-}R$	Dilute mineral acid
Acetoacetyl group	$CH_3\text{-}CO\text{-}CH_2\text{-}CO\text{-}NH\text{-}R$	Phenylhydrazine in acetic acid

Table 14. (cont.)

Protecting group	Formula (R = amino acid residue)	Cleavage
5,5-Dimethylcyclohexandione group	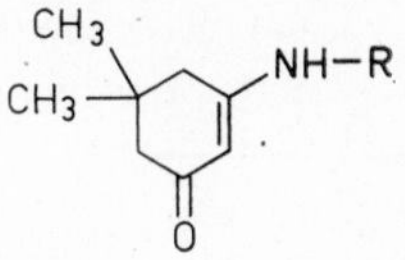 [structure: 5,5-dimethylcyclohexenone with NH–R]	Bromine water
Benzyl group	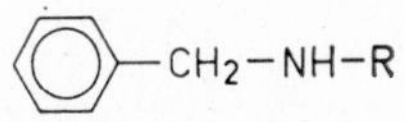 C_6H_5–CH_2–NH–R	H_2 / Pd; resistant to acids
Schiff bases (enamine structure)	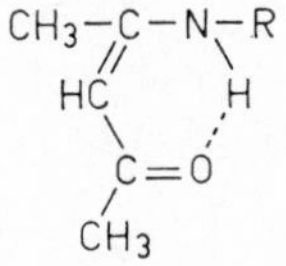 CH_3–C(=CH–C(=O)–CH_3)–N(H)–R	Weak acids
o-Nitrophenoxyacetyl group	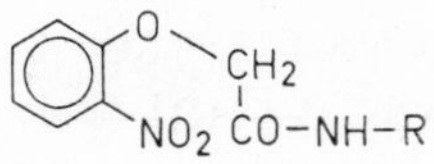 o-NO_2–C_6H_4–O–CH_2–CO–NH–R	Partial reduction to the hydroxylamine derivative and solvolytic cleavage

2.2.2.2. Carboxyl protecting Groups

During peptide synthesis it is also necessary to block the carboxyl function of the amino component. The reason for this has already been explained in Section 2.2.1. Protection of the carboxyl group is achieved most simply by salt formation. Besides alkali salts the salts of organic bases, such as triethylamine, tributylamine, *N*-ethylpiperidine or dicyclohexylamine, can be used, since they are more soluble in organic solvents.

In the usual procedures of peptide synthesis various esters and substituted hydrazides are used for the protection of the carboxyl group. These are briefly described below.

2.2.2.2.1. Methyl and Ethyl Esters

Methyl esters (—OMe) and ethyl esters (—OEt) were used in peptide synthesis by those pioneers of synthetic peptide chemistry Th. Curtius and E. Fischer. Besides the well-known organic chemical procedures, these alkyl esters can be formed elegantly by the reaction of free amino acids with a solution of thionyl chloride in cold methanol or ethanol[32]. The conversion of amino acids into their ethyl esters using azeotropic distillation has been reported by Dymocky and co-workers[33].

To liberate the C-terminal carboxyl function after the coupling reactions are complete alkaline hydrolysis is used in most cases, although dilute hydrochloric acid in aqueous acetone is sometimes used. Alkaline saponification is accomplished with sodium hydroxide in acetone, methanol, dioxane or dimethylformamide at, or below, room temperature. An excess of sodium hydroxide should be avoided, or racemisation may occur. The carboxamide groupings of glutamine and asparagine in peptides are especially sensitive to alkali. In longer peptide sequences alkaline saponification can fail or, as in some serine-containing peptides, be accompanied by cleavage of peptide bonds.

Ammonolysis in absolute ethanol is convenient if the C-terminal amino acid is to bear an amide grouping.

2.2.2.2.2. Benzyl Esters

Benzyl esters (—OBzl) of free amino acids are obtained by direct esterification with benzyl alcohol in presence of acidic catalysts (hydrochloric acid, benzenesulphonic acid, *p*-toluenesulphonic acid, polyphosphoric acid, etc.); the water produced during the esterification is eliminated by azeotropic distillation. For the preparation of benzyl esters of N-protected amino acids various transesterification methods are available—for example, by means of boric acid benzyl esters.

In addition to catalytic hydrogenation, hydrogen bromide in acetic acid (2 h at 50–70°), hydrogen bromide in trifluoroacetic acid, sodium in liquid ammonia and alkaline saponification are suitable for the removal of a C-terminal benzyl ester group.

2.2.2.2.3. p-Nitrobenzyl Esters

The methods described for benzyl esters are also suitable for the preparation of *p*-nitrobenzyl esters (—ONb). N-protected amino acid and peptide derivatives react easily with *p*-nitrobenzyl chloride or *p*-nitrobenzyl tosylate in presence of triethylamine to give the corresponding esters. *p*-Nitrobenzyl esters are resistant to hydrogen bromide in acetic acid. On the other hand, they can be removed successfully by catalytic hydrogenation or alkaline saponification.

2.2.2.2.4. Benzhydryl Esters

Benzhydryl esters (—OBzh) of amino acids and peptides (also called diphenylmethyl esters, —ODpm) can be prepared in good yield by the reaction between their *p*-toluenesulphonic salts and diphenyldiazomethane[34]:

$$(C_6H_5)_2CN_2 + HOOC-R \xrightarrow{-N_2} (C_6H_5)_2CHOOC-R$$

The ester grouping can be removed together with the Z-group by catalytic hydrogenation or selectively by CF_3COOH. Other acidolytic cleavage procedures are also feasible. In combination with the Nps-, Trt- or For-residue for N-protection diverse strategies are possible[35].

The 2,2-dinitrodiphenylmethyl ester grouping can also be cleaved by photolysis[17].

2.2.2.2.5. *tert-Butyl Esters*

tert-Butyl esters (—OBu) are very easily obtained by esterification with isobutylene and sulphuric acid under pressure or by transesterification with *tert*-butyl acetate in the presence of acidic catalysts—for example, perchloric acid.

$$R-\overset{\overset{\displaystyle O}{\|}}{C}-OH + \overset{\overset{\displaystyle CH_3}{\|}}{\underset{\underset{\displaystyle CH_2}{|}}{C}}-CH_3 \xrightarrow{(H^{\oplus})} R-\overset{\overset{\displaystyle O}{\|}}{C}-O-\overset{\overset{\displaystyle CH_3}{|}}{\underset{\underset{\displaystyle CH_3}{|}}{C}}-CH_3$$

They are removable under very mild conditions by acid catalysis. The stability of the *tert*-butyl ester group to bases allows the selective hydrolysis and hydrazinolysis of ethyl, methyl or benzyl esters present in the same peptide. The strong +I-effect of the *tert*-butyl grouping lowers the electrophilicity of the carbonyl carbon atom somewhat and this, combined with the steric hindrance of the three methyl groups, explains the stability to nucleophilic reagents. For this reason free *tert*-butyl esters can be distilled without danger of dioxopiperazine formation or polymerisation.

The *tert*-butyl ester group is not attacked by hydrogenolysis. For its removal hydrogen bromide in acetic acid, hydrogen chloride in acetic acid (or dioxane) or trifluoroacetic acid may be used. With trifluoroacetic acid in the cold the cleavage of the ester grouping can be performed selectively in the presence of an *N*-benzyloxycarbonyl group.

2.2.2.2.6. *Further Carboxylprotecting Groups*

In Table 15 further ester groups suitable for blocking of carboxyl function are listed.

Owing to their basic centre 4-picolyl esters[41] of peptides can be reversibly bound to an ion exchange resin. This allows the separation of by-products after each coupling step of the peptide synthesis. This method has the advantage that all the reactions proceed in homogeneous solution. A similar technique has been described by Wieland and Racky, using *p-dimethylaminobenzyl esters*[46].

The use by Merrifield of *insoluble polymeric esters* in peptide synthesis is a methodologically interesting method of carboxyl protection. Because of its great importance in modern peptide chemistry the whole problem of synthesis on polymeric supports will be discussed in a separate section (2.2.4), even though this is a deviation from our classification.

Table 15. Further carboxylprotecting groups $R-C(=O)OR'$

Protecting group	Formula of R′	Cleavage
p-Methoxybenzyl ester 36, 37	$-CH_2-C_6H_4-OCH_3$	HBr/CH_3COOH; CF_3COOH
Methylsubstituted benzyl ester 38	$-CH_2-C_6H_{5-n}(CH_3)_n$ (2, 4, 6-trimethyl- and pentamethylbenzyl ester)	2 N HBr/CH_3COOH; CF_3COOH
Phenacyl ester 39	$-CH_2-CO-C_6H_5$	H_2/Pd; sodium thiophenate; Zn/CH_3COOH
p-Methoxyphenacyl ester	$-CH_2-CO-C_6H_4-OCH_3$	At 20° by UV light
Phenyl ester 40	$-C_6H_5$	H_2O_2 (pH 10. 5)
4-Picolyl ester 41	$-CH_2-C_5H_4N$	H_2/Pd; $Na/liq. NH_3$; electrolytic reduction; alkaline hydrolysis
4-Methylthioethyl ester 42	$-CH_2-CH_2-S-CH_3$	Oxidation to the sulphone and subsequent hydrolysis under mild alkaline conditions
4- Methylthio phenyl ester 43	$-C_6H_4-S-CH_3$	Oxidation to sulphoxide or sulphone and use as activated ester
Trimethylsilyl ester	$-Si(CH_3)_3$	Hydrolysis

Table 15 (cont.)

Protecting Group	Formula of R′	Cleavage
Phthalimidomethyl ester	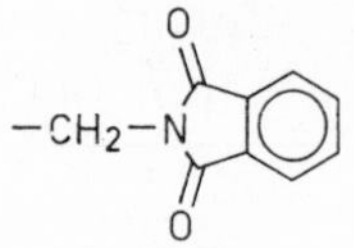 $-CH_2-N$ (phthalimido)	HCl/organic solvent; diethylamine/ethanol; N_2H_4; dilute NaOH; Zn/CH_3COOH 44
Piperonyl ester	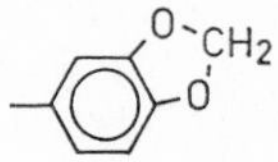 (3,4-methylenedioxyphenyl, $-O-CH_2-O-$)	Brief treatment with CF_3COOH; 2 M hydrogen bromide in acetic acid 45

2.2.2.2.7. Substituted Hydrazides

The chief advantage of substituted hydrazides for carboxyl protection lies in the fact that cleavage of substituents (Z-, Boc- or Trt-) produces the starting materials for azide coupling (see p. 97).

In principle, however, acylamino acid hydrazides can also be converted into acylamino acids by oxidation with *N*-bromosuccinimide, iodine or other oxidants. For the preparation of substituted hydrazides of acylamino acids, benzyloxycarbonylhydrazine, *tert*-butyloxycarbonylhydrazine or tritylhydrazine is coupled to acylamino acids by use of the common coupling methods (see also p. 112).

A new protected hydrazine, namely trichloroethoxycarbonylhydrazine, $Cl_3C-CH_2-O-CO-NH-NH_2$, has been introduced by Kiso[47]. Acylamino acids are coupled with the hydrazine using the mixed-anhydride method or dicyclohexylcarbodiimide. Removal of the trichlorethoxylcarbonyl group is then effected with zinc dust in acetic acid.

2.2.2.3. Special Side-chain protecting Groups

If polyfunctional amino acids, such as glutamic acid, aspartic acid, lysine, arginine, cysteine, serine, tyrosine, etc., are employed in peptide synthesis, their side-chain functional groups must be selectively blocked.

2.2.2.3.1. ω-Amino protecting Groups[48]

Protection of the ω-function of diaminocarboxylic acids—especially of lysine and ornithine— is often necessary. There is no difficulty if lysine or ornithine is incorporated at the N-terminal end. In this case both amino groups (α- and ω-) can be blocked with the same N-protecting group, since both will be removed together after the synthesis is finished. α,ω-Dibenzyl-

Benzyloxycarbonylhydrazide — R–CO–NH–NH–CO–O–CH$_2$–C$_6$H$_5$

Tritylhydrazide — R–CO–NH–NH–C(C$_6$H$_5$)$_3$

tert-Butyloxycarbonylhydrazide — R–CO–NH–NH–CO–O–C(CH$_3$)$_3$

-Z, -Trt, -Boc → R–CO–NH–NH$_2$

R–CO–NH–NH$_2$ $\xrightarrow{\text{oxidation}}$ R–CO–N=NH

R–CO–NH–NH$_2$ ← Azide procedure

$\xrightarrow{+H_2O}$ R–COOH

oxycarbonyl derivatives are usual in such cases. Masking the ε-amino group of lysine and especially the δ-amino group of ornithine with a tosyl residue does not give complete protection. Carboxyl-activated δ-Tos-ornithine derivatives have a remarkable tendency towards lactam formation:

$$\mathrm{Y{-}NH{-}\underset{\underset{CH_2{-}CH_2}{|}}{\overset{}{CH}}{-}\overset{\overset{O}{\|}}{C}{-}X \quad NH{-}SO_2{-}C_6H_4{-}CH_3 \xrightarrow{-HX} Y{-}NH{-}CH{-}CO \ (CH_2{-}CH_2{-}CH_2 \text{ ring}) \ N{-}SO_2{-}C_6H_4{-}CH_3}$$

In ε-tosyl derivatives lactam formation is not observed because of the difficulty of forming seven-membered rings. However, the N-atom in all tosyl amides possesses some nucleophilic properties and so undesirable side reactions cannot be completely excluded.

For the synthesis of other lysine and ornithine peptides N^{ω}-protected derivatives are needed. Because of the difference in reactivity between α- and ω-amino groups there are several possibilities. Either monoprotection is accomplished directly or one protecting group is removed from a di-substituted derivative. Selective trifluoroacetylation of the ω-amino groups of lysine and ornithine is obtained on the reaction between the amino acid and trifluoroacetic acid thioethyl ester in aqueous media at pH 8–9. The conversion of either of the amino acids into its copper complex leads to blocking of the α-amino function, and allows ω-acylation to be performed.

The combination of N^{α}-Boc-group and N^{ε}-Z-group often used in MERRIFIELD synthesis often results in a partial deacylation of the N^{ε}-protecting function during the repeated α-protection removal at each cycle. Therefore, more acid-stable modified Z-residues, such as Z(*p*-Cl)-, Z(*m*-Cl)-, Z(*p*-CN)-, Z(*p*-NO_2)-groups[49], have been suggested as alternatives. Of these, however, only the Z(*p*-NO_2)-group displays sufficient stability. It can be cleaved by hydrogenolysis, but not by means of anhydrous HF. More suitable may be the *diisopropylmethoxycarbonyl* (*Dipmoc-*) *group*[50], an ω-amino-protecting group which is stable to prolonged action of N HCl/acetic acid and can be cleaved by HF in presence of anisol.

2.2.2.3.2. Guanidine protecting Groups

In spite of numerous protecting groups known for the guanidine function of arginine, the synthesis of peptides containing arginine is still a difficult problem. The guanidine function is only incompletely protected by N^G-tosyl, N^G-Boc, N^G-nitro, and N^G-Z(*p*-NO_2) residues. Apart from the potential danger of undesired lactam formation that this presents, the cleavage of the tosyl and nitro groups is often accompanied by side reactions, unless the rather laborious *HF* method[51] can be used.

The simplest way to block the strongly basic guanidine function is certainly protonation, although protonated arginine peptides can only be purified with difficulty because of their poor solubility in organic solvents. Complete protection of the guanidine function is also achieved by the introduction of two Z-residues (which does not proceed in quantitative yield!), two *adamantyloxycarbonyl*[52] or two *isobornyloxycarbonyl groups*[53]. In the last two cases the derivatives have better solubility in organic solvents.

Further possibilities for guanidine protection have been investigated by PLESS and GUTTMAN[54].

Another way to synthesise arginine peptides is by introduction of the guanidine group into N^{α}-acylated but N^{ω}-unsubstituted ornithine peptides. The guanylation can be accomplished very simply with 1-guanyl-3,5-dimethylpyrazole:

$$\text{-CO-NH-CH(}(CH_2)_3\text{-}NH_2\text{)-CO-NH-} + \text{1-guanyl-3,5-dimethylpyrazole } (H_3C,\ CH_3\text{-pyrazole-N-C(=NH)-}NH_2) \xrightarrow[\text{pyrazole}]{-3,5\text{-dimethyl}} \text{-CO-NH-CH(}(CH_2)_3\text{-NH-C(=NH)-}NH_2\text{)-CO-NH-}$$

N^{α}-benzyloxycarbonyl-N^{ω}-phthalyl-L-ornithine *p*-nitrophenyl ester is an important starting material for a strategy in which the N^{α}-protecting group can be removed after coupling stages by means of hydrogen bromide in glacial acetic acid: when the lengthening of the peptide chain is finished, the N^{ω}-phthalyl group is cleaved by hydrazinolysis at room temperature, and the guanidine group is introduced as described above.

2.2.2.3.3. Imidazole protecting Groups

Although the protection of the secondary amino function in the imidazole ring of histidine is not necessary if histidine is the carboxyl component of coupling, in general, the synthesis of peptides containing histidine presents difficulties caused by the basicity and easy acylation of the imidazole residue, as well as by their usually low solubility.

If histidine is the carboxyl component, blocking of the imino group is required. Convenient protecting groups are *benzyl*, *benzyloxycarbonyl*, *adamantyloxycarbonyl*[55], *trityl residues*, etc. The trityl residue cannot be combined with the usual urethane-protecting groups because of its acid lability. However, the cleavage of the benzyl residue by sodium in liquid ammonia or by catalytic hydrogenation does not proceed smoothly.

The N^{im}-urethane-protecting groups mentioned are activated amides, and their lability towards aminolysis and alkali presents difficulties if, in the course of the subsequent synthesis, alkaline saponification or hydrazinolysis is required.

The *2,2,2-trifluoro-1-acylaminoethyl-protecting groups*—for example, the *2,2,2-trifluoro-1-benzyloxycarbonylethyl* (N^{im}-Z–TF) *group*:

$$C_6H_5-CH_2-O-\underset{\underset{O}{\|}}{C}-NH-\overset{F_3C-CH-N<}{}$$

suggested by WEYGAND and co-workers[56] avoid these disadvantages. The latter is introduced by means of 2,2,2-trifluoro-1-chloro-N-benzyloxycarbonylethylamine, and may be removed by catalytic hydrogenation in methanol.

An N^{im}-carbamoyl group[58], such as the piperidinocarbonyl residue, is resistant to acidolysis and hydrogenolysis but can be removed by hydrazinolysis.

The reaction of α-N-protected histidine derivatives in aqueous sodium bicarbonate solution with 2,4-dinitrofluorobenzene yields N^{im}-2,4-dinitrophenyl histidine derivatives, from which the N^{im}-protecting group can be removed by thiolysis with ethanethiol at pH of 8 [57].

2.2.2.3.4. *Protection of the Thioether Bond in Methionine*

So far, only one method for the reversible protection of the thioether bond of methionine exists. By oxidation with hydrogen peroxide the methionine derivative can be converted into the corresponding sulphoxide, from which it may be regenerated by reduction with thioglycolic acid or by CLELAND's reagent. This formation of an additional asymmetric centre results in a mixture of diastereoisomers which may complicate purification. On deacylation of N-benzyloxycarbonylmethionine peptides with hydrogen bromide in glacial acetic acid the benzylbromide produced in the reaction can form sulphonium salts which may lead to undesired alkyl exchange reactions:

$$-HN-CH(CH_2-CH_2-S-CH_3)-CO- \;+\; C_6H_5-CH_2-Br \longrightarrow \left[C_6H_5-CH_2-\overset{\oplus}{S}(CH_3)-CH_2-CH_2-CH(-HN)-CO-\right]^{\oplus} Br^{\ominus}$$

Adding methyl ethyl sulphide to the reaction mixture suppresses this unwanted side reaction. Catalytic hydrogenation in presence of triethylamine or cyclohexylamine (Medzihradszky, 1966) or boron trifluoride etherate (Okamoto, 1967) can be used to remove a benzyloxycarbonyl group from methionine-containing peptides.

2.2.2.3.5. *Protection of Aliphatic Hydroxyl Groups*

When serine, threonine or hydroxyproline is used as the amino component for a coupling reaction, blocking of the hydroxyl function is not absolutely necessary. In some cases, however, undesired side reactions have been reported. Various alternatives are available for selective protection of the hydroxyl group. The use of the *O*-acetyl derivatives of serine and threonine carries the danger of possible racemisation due to reversible β-elimination. More important protecting groups are the benzyl and *tert*-butyl residues. But on cleavage of *O*-benzyl ethers by hydrogen bromide in glacial acetic acid partial or even complete acetylation is possible and so the *O-tert*-butyl ethers are more important. They can be synthesised, for instance, from the corresponding *N*-benzyloxycarbonylamino acid *p*-nitrobenzyl esters by reaction with isobutylene in presence of catalytic amounts of sulphuric acid and subsequent catalytic hydrogenation. Weygand and co-workers[59] have also suggested the 2,2,2-trifluoro-1-benzyloxycarbonylamino ethyl residue as a protecting group.

2.2.2.3.6. *Protection of Aromatic Hydroxyl Groups*

The necessity to protect the phenolic hydroxyl group in tyrosine essentially parallels the case of the aliphatic hydroxyl groups discussed above.

Tyrosine can be converted selectively to the *O*-benzyloxycarbonyl or the benzyl derivative via its copper complex. In addition, *O*-tosyl and *O*-benzoyl groups have been used as protection. However, it must be borne in mind that *O*-acyl tyrosine derivatives may suffer nucleophilic attack by the amino component.

The *tert*-butyl and the benzyl groups can be removed by acidolysis; the latter can also be eliminated by catalytic hydrogenation.

O-Alkyloxycarbonyl tyrosine derivatives[60] are stable under the conditions of peptide synthesis and also towards hydrogenolysis and acidolysis, whereas they can be cleaved by alkaline saponification or hydrazinolysis.

2.2.2.3.7. *SH-protecting Groups*

The nucleophilicity, ease of oxidation and acidic character of the SH group in cysteine render a selective blocking of this functional group essential in all synthetic operations. As early as 1930, V. du Vigneaud used *S*-benzylated cysteine derivatives and opened the way for synthetic work in the cystine–cysteine field.

Table 16 lists the most important SH-protecting groups.

Table 16. Important SH-protecting groups

Protecting group (abbreviation)	Formula	Cleavage
Benzyl (Bzl-)	$C_6H_5-CH_2-S-$	Na/liq. NH_3; HF; resistant to HBr/CH_3COOH
p-Methoxybenzyl (Mbzl-)	$CH_3-O-C_6H_4-CH_2-S-$	Boiling CF_3COOH; HF; resistant to HBr/CH_3COOH
p-Nitrobenzyl (Nbzl-)	$O_2N-C_6H_4-CH_2-S-$	Catalytic hydrogenation (by-product: *p*-aminobenzyl cysteine derivatives)
Trityl (Trt-)	$(C_6H_5)_3C-S-$	Ag and Hg salts at 0°; HCl in chloroform; CF_3COOH; Na/liq. NH_3
tert-Butyl (Bu^t-)	$(CH_3)_3C-S-$	Anhydrous HF; oxidative sulphitolysis followed by reduction; CLELAND's reagent
Diphenylmethyl (Dpm-)	$(C_6H_5)_2CH-S-$	CF_3COOH (anisole); HF; Na/liq. NH_3; resistant to $AgNO_3$ and $HgCl_2$
4,4′-dimethoxy-diphenylmethyl (Dmb-)	$(CH_3O-C_6H_4)_2CH-S-$	Na/liq. NH_3; acidolysis
Acetyl (Ac-)	CH_3CO-S-	Alkaline hydrolysis or ammonolysis
Benzoyl (Bz-)	$C_6H_5-CO-S-$	Alkaline hydrolysis or ammonolysis

Protecting group (abbreviation)	Formula	Cleavage
Benzyloxycarbonyl (Z-)	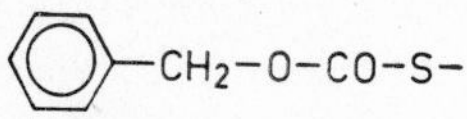 $C_6H_5-CH_2-O-CO-S-$	Alkaline hydrolysis or ammonolysis; CF_3COOH; resistant to 2N HBr/CH_3COOH
Benzylthiomethyl (Btm-)	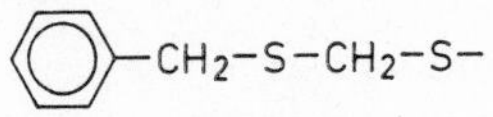 $C_6H_5-CH_2-S-CH_2-S-$	$(CH_3COO)_2Hg$ in 80% HCOOH; resistant to HBr/CH_3COOH
Tetrahydropyranyl (Thp-)	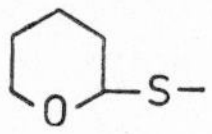 (tetrahydropyran-2-yl)$-S-$	Na/liq. NH_3; $AgNO_3$; acidolysis
Ethylthio (Et-)	H_5C_2-S-S-	Thiophenol or thioglycolic acid
Acetamidomethyl (Acm-)	$CH_3-C(=O)-NH-CH_2-S-$	Hg(II) salts at pH 4; resistant to HF
Isobutoxymethyl (iBm-)	$(CH_3)_2CH-CH_2-O-CH_2-S-$	$(SCN)_2$
β,β-Diethoxycarbonylethyl (Dce-)[61]	$(H_5C_2O_2C)_2CH-CH_2-S-$	β-Elimination by 1N ethanolic KOH at 20°
p-Methoxybenzyloxycarbonyl (Z(OMe)-)[62]	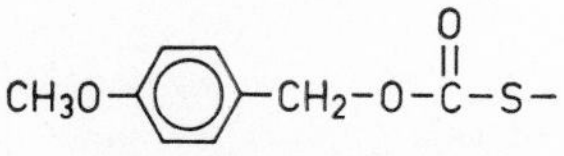 $CH_3O-C_6H_4-CH_2-O-C(=O)-S-$	1.5 N HCl/CH_3COOH; 0.2 N HBr/CH_3COOH; methanolysis

S-Benzylcysteine can easily be prepared by reduction of cysteine with sodium in liquid ammonia benzylation *in situ*. Although the benzyl ether procedure has proved convenient in syntheses of glutathione, oxytocin, vasopressin, insulin, etc., the drastic conditions for its removal frequently result in undesired side reactions, such as cleavage of bonds or desulphuration of cysteine, etc. Removal of the *S*-benzyl group by HF is possible but not quantitative.

SH-protecting groups that can be removed selectively under mild conditions are more important. Thus, the *S-trityl residue* has been often used because of its easy removability by acids, iodine, mercury and silver salts. However, owing to its high lability towards acids, the trityl residue cannot be used in conjunction with Boc- and Nps-amino-protecting groups in the MERRIFIELD synthesis.

The *S-ethylthio group*[63] and *S-acetamidomethyl group*[64] described recently seem to be more favourable. Both of these are fairly stable under the conditions normally used to remove the common amino-protecting groups. The *S*-ethylthio group can be selectively removed by thiophenol or thioglycolic acid in an appropriate solvent below 50° and the *S*-acetamidomethyl residue by mercury(II) salts at pH 4.

The alkali-labile *S-ethylcarbamoyl* group (Ec-) recommended by St. GUTTMAN in 1966 and the *β-(N-acyl-N-methyl-aminoethyl)-carbamoyl* group[65] have both been employed as S-protecting groups: the latter can be removed by an intramolecular displacement.

2.2.2.3.8. ω-Carboxyl protecting Groups

Peptides with C-terminal aminodicarboxylic acid residues can be handled simply and without special blocking procedures, whereas difficulties are encountered in the synthesis of peptides with N-terminal or in-chain aminodicarboxylic acid residues since only one carboxyl function may take part in coupling. Both the carboxyl groups could take part in peptide formation, so simultaneous formation of both the α- and ω-isomers will occur unless the reactivity of one of the carboxyl functions is diminished or concealed.

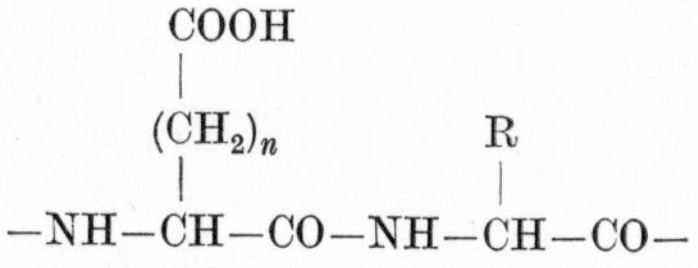

$$\begin{array}{ll}
\begin{array}{l}
\mathrm{COOH} \\
| \\
\mathrm{(CH_2)}_n \qquad\quad \mathrm{R} \\
| \qquad\qquad\qquad | \\
-\mathrm{NH-CH-CO-NH-CH-CO-}
\end{array}
&
\begin{array}{l}
\qquad\qquad\qquad \mathrm{R} \\
\qquad\qquad\qquad | \\
\mathrm{CO-NH-CH-CO-} \\
| \\
\mathrm{(CH_2)}_n \\
| \\
-\mathrm{NH-CH-COOH}
\end{array}
\end{array}$$

α-Aminodicarboxylic peptide 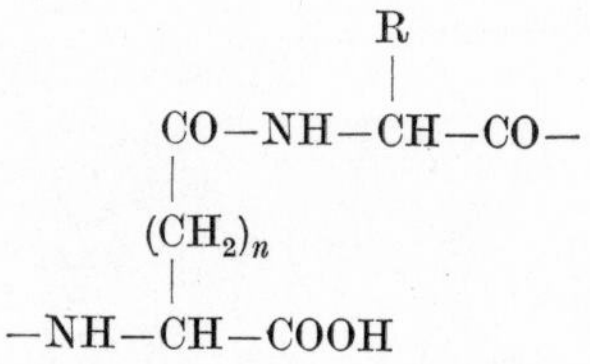 ω-Aminodicarboxylic peptide

For an unambiguous synthesis of α- or ω-peptides of aminodicarboxylic acids selective protection of one carboxyl group is necessary. Because of the complexity of this problem only some examples can be mentioned here: there are reviews on the topic[66,67].

Starting from anhydrides of N-protected aminodicarboxylic acids (I), it is possible to form both α- and ω-derivatives by the attack of nucleophilic reagents.

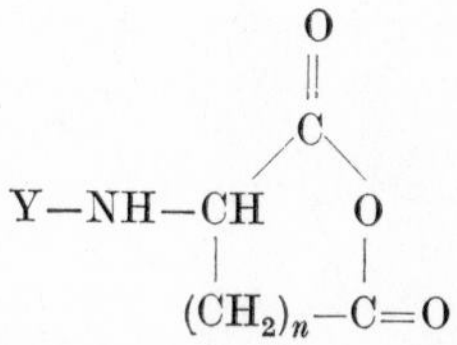

$$\begin{array}{l}
\qquad\qquad\qquad \mathrm{O} \\
\qquad\qquad\qquad \| \\
\qquad\qquad\qquad \mathrm{C} \\
\qquad\qquad\quad / \quad \backslash \\
\mathrm{Y-NH-CH} \qquad \mathrm{O} \\
\qquad\qquad | \qquad\quad | \\
\qquad\quad \mathrm{(CH_2)}_n\mathrm{-C{=}O}
\end{array} \qquad \mathrm{(I)}$$

Apart from steric effects two factors determine the position of cleavage of the anhydride ring. On the one hand, nucleophilic attack on the ω-carbonyl group of the anhydride ring should be energetically favoured, since the more acidic α-carboxyl group is thereby converted into the thermodynamically more stable carboxylate. On the other hand, the electron-attracting effect of most N-protecting groups (Y) accentuates the positive character of the α-carboxyl C-atom and makes it the preferred site for nucleophilic attack. The product composition therefore depends on the N-protecting group and the nucleophile; and, in addition, it is affected by the choice of solvents and other reaction conditions. In any case, separation of an isomeric mixture, whose composition is not predictable, is inevitable in preparative work. Such separations are best achieved by fractional alkaline extraction or countercurrent distribution.

The protected aminodicarboxylic acid anhydride can be ring-opened by aminolysis or, alternatively, reaction with alcohols will give the α-semi-esters. The preparatively important α-esters can be obtained as the easily crystallised dicyclohexylammonium salts if the alcoholysis is performed in presence of dicyclohexylamine. The selective esterification of the ω-carboxyl group of the free aminodicarboxylic acid is accomplished by reaction with alcohols in presence of anhydrous hydrochloric acid, sulphuric acid, *p*-toluenesulphonic acid or cation exchangers. However, in no case is the selectivity complete. The ω-esters of glutamic acid and aspartic acid are obtainable without isomeric by transesterification contaminants with the corresponding acetic acid esters, using perchloric acid as catalyst.

The preparation of the very important *tert*-butyl esters starts from other semi-esters which are converted into unsymmetrical diesters by standard methods (see p. 94), followed by the selective cleavage of the starting ester grouping.

Intramolecular reaction between the ω-carboxyl group of glutamic acid or aminoadipic acid and the α-amino group gives the five- or six-membered lactams, in which the ω-carboxylic function is blocked. In the case of aspartic acid the formation of a four-membered ring system is prevented by ring strain. The N-protected derivatives are necessary for the selective formation of α- and ω-peptides of glutamic and aspartic acids. For the preparation of *N*-acyl pyrrolidone carboxylic acid the rearrangement of *N*-acyl glutamic acid anhydride in presence of dicyclohexylamine proved to be useful. For example, the formation of benzyloxycarbonylpyrrolidone carboxylic acid (benzyloxycarbonylpyroglutamic acid):

$$\begin{array}{l} Y{-}NH{-}CH{-}COOH \\ \quad\quad\;\; | \\ \quad\quad\; CH_2 \\ \quad\quad\;\; | \\ \quad\quad\; CH_2{-}COOH \end{array} \xrightarrow{-H_2O} \begin{array}{l} Y{-}NH{-}CH{-}CO \\ \quad\quad\;\; | \quad\quad\;\; \backslash \\ \quad\quad\; CH_2 \quad\quad O \\ \quad\quad\;\; | \quad\quad\;\; / \\ \quad\quad\; CH_2{-}CO \end{array} \xrightarrow{(C_6H_{11})_2NH}$$

$$\left[\begin{array}{l} Y{-}N{-}CH{-}COO^{\ominus} \\ \;\;\;\, | \quad\;\; | \\ \;\;\;\, | \quad CH_2 \\ \;\;\;\, | \quad\;\; | \\ O{=}C{-}CH_2 \end{array}\right]^{\ominus} \left[(C_6H_{11})_2NH_2^{\oplus}\right]^{\oplus}$$

The tosylpyrrolidone carboxylic acid was first synthesised by reaction of tosylglutamic acid with acetic anhydride. However, this procedure has attained significance only in studies with N-protecting groups of the urethane type, accomplished independently by J. RUDINGER and V. DU VIGNEAUD as well as E. SCHROEDER and co-workers. For the selective synthesis of α- and δ-aminoadipic acid the tosyl-L-piperidone carboxylic acid and its acid chloride is an important key compound.

Also of interest are the α- and γ-phenacyl esters and the corresponding benzhydryl esters of benzyloxycarbonylglutamic acid, which were introduced into peptide chemistry by L. ZERVAS in **1967**. In the reaction of phenacyl bromide and triethylamine with benzyloxycarbonylglutamic acid, for instance, the α-ester is formed preferentially. For the preparation of the isomeric γ-ester, benzyloxycarbonylglutamic acid is treated with trityl chloride and triethylamine, and then phenacyl bromide is added. The benzyloxycarbonyl-L-glutamic acid α-trityl-γ-phenacyl ester produced is heated in alcoholic solution without being isolated when the trityl group is cleaved selectively.

References

1 WOLMAN, Y. (1968). In *Chemistry of the Amino Group*, ed. S. Patai, 669

2 BERGMANN, M. and ZERVAS, L. (1932). *Ber. Dtsch. Chem. Ges.*, **65**, 1192

3 MCKAY, F. C. and ALBERTSON, N. F. (1957). *J. Amer. Chem. Soc.*, **79**, 4686; ANDERSON, G. W. and MCGREGOR, A. C. (1957). *J. Amer. Chem. Soc.*, **79**, 6180

4 CARPINO, L. A. (1957). *J. Amer. Chem. Soc.*, **79**, 4427

5 SCHNABEL, E. (1967). *Annalen*, **702**, 188

6 SCHNABEL, E. *et al.* (1968). *Annalen*, **716**, 175; (1971). *Annalen*, **743**, 57

7 GUIBE-JAMPEL, E. and WAKSELMAN, M. (1971). *Chem. Comm.*, 267

8 HALPERN, B. and NITEKI, D. E. (1967). *Tetrahedron Letters*, 3031

9 NITEKI, D. E. and HALPERN, B. (1969). *Austral. J. Chem.*, **22**, 871

10 KARLSON, S. *et al.* (1970). *Acta Chem. Scand.*, **24**, 1010

11 SCHNABEL, E., KLOSTERMEYER, H. and BERNDT, H. (1972). *Proc. 11th Europ. Peptide Symp.*, Vienna, 1971, p. 69, North-Holland, Amsterdam

12 MEIENHOFER, J. (1970). *J. Amer. Chem. Soc.*, **92**, 3771

13 HISKEY, R. G. *et al.* (1971). *J. Org. Chem.*, **36**, 488

14 SCHNABEL, E. *et al.* (1971). *Annalen*, **743**, 69

15 CHABERLIN, J. W. (1966). *J. Org. Chem.*, **31**, 1658

16 WIELAND, TH. and BIRR, CH. (1967). *Proc. 8th Europ. Peptide Symp.*, Noordwijk, 1966, p. 144, North-Holland, Amsterdam

17 PATCHORNIK, A., AMIT, B. and WOODWARD, R. B. (1970). *J. Amer. Chem. Soc.*, **92**, 6333

18 BIRR, CH., WIELAND, TH. *et al.* (1972). *Proc. 11th Europ. Peptide Symp.*, Vienna, 1971, p. 175, North-Holland, Amsterdam

19 JAEGER, G. and GEIGER, R. (1972). *Proc. 11th Europ. Peptide Symp.*, Vienna, 1971, p. 78, North-Holland, Amsterdam

20 Carpino, L. A. and Han, G. Y. (1970). *J. Amer. Chem. Soc.*, **92**, 5748

21 Horner, L. and Neumann, H. (1965). *Chem. Ber.*, **98**, 3462

22 Okamura, K. *et al.* (1971). *Chem. Ind.*, 929

23 Banks, G. R. *et al.* (1967). *J. Chem. Soc.*, 126

24 Nefkens, G. H. L. (1960). *Nature*, **185**, 309

25 Beyerman, H. C. *et al.* (1971). *Proc. 10th Europ. Peptide Symp.*, Padova, 1969, p. 7, North-Holland, Amsterdam

26 Weygand, F. and Csendes, E. (1952). *Angew. Chem.*, **64**, 136

27 Weygand, F. and Frauenhofer, E. (1970). *Chem. Ber.*, **103**, 2437

28 Panetta, C. A. and Casanova, T. G. (1970). *J. Org. Chem.*, **35**, 4270

29 Helferich, B. *et al.* (1925). *Chem Ber.*, **58**, 872

30 Tamaki, T., Kudo, G. *et al.* (1971). *Yuki Gosei Kagaku Kyokai Shi*, **29**, 599; *Chem. Abstracts*, 1971, 75, 98 808r

31 Zervas, L. *et al.* (1963). *J. Amer. Chem. Soc.*, **85**, 3660

32 Brenner, M. and Huber, W. (1953). *Helv. Chim. Acta*, **36**, 1109

33 Dymocky, M. *et. al.* (1971). *Anal. Biochem.*, **41**, 487

34 Fruton, J. S. *et al.* (1965). *J. Amer. Chem. Soc.*, **87**, 5469

35 Zervas, L. *et al.* (1966). *J. Chem. Soc.*, 1191

36 Weygand, F. and Hunger, K. (1962). *Chem. Ber.*, **95**, 1

37 Stewart, F. H. C. (1968). *Austral. J. Chem.*, **21**, 2543

38 Stewart, F. H. C. (1967). *Austral. J. Chem.*, **20**, 2243

39 Sheehan, J. C. and Davis, G. D. (1964). *J. Org. Chem.*, **29**, 2006

40 Kenner, G. W. and Seely, J. H. (1972). *J. Amer. Chem. Soc.*, **94**, 3259

41 Young, G. T. *et al.* (1968). *Nature*, **217**, 247; (1969). *J. Chem. Soc.*, 1911; (1972). *Proc. 11th Europ. Peptide Symp.*, Vienna, 1971, p. 82, North-Holland, Amsterdam

42 Rydon, H. N. *et al.* (1966). *J. Chem. Soc.*, 807; (1968). *Tetrahedron Letters*, 2525

43 Johnson, B. J. and Jakobs, P. M. (1968). *Chem. Comm.*, 73; (1968). *J. Org. Chem.*, **33**, 452

44 Nefkens, G. H. (1962). *Nature*, **193**, 974

45 Stewart, F. H. C. (1971). *Austral. J. Chem.*, **24**, 2193

46 Wieland, Th. and Racky, W. (1968). *Chimia*, **22**, 375

47 Yajima, H. and Kiso, Y. (1971). *Chem. Pharm. Bull. (Japan)*, **19**, 420

48 Rudinger, J. (1959). *Coll. Czech. Chem. Comm.*, **24**, 95

49 Izumiya, N. *et al.* (1970). *Bull. Chem. Soc. Japan*, **43**, 1883

50 Sakakibara, S. *et al.* (1970). *Bull. Chem. Soc. Japan*, **43**, 3322

51 Sakakibara, S. *et al.* (1967). *Bull. Chem. Soc. Japan*, **40**, 2164

52 Jaeger, G. and Geiger, R. (1970). *Chem. Ber.*, **103**, 1727

53 Jaeger, G. and Geiger, R. (1972). *Proc. 11th Europ. Pepide Symp.*, Vienna, 1971, p. 78, North-Holland, Amsterdam

54 Pless, J. and Guttman, St. (1967). *Proc. 8th Europ. Peptide Symp.*, Noordwijk, 1966, p. 50, North-Holland, Amsterdam

55 Wuensch, E. (1967). *Z. Naturforsch.*, **22b**, 1269

56 Weygand, F., Steglich, W. and Pietta, P. (1967). *Chem. Ber.*, **100**, 3841

57 Chillemi, F. and Merrifield, R. B. (1969). *Biochemistry*, **8**, 4344

58 JAEGER, G., GEIGER, R. and SIEDEL, W. (1968). *Chem. Ber.*, **101**, 3537
59 WEYGAND, F. *et al.* (1968). *Chem. Ber.*, **101**, 923
60 GEIGER, R. *et al.* (1968). *Chem. Ber.* **101**, 2189
61 WIELAND, Th. and SIEBER, A. (1969). *Annalen*, **722**, 222
62 PHOTAKI, I. (1970). *J. Chem. Soc.*, 2687
63 INUKAI, N. *et al.* (1967). *Bull. Chem. Soc. Japan*, **40**, 2913
64 HIRSCHMANN, R. *et al.* (1968). *Tetrahedron Letters*, 3057
65 JAEGER, G. and GEIGER, R. (1973). *Proc. 12th Europ. Peptide Symp.*, Reinhardsbrunn, 1972, p. 90, North-Holland, Amsterdam
66 RUDINGER, J. (1962). *Record Chem. Progr.*, **23**
67 JESCHKEIT, H. and LOSSE, G. (1965). *Z. Chem.*, **5**, 81

2.2.3. Methods for the Formation of Peptide Bond

At present about 50 different coupling methods are known. For the formation of larger peptides, however, only a few variants have proved to be suitable, as is revealed by the statistical analysis of H. N. RYDON (1960 to 1962) and J. H. JONES (1968), respectively.

Table 17. Statistical studies on the practical application of coupling methods (referred to 91 publications)

Method	Percentage of application	
	1960–1962	1968
DCC	68	48
DCC/HONSu	—	10
Azide	43	44
Anhydride	42	24
p-Nitrophenyl ester	24	50

The increasing popularity of *p*-nitrophenyl esters (and other activated esters) is evident. Meanwhile there is a marked increase in the use of DCC-combination procedures and the anhydride method, which is practically free of racemisation.

A comprehensive review of coupling reagents in peptide synthesis has appeared[1]. A list, with physical constants, of some 1000 protected amino acid derivatives which are useful in peptide synthesis has been published[2].

2.2.3.1. Carbodiimide Method

The usefulness of this procedure for the formation of peptide bonds was first recognised by J. C. SHEEHAN and G. P. HESS[3]. Most frequently *N,N′*-dicyclohexylcarbodiimide (DCC) is used. Acylamino acids and amino components are coupled after adding an equimolar quantity of DCC in solutions as concentrated as possible at 0° to give the corresponding peptide derivatives in good yields. According to H. G. KHORANA, the first reaction step proceeds via an addition of the carboxyl group to the diimide, giving an *O*-acyllactim (isourea derivative) which reacts to give the peptide derivative and *N*,*N*′-dicyclohexylurea (route A) or the symmetrical anhydride and *N*,*N*′-dicyclohexylurea (route B). The anhydride is cleaved by aminolysis with the amino component, whereby besides the peptide derivative one mole of the acylamino acid is regenerated (route C). An undesired side reaction is the formation of acylurea derivatives (route D) by rearrangement of the *O*-acyllactim. Acylurea derivatives are not acylating agents and are, moreover, very difficult to separate. In the following reaction scheme the several different pathways are shown.

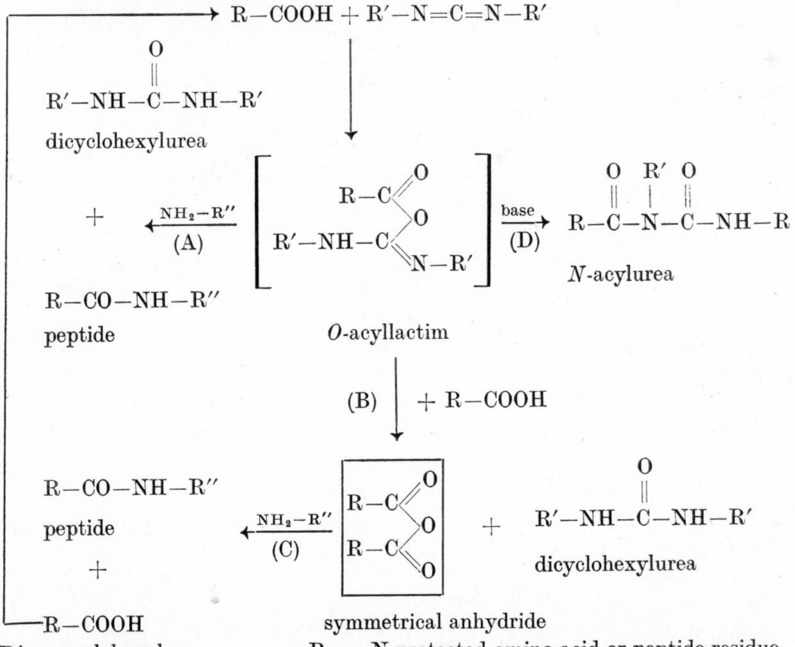

R–COOH + R′–N=C=N–R′

↓

[R–C(=O)–O–C(–NH–R′)=N–R′] *O*-acyllactim

(A) + NH_2–R″ → R–CO–NH–R″ (peptide) + R′–NH–C(=O)–NH–R′ (dicyclohexylurea)

(D) base → R–C(=O)–N(R′)–C(=O)–NH–R′ (*N*-acylurea)

(B) + R–COOH → (R–CO)$_2$O (symmetrical anhydride) + R′–NH–C(=O)–NH–R′ (dicyclohexylurea)

(C) + NH_2–R″ → R–CO–NH–R″ (peptide) + R–COOH (returned to the start)

R′ = cyclohexyl   R = N-protected amino acid or peptide residue
R″ = C-protected amino acid or peptide residue

Using acetonitrile as solvent the acylurea formation is suppressed. The coupling with DCC in dimethylformamide is extremely slow. Dimethylsulphoxide is not advisable as solvent owing to side reactions.

For the activation of asparagine and glutamine derivatives the DCC procedure is not applicable, since the amide group may be dehydrated to a nitrile.

In peptide couplings by means of DCC, racemisation has often been detected. Thus, on the preparation of Z-Leu–Phe–ValOBu^t from Z-Leu–PheOH and ValOBu^t in tetrahydrofuran at 20° F. WEYGAND and coworkers found 12.5 per cent D-Phe and at $-10°$ 6 per cent D-Phe in the protected tripeptide. Gas chromatographic studies on racemisation in peptide synthesis (see p. 170) later made by the same authors showed that the *simultaneous use of dicyclohexylcarbodiimide and 2 equivalents of N-hydroxysuccinimide*[4] enables coupling reactions practically free of racemisation to be carried out. This modified procedure gives high yields and no detectable *N*-acylurea formation.

In peptide syntheses carried out according to the WUENSCH–WEYGAND procedure[4] D. S. KEMP detected between 0.20 and 0.62 per cent racemisation in dimethylformamide but only between 0.024 and 0.12 per cent in tetrahydrofuran, using the highly sensitive racemisation detection method developed by him[5]. One equivalent of *N*-hydroxysuccinimide was used (DCC was added last!)[6]. These results indicate that the WUENSCH–WEYGAND procedure is a satisfactory alternative to the azide method for fragment coupling. However, in some cases side reactions were noted. The origin of these was explained by GROSS and BILK[7]. Nucleophilic attack of *N*-hydroxysuccinimide on the diimide leads to ring-opening and formation of an acylnitrene which undergoes a LOSSEN-type rearrangement to yield the isocyanate. Subsequent addition of a further mole of *N*-hydroxysuccinimide gives succinimidooxycarbonyl-β-alanine-*N*-hydroxysuccinimide ester, the structure of which was proved by its NMR spectrum. This mechanism was confirmed by the reaction of DCC with *N*-hydroxyglutarimide which led to the corresponding aminobutyric acid derivative[8]. The employment of the WUENSCH–WEYGAND procedure seems to give problems in cases such as these when, for steric reasons, the synthesis of the intermediate *N*-hydroxysuccinimide ester proceeds slowly[9]. In this connection the development of additional components whose structural properties exclude such side reactions is important. *N*-hydroxycarbamates[10] and *N*-hydroxybenztriazole[11], etc., have been suggested as suitable compounds in this respect. The KOENIG–GEIGER procedure[11] has already been tried in practice with success.

Generally, in the classical DCC method, complete removal of dicyclohexylurea is difficult. With small peptides the insoluble dicyclohexylurea crystallises nearly quantitatively, whereas poorly soluble peptides are most appropriately purified by repeated boiling with methanol. Because the latter procedure is rarely completely satisfactory, several *modified carbodiimides*, which form soluble urea derivatives, have been developed[12]. Hydrophilic or basic substituents on carbodiimides allow an easy separation by extraction with water or acids: for example, *N*-ethyl, *N'*-(γ-dimethylaminopropyl) carbodiimide.

In order to avoid the difficulties of quantitative separation of the coproduct, the use of a polymeric carbodiimide was proposed by FRANKEL[13]. While the acylamino acid and the growing peptide chain are in solution, the reagent, polyhexamethylenecarbodiimide (I), is bound to an insoluble support.

$$-(CH_2)_6-N=C=N-[-(CH_2)_6-N=C=N-]_n-(CH_2)_6-$$

(I)

2.2.3.2. Azide Method

The azide method[14], introduced into peptide chemistry by Th. CURTIUS in 1902, is still one of the most efficient coupling methods available. CURTIUS synthesised a series of *N*-benzoylated di- up to hexapeptides by this procedure. Both amino acids and peptides in aqueous alkaline solution and amino acid esters in organic solvents have been used as amino components. In contrast to the acid chloride method, a second equivalent of amino component is not required for binding the acid which is formed during the reaction since the liberated hydrazoic acid escapes as a gas from the reaction mixture:

$$R-CO-NH-NH_2 + HNO_2 \xrightarrow[-2H_2O]{} \left[R-C\begin{matrix}\nearrow \overline{O} \\ \searrow \underline{N}=\overset{\oplus}{N}=\overset{\ominus}{\overline{N}}\end{matrix} \longleftrightarrow R-C\begin{matrix}\nearrow \overline{O} \\ \searrow \overset{\ominus}{\overline{\underline{N}}}-\overset{\oplus}{N}\equiv N|\end{matrix} \right.$$

$$\left. \longleftrightarrow R-C\begin{matrix}\nearrow \overline{\underline{O}}|^{\ominus} \\ \searrow \underline{N}-\overset{\oplus}{N}\equiv N|\end{matrix} \right] \xrightarrow{+ NH_2-R'} R-CO-NH-R' + HN_3$$

(R = N-protected amino acid or peptide residue;
R′ = C-protected amino acid or peptide residue)

With the introduction of easily removable N-protecting groups the azide method, which until recently was thought to be the only coupling method free of racemisation, has become the method of choice.

The starting compounds are the well-crystalline N-protected amino acid or peptide hydrazides obtainable from the corresponding esters with hydrazine hydrate in alcohol or dimethylformamide. For conversion to the azides the hydrazides are dissolved in dilute hydrochloric acid, and after cooling to $-10°$ the calculated amount of sodium nitrite in concentrated aqueous solution is added. Convenient reaction media include mixtures of acetic acid and hydrochloric acid, tetrahydrofuran and hydrochloric acid, dimethylformamide and hydrochloric acid or trifluoroacetic acid, etc. The azide formed is extracted with ethyl acetate in the cold, washed and dried and caused to react with the amino component. Some azides can also be precipitated by dilution with ice water and after isolation treated with the amino component in dimethylformamide. Owing to the instability of azides such operations have to be performed at low temperatures. However, with larger N-protected peptide azides the reaction time at room temperature often extends to several days.

In azide syntheses any contact with bases is avoided, since bases promote the racemisation of acyl azides[15]. Instead of triethylamine, *N,N*-diisopropylethylamine, *N*-methyl- or *N*-ethylmorpholine should be used[16].

N-protected amino acid or peptide esters bearing a phthaloyl or a trifluoroacetyl group cannot be converted into the corresponding hydrazides via hydrazinolysis, since under these conditions simultaneous cleavage of the N-protecting group occurs. As an alternative, the reaction of phthaloylamino acid chloride or phthaloylpeptide acid chloride with sodium azide is suitable. F. Weygand and W. Steglich have suggested trityl hydrazine for introducing the azide residue. Trifluoroacetylamino acids are coupled with tritylhydrazine by the DCC method, followed by alkaline removal of the amino-protecting group and acidolytic cleavage of the hydrazine-protecting group:

$$\mathrm{CF_3{-}CO{-}NH{-}CH(R){-}COOH} + \mathrm{H_2N{-}NH{-}C(C_6H_5)_3} \xrightarrow{+\,\mathrm{DCC}}$$

$$\mathrm{CF_3{-}CO{-}NH{-}CH(R){-}CO{-}NH{-}NH{-}C(C_6H_5)_3}$$

$$\xrightarrow{+\,\mathrm{OH^{\ominus}}} \mathrm{H_2N{-}CH(R){-}CO{-}NH{-}NH{-}C(C_6H_5)_3}$$

$$\xrightarrow{+\,\mathrm{H^{\oplus}}} \mathrm{CF_3{-}CO{-}NH{-}CH(R){-}CO{-}NH{-}NH_2}$$

An analogous technique had been previously applied by K. Hofmann to the synthesis of phthaloyl peptide benzyloxycarbonylhydrazides:

$$\text{Pht}{>}N-\underset{\underset{}{}}{\overset{R}{\overset{|}{CH}}}-CO-(NH-\overset{R}{\overset{|}{CH}}-CO)_n-NH-NH-CO-O-CH_2-C_6H_5$$

$$\downarrow + H_2/Pd \qquad\qquad \downarrow + H_2N-NH_2$$

$$\text{Pht}{>}N-\overset{R}{\overset{|}{CH}}-CO-(NH-\overset{R}{\overset{|}{CH}}-CO)_n-NH-NH_2$$

$$H_2N-\overset{R}{\overset{|}{CH}}-CO-(NH-\overset{R}{\overset{|}{CH}}-CO)_n-NH-NH-CO-O-CH_2-C_6H_5$$

(R = side chain residues)

The use of the *tert*-butyloxycarbonylhydrazide residue in combination with the *N*-benzyloxycarbonyl group has since been reported by R. SCHWYZER. These studies were aimed at avoiding the hydrazinolysis of longer peptide esters, which is a difficult procedure.

Since in a quantitative azide coupling the only by-product is gaseous hydrazoic acid, the work-up of the reaction products should not present difficulty in principle. Unfortunately, undesired products generated by the numerous side reactions reduce the practical value of this coupling approach. This position is not changed by the fact that many competitive reactions are greatly suppressed if temperatures between $-10°$ and $+5°$ are used. On the other hand, N-protected aminodicarboxylic acids, hydroxyamino acids and amino acid amides, such as glutamine and asparagine, can be converted into the reactive azides via the hydrazides by means of nitrous acid without special protection of side chain functions. Thus, from benzyloxycarbonyl serine hydrazide the azide is obtained in good yields at low temperature.

At elevated temperatures or on longer standing of the azide the same reaction leads to the formation of a heterocyclic derivative. This and other side reactions can be interpreted as consequences of the CURTIUS rearrangement. The intermediate isocyanate reacts intramolecularly with the β-hydroxyl group:

$$C_6H_5-CH_2-O-CO-NH-\underset{\underset{CH_2-OH}{|}}{CH}-N{=}C{=}O \longrightarrow C_6H_5-CH_2-O-CO-NH-HC\overset{NH}{\frown}C{=}O \;(\text{ring: } HC-CH_2-O-C{=}O)$$

Moreover, isocyanates are capable of reacting with the amino component to form urea derivatives:

$$R-\underset{\underset{O}{\|}}{C}-NH-\underset{\underset{R'}{|}}{CH}-N=C=O + NH_2-R'' \longrightarrow R-\underset{\underset{O}{\|}}{C}-NH-\underset{\underset{R'}{|}}{CH}-NH-\underset{\underset{O}{\|}}{C}-NH-R''$$

This is easily verified since, on hydrolysis of the urea derivative, the amino acid is found to be lost:

$$R-\underset{\underset{O}{\|}}{C}-NH-\underset{\underset{R'}{|}}{CH}-NH-\underset{\underset{O}{\|}}{C}-NH-R'' \xrightarrow[-2\,NH_3;\;-CO_2]{+H_2O} R-COOH + NH_2-R'' + R'-\underset{\underset{O}{\|}}{C}-H$$

Furthermore, nitrosation of the benzene ring in tyrosine, oxidation of *S*-benzylcysteine derivatives to sulphoxides, formation of *N*-nitroso derivatives of tryptophan, decomposition of α-tosylamino acid azides, etc., have been observed. 1,2-Bisacylhydrazines are rapidly formed if the hydrazide not yet converted reacts with the azide already formed:

$$Y-NH-\underset{\underset{R}{|}}{CH}-CON_3 + H_2N-NH-CO-\underset{\underset{R}{|}}{CH}-NH-Y$$

$$\rightarrow Y-NH-\underset{\underset{R}{|}}{CH}-CO-NH-NH-CO-\underset{\underset{R}{|}}{CH}-NH-Y$$

The acyl hydrazide can be converted into an amide, if N_2O is eliminated from the initially formed *N*-nitrosohydrazide:

$$R-C\!\begin{smallmatrix}\nearrow O\\ \searrow NH-NH-NO\end{smallmatrix} \xrightarrow{+H^{\oplus}} R-C\!\begin{smallmatrix}\nearrow O\\ \searrow \overset{\oplus}{N}(H)-\underset{H}{\overset{|}{}}N(H)-NO\end{smallmatrix} \xrightarrow{-H^{\oplus}} R-C\!\begin{smallmatrix}\nearrow O\\ \searrow NH_2\end{smallmatrix} + N_2O$$

According to J. Rudinger, this side reaction can be largely excluded if high nitrite and proton concentrations are used or if the nitrosation is executed in an organic phase with alkyl nitrites and nitrosyl chloride.

A detailed survey of such side reactions is given in the literature[17].

2.2.3.3. Anhydride Method

The foundations for the development of the mixed anhydride procedure were laid by Th. Curtius. During the reaction of benzoyl chloride with silver glycinate, which was expected to result in the benzoylation of the amino acid, in addition to hippuric acid, benzoyl diglycine, triglycine ... up to benzoyl hexaglycine were isolated. Already at that time, anhydrides of *N*-benzoylated amino acids or peptides were presumed to be the reactive intermediates:

$$C_6H_5-COCl + H_2N-CH_2-COOAg \xrightarrow{-AgCl} \begin{cases} C_6H_5-CO-NH-CH_2-COOH \\ C_6H_5-CO-O-CO-CH_2-NH_2 \end{cases}$$

$$C_6H_5-CO-O-CO-CH_2-NH_2 \xrightarrow{C_6H_5COCl} C_6H_5-CO-O-CO-CH_2-NH-CO-C_6H_5$$

$$C_6H_5-CO-O-CO-CH_2-NH-CO-C_6H_5 \xrightarrow[-C_6H_5COOAg]{+\ H_2N-CH_2-COOAg} C_6H_5-CO-NH-CH_2-CO-NH-CH_2-COOH$$

In the same way the reaction can result in hexaglycine derivatives via tri-, tetra-, pentapeptides.

About 70 years after these studies Th. Wieland and his co-workers, starting from these results, utilised the mixed anhydride method for controlled peptide synthesis.

The most useful reagents for forming anhydrides are the carbonic acid half-ester chlorides. Ethyl chloroformate proposed independently by Th. Wieland and R. A. Boissonnas and also isobutyl chloroformate (J. R. Vaughan, 1952) are the most widely used. For the preparation of mixed anhydrides the N-protected amino acids or peptides are dissolved in tetrahydrofuran, dioxane, chloroform or toluene, and after adding an equimolar amount of tertiary base the reaction with the corresponding alkyl chloroformate is accomplished at $-5°$ to $-15°$. In order to prevent racemisation, it is advisable to choose a short 'activating time' (4 to maximum 10 min) and to avoid an excess of tertiary amine.

According to investigations by Anderson and co-workers[18], the best results regarding yield and prevention of racemisation are achieved by use of isobutyl chloroformate as the anhydride-forming reagent and *N*-methylmorpholine as base. These results were confirmed by Kemp and co-workers[15] using their sensitive racemisation test. After an activating time of 1 min with an exactly equivalent amount of *N*-methylmorpholine only 0.01 per cent racemate was detected. If these conditions are adhered to, the mixed anhydride method can be considered to be practically free of racemisation. The preparation of mixed anhydrides must take place with exclusion of moisture, whereas the acylation reaction itself can be performed even in aqueous media. Frequently, it is advantageous to use the amino components as esters in indifferent solvents instead of the sodium salts.

The aminolytic cleavage of the unsymmetrical (mixed) anhydride formed from the carboxyl component and the alkyl chloroformate (see scheme) usually occurs on the carboxyl C-atom of the acyl amino acid, to form the desired peptide derivative and liberating the other acid (route a). In the case of alkyl chloroformates the latter is very unstable and decomposes immediately to CO_2 and ethanol (or isobutanol). However, some examples of nucleophilic attack of the amino component to the wrong carboxyl C-atom have been reported as well, where the acyl amino acid used is eliminated and urethanes (route b) are formed as by-products. This side reaction has been observed especially with tosyl and trityl amino acids. The theoretically possible disproportionation of unsymmetrical anhydrides is suppressed at temperatures below 0°.

$$\text{Y—NH—CH(R)—COOH} + \text{Cl—C(=O)—O—R''} \xrightarrow[-[\text{BH}]^+\text{Cl}^-]{+\overline{B}} \left[\begin{array}{c}\text{Y—NH—CH(R)—C(=O)}\\ \text{O}\\ \text{R''—O—C(=O)}\end{array}\right]$$

$$\left[\ \text{mixed anhydride}\ \right] \xrightarrow[\text{(a)}]{\text{NH}_2\text{—R'}} \text{Y-NH-CH(R)-C(=O)-NH-R'} + \boxed{\text{R''—O—C(=O)OH}} \longrightarrow \text{R''—OH} + CO_2$$

$$\left[\ \text{mixed anhydride}\ \right] \xrightarrow[\text{(b)}]{\text{NH}_2\text{—R'}} \text{Y—NH—CH(R)—C(=O)—OH} + \text{R''—O—C(=O)—NH—R'}$$

Mixed anhydrides from acylamino acids and isovaleric and trimethylacetic (pivalic) acid, respectively, are obtainable in an analogous way and can be coupled in excellent yield (M. ZAORAL, 1962):

$$\text{R—COOH} + \text{Cl—C(=O)—C(CH}_3)_2\text{—CH}_3 \xrightarrow{-\text{HCl}} \left[\begin{array}{c}\text{R—C(=O)}\\ \text{O}\\ \text{CH}_3\text{—C(CH}_3)_2\text{—C(=O)}\end{array}\right] \xrightarrow{+\text{NH}_2\text{—R'}} \text{R—C(=O)—NH—R'} + \text{CH}_3\text{—C(CH}_3)_2\text{—COOH}$$

Owing to the strong +I-effect of the *tert*-butyl grouping the electrophilic potential of the carbonyl C-atom of the pivaloyl group is somewhat depressed, and this with the steric hindrance directs nucleophilic attack to the carbonyl group of the carboxyl component.

It is possible in principle to form peptides via anhydrides of inorganic acids (phosphoric acid, phosphorous acid, arsenous acid, sulphuric acid, sulphurous acid, etc.) as well as other aliphatic and aromatic carboxylic acids and also thio acids. A detailed survey of the numerous variants is given by ALBERTSON[19]. However, most of these procedures, although theoretically of considerable interest, have not found wide practical application.

A rapid quantitative synthesis without purification of intermediates or final products, in which the acylation steps are performed with an excess of 0.6 equivalents of mixed anhydrides, has been published by TILAK[20].

BELLEAU and MALEK[21] suggested *N-ethoxycarbonyl-2-ethoxy-1,2-dihydroquinoline* (EEDQ) as coupling agent. The activation proceeds via mixed anhydrides which are formed after the exchange of the 2-ethoxy substituent on the carboxyl component according to the following mechanism:

$$\text{EEDQ (N-}COOC_2H_5\text{, 2-}OC_2H_5) \xrightarrow{R-COOH} \text{(N-}COOC_2H_5\text{, 2-}O-CO-R) \longrightarrow \text{quinoline} + R-CO-O-CO-OC_2H_5$$

The slow formation of the mixed anhydride and its fast reaction with the amino components prevent undesired side reactions. This coupling reagent seems to be of principal interest for fragment condensations.

As was first shown by WEYGAND[22], stepwise peptide syntheses can be accomplished by means of *symmetrical anhydrides*. Symmetrical anhydrides are readily formed from acylamino acids and ynamines (methylethinyldiethylamine):

$$2\,Y-NH-\underset{\displaystyle R}{\overset{}{CH}}-COOH + CH_3-C\equiv C-N(C_2H_5)_2 \longrightarrow CH_3-CH_2-\overset{\displaystyle O}{\overset{\|}{C}}-N(C_2H_5)_2 + (Y-NH-\underset{\displaystyle R}{CH}-CO)_2O$$

The amino component is added to the solution of the symmetrical anhydride, and the acyl dipeptide usually forms within a few minutes. By extraction with sodium bicarbonate the portion of the acylamino acid which is regenerated in the aminolytic cleavage of the symmetrical anhydride may be largely recovered.

The preparation of symmetrical anhydrides can also be accomplished by reacting 2 equivalents of the sodium salt of a Boc-amino acid with one equivalent of phosgene in tetrahydrofuran at $-40°$ for about five minutes[23]:

$$2\ \text{Boc-NH-CH(R)-C(=O)-}\overline{\underline{O}}|^{\ominus}\text{Na}^{\oplus} + \text{Cl}_2\text{C=O} \xrightarrow[-2\,\text{NaCl};\ -\text{CO}_2]{} \text{Boc-NH-CH(R)-C(=O)-O-C(=O)-CH(R)-NH-Boc}$$

2.2.3.4. *Method of Activated Esters*

From the very beginning of synthetic peptide chemistry amino acid methyl and ethyl esters have been used as acylating agents in the formation of peptide bonds. Even if the pioneering work of Th. CURTIUS and E. FISCHER has found no further use, it resulted in the knowledge that esters of acylated amino acids and peptides represent 'energy-rich' compounds.

Eighty years after these initial studies the coupling method of activated esters was applied by WIELAND[24], who used N-protected amino acid thiophenyl esters for peptide coupling.

A little later R. SCHWYZER succeeded in converting the relatively unreactive acylamino acid methyl esters into intermediates sufficiently reactive for peptide coupling by the introduction of —I-substituents into the alcohol component.

Since then, a great number of activated esters has found successful application in peptide synthesis.

The formation of peptide bonds by aminolysis of esters is classified, by analogy with the saponification of esters, as a bimolecular base-catalysed cleavage of the ester group between acyl residue and oxygen with the symbol $B_{AC}2$:

$$\text{R-}\overline{\text{N}}\text{H}_2 + \text{R}'\text{-C}(=\overline{\underline{O}})\text{-XR}'' \underset{\text{fast}}{\overset{\text{slow}}{\rightleftharpoons}} \text{R-}\overset{\oplus}{\text{N}}\text{H}_2\text{-C}(\text{R}')(\text{-}\overline{\underline{O}}|^{\ominus})\text{-XR}'' \underset{\text{slow}}{\overset{\text{fast}}{\rightleftharpoons}} \text{R-}\underline{\text{N}}\text{H-C}(=\overline{\underline{O}})\text{-R}' + \text{R}''\text{XH}$$

X = O, S or Se; R = residue of the amino component; R′ = residue of the carboxyl component; R″ = substituted or unsubstituted alkyl or aryl residue.

Table 18. Some activated esters used for peptide synthesis $R{-}C{\langle}^{O}_{X}$

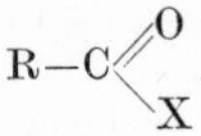

Activated ester (abbreviation)	Formula of the activating component X	Authors
p-Nitrophenyl ester (-ONp)	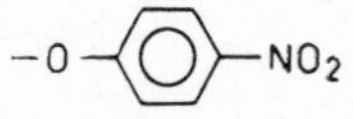 $-O-C_6H_4-NO_2$ (also *o*- and 2,4-dinitrophenyl ester)	M. Bodanszky
Chlorine-substituted phenyl ester	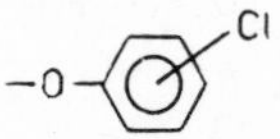 $-O-C_6H_4-Cl$ (2,4-di-, 2,4,6-tri, and pentachlorophenyl ester)	G. Kupryszewski; J. Kovacs
Pentafluorophenyl ester	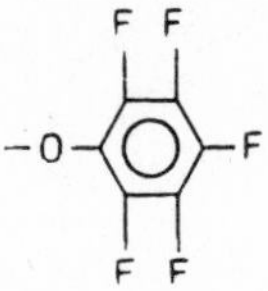 $-O-C_6F_5$	J. Kovacs
Thiophenyl ester (-SPh)	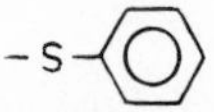 $-S-C_6H_5$ (also *p*-nitro- and *p*-methylthiophenyl ester)	Th. Wieland; G. W. Kenner
Selenophenyl ester (-SePh)	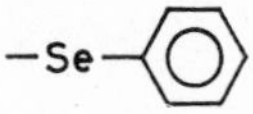 $-Se-C_6H_5$	H.-D. Jakubke
N-Hydroxypiperidyl ester (-Opip)	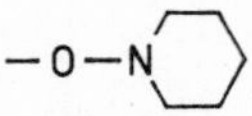 $-O-N$ (piperidyl)	G. T. Young
Quinolyl-(8) ester (-OQ or -OQ(Cl))	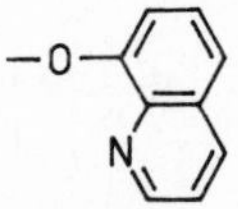 $-O-$ quinolyl-(8) (also 5-chloroquinolyl-(8) ester)	H.-D. Jakubke; A. Voigt

Table 18 (cont.)

Activated ester (abbreviation)	Formula of the activating component X	Authors
o-Hydroxyphenyl ester	$-O-C_6H_4-OH$ (o)	G. T. Young; J. H. Jones
2-Pyridyl ester (-OPyr)	$-O-C_5H_4N$ (2-pyridyl)	G. T. Young; J. S. Morley
2-Thiopyridyl ester (-SPyr)	$-S-C_5H_4N$ (2-pyridyl)	G. T. Young
N-Hydroxysuccinimide ester (-OSu)	—O—N (succinimide ring)	G. W. Anderson
N-Hydroxyphthalimide ester	—O—N (phthalimide ring)	H. G. L. Nefkens; G. I. Tesser
N-Hydroxyglutarimide ester	—O—N (glutarimide ring)	H. Jeschkeit
N-Hydroxyurethane ester	$-O-NH-COOC_2H_5$	H. Jeschkeit
N-Hydroxypyridine-(2,3)-dicarboxylic acid imide ester	—O—N (pyridine-2,3-dicarboximide ring)	E. Taschner
Vinyl ester	$-O-CH=CH_2$	F. Weygand; W. Steglich
Cyanomethyl ester	$-O-CH_2-C\equiv N$	R. Schwyzer

Table 18 (cont.)

Activated ester (abbreviation)	Formula of the activating component X	Authors
Phenylazophenyl ester	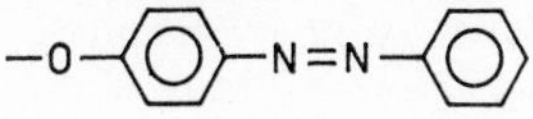 $-O-C_6H_4-N=N-C_6H_5$	G. Kupryszewski; A. Barth
Hydroxamic acid ester	$-O-NH-C(=O)-R$	E. Hoffmann; T. R. Govin-Dachari
O-(Aminoacyl)-oxime	$-O-N=CR_2$	G. Losse
3-Acyloxy-2-hydroxy-*N*-ethyl benzamide	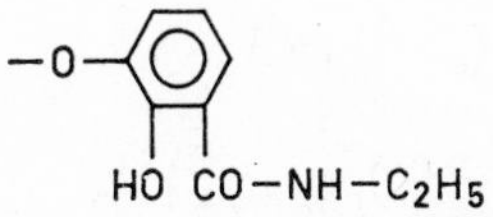 $-O-C_6H_3(OH)-CO-NH-C_2H_5$	D. S. Kemp

The rate-determining step of this reaction is the nucleophilic attack of the amino component on the carbonyl C-atom leading to a tetrahedral adduct. Strongly electron-attracting groupings (—XR″) promote the formation of the adduct. Hence, the formation of the peptide bond will take place most quickly if R″XH is as weak a base as possible, or—in other words—if R″XH is the conjugated base of a relatively strong acid. The cause of the great reactivity lies in the electron-attraction in the anion, which facilitates cleavage of the C—XR″ bond. Differentiation between active and activated esters seems to be unnecessary, since the increase of the acyl potential can always be attributed to electrometric effects.

In the strictest sense of the definition only energy-rich oxygen esters (X=O) should be classed as activated esters, but the term is usually extended to include the analogous sulphur (X=S) and selenium compounds (X=Se), and other activated acyl groupings which enhance the electrophilic character of the carboxyl C-atom, such as *O*-acyl semi-acetals (-ketals), derivatives of hydroxylamine, etc., are also usually called activated esters.

Since the reactivity of an ester directly depends on the acidity of the group R″XH eliminated during the aminolysis, the classification of activated esters as unsymmetrical anhydrides is not actually incorrect, and the thiophenyl esters used for peptide synthesis by Wieland[24] were in fact first thought of as mixed anhydrides. Mixed anhydrides are much more 'energy-rich' than activated esters. Thus, mixed anhydrides react under mild conditions with amino components and also with the weaker nucleophilic hydroxyl groups of alcohols and phenols or the sulphhydryl groups of mercaptans. The reaction of an N-protected activated ester with an amino component leads to an only acylation product, whereas an unsymmetrical anhydride may be cleaved by aminolysis at two sites (p. 116).

Some important activated esters are listed in Table 18, which does not, however, claim to be complete.

Detailed references can be found in the relevant monographs on peptide synthesis as well as in two reviews on activated esters[25,26].

p-Nitrophenyl esters, first described by M. BODANSZKY, and to a lesser extent *chlorine-substituted phenyl esters*, to date have been most popular in the synthesis of biologically active peptides. The activated esters are usually prepared by the DCC method or the mixed anhydride procedure. In addition, many other methods of preparation have been described, of which one deserves special mention. According to M. FRANKEL, in the reaction of benzyl-(*p*-nitrophenyl) carbonate with sodium salts of amino acids the corresponding benzyloxycarbonylamino acid salts are obtained. After acidification and addition of DCC they couple with the *p*-nitrophenol co-product of the first stage to give the N-protected amino acid *p*-nitrophenyl esters in yields between 65 and 80 per cent:

$$C_6H_5\text{-}CH_2\text{-}O\text{-}\overset{\overset{\displaystyle O}{\|}}{C}\text{-}O\text{-}C_6H_4\text{-}NO_2 \;+\; H_2N\text{-}\overset{\overset{\displaystyle R}{|}}{CH}\text{-}COO^{\ominus} \xrightarrow{\Delta}$$

$$C_6H_5\text{-}CH_2\text{-}O\text{-}\overset{\overset{\displaystyle O}{\|}}{C}\text{-}NH\text{-}\overset{\overset{\displaystyle R}{|}}{CH}\text{-}COO^{\ominus} \;+\; O_2N\text{-}C_6H_4\text{-}OH \xrightarrow[\text{2) }+DCC]{\text{1) }+H^{\oplus}}$$

$$C_6H_5\text{-}CH_2\text{-}O\text{-}\overset{\overset{\displaystyle O}{\|}}{C}\text{-}NH\text{-}\overset{\overset{\displaystyle R}{|}}{CH}\text{-}\overset{\overset{\displaystyle O}{\|}}{C}\text{-}O\text{-}C_6H_4\text{-}NO_2$$

By appropriate choice of the benzyl (aryl) carbonates this 'one-pot' procedure is also applicable to the preparation of other activated esters of aralkyloxycarbonyl- and alkoxycarbonyl-amino acids. Thus, in addition, the *p*-nitrophenyl esters, thiophenyl esters and different substituted chlorophenyl esters of benzyloxycarbonylamino acids can also be synthesised this way. *tert*-Butyl (*p*-nitrophenyl) carbonate can be used similarly for the preparation of *tert*-butyloxylcarbonylamino acid *p*-nitrophenyl esters.

The *N-hydroxysuccinimide esters* have attained increasing practical importance. They have the great advantages of (i) giving a high rate of aminolysis and (ii) the water-solubility of the *N*-hydroxysuccinimide liberated during the coupling reaction which enables it to be easily removed.

Pentafluorophenyl esters, which are eminently suitable for the synthesis of sequence polymers (Section 2.4) can be obtained by the mixed anhydride method and via so-called 'pentafluorophenol complexes'[28]. The 4-(methylthio)-phenyl esters can act as carboxyl-protecting groups (see Table 15) but can be converted into the activated *4-methylsulphonyl-phenyl esters* by oxidation, and, hence, offer interesting possibilities for application to fragment condensation[29]. Amino acids sensitive to oxidation, such as methionine, cysteine, cystine and tryptophan, however, preclude this manoeuvre.

In 1965 two most interesting esters, the *N-hydroxypiperidine esters*[30] and the *quinolyl-(8) esters*[31], were proposed for peptide coupling. They are distinguished by an unusually high aminolysis rate, although, on the basis of the high pK_a values of the hydroxy components (12–13 and 9.89, respectively), only a slightly enhanced electrophilic character of the carbonyl C-atom might be expected. This was contrary to the then current view of activated esters, which supposed that the enhanced electrophilic character of the carbonyl carbon arose exclusively from the electromeric effect of the ester component. The pK_a value of *p*-nitrophenol is 7.21 and, therefore, the strong direct activation of the carbonyl C-atom is understandable, because both the acyl cleavage and the stabilisation of the anion expelled on aminolysis are caused by the same electronic interaction. The mechanism[32] first postulated and later verified by kinetic studies on quinolyl-(8) ester aminolysis demonstrates a new principle of carboxyl activation by synchronous proton abstraction from the amino function of the nucleophile:

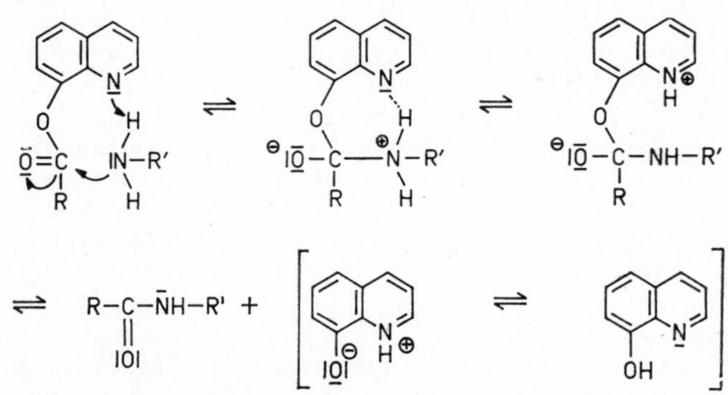

(R = residue of the carboxyl component;
(R' = residue of the amino component)

The rate-determining addition of the attacking amino component is accelerated by intramolecular general base catalysis and the resulting tetrahedral adduct is stabilised by intramolecular H-bonds. In the subsequent fast steps the proton transfer to the tertiary N-atom of the activating component is followed by acyl cleavage.

Conclusive support for this mechanism has been provided by the extensive kinetic investigations of Felton and Bruice[33] on the 6- and 8-acetoxy-quinoline and other model compounds.

The class of activated esters containing an aminolysis-accelerating basic centre includes *o-hydroxyphenyl*, *2-hydroxy-* and *2-thio-pyridine esters* as well as *3-acyloxy-2-hydroxy-N-ethyl benzamides*[34], etc. Only the esters last mentioned are readily obtainable from N-protected peptides:

H, C⊕, N−C_2H_5, O, OH $\xrightarrow[\text{pH 5}]{R-COO^{\ominus}}$ HN−C_2H_5, C=O, OH, O−C(=O)−R

The coupling rate of the *p*-nitrophenyl esters is strongly dependent on the solvent, the highest rate of aminolysis being reached in DMF and DMSO, whereas the reactivity of the 3-acyloxy-2-hydroxy-*N*-ethyl benzamides and also of the other esters of this type is independent of solvent because of the intramolecular general base catalysis[6]. It has also been shown that polar solvents have an inhibiting effect on the aminolysis rate of *N*-hydroxy-succinimide esters. But a mechanistic interpretation of the aminolysis of *N*-hydroxysuccinimide esters with the currently available results is difficult.

The value of activated esters for peptide synthesis was doubted at first, since an additional acylation reaction is required. Apart from the 3-acyloxy-2-hydroxy-*N*-ethyl benzamides already mentioned only the cyanomethyl and vinyl esters are readily preparable by the direct reaction of acylamino acids without carboxyl activation (by treatment with chloroacetonitrile in presence of tertiary amine or by transvinylation with vinyl acetate in presence of a $PdCl_2$–NaCl catalyst). According to Taschner and co-workers[35], bromoacetonitrile is advantageous for the preparation of cyanomethyl esters, whereas Ichimura[36] proposed *p*-bromobenzenesulphonyloxyaceto-nitrile as cyanomethylating agent.

In spite of the labour involved in the preparation of the activated esters, they have some advantages which have resulted in their attaining wide application within a short time. For instance, it is very useful in peptide syntheses to use definite 'energy-rich' intermediates whose chemical and optical purity can be tested beforehand. A collection of activated ester derivatives prepared and in store is the basis for the continuous and rapid synthesis of peptide sequences in the modern peptide laboratory. Moreover, in the stepwise building up of repeating units of amino acids it is possible to accomplish various acylating reactions with one activated ester. Incidentally, it should be borne in mind that among all the 'energy-rich' intermediates used in synthetic peptide chemistry for the formation of peptide bonds, only the activated esters show a parallel to the natural activation of amino acids in protein biosynthesis (see p. 317).

The *isoxazolium salt method* recommended by WOODWARD and coworkers[37,38] for the formation of peptide bonds uses what is—considered from the mechanistic point of view—an 'activated ester' reagent since *N*-ethyl-5-phenyl-isoxazolium-3′-sulphonate reacts with N-protected amino acids or peptides to form an 'energy-rich' enol ester which couples with the amino component without isolation:

$$\text{(isoxazolium salt, } SO_3^{\ominus}\text{, } N\text{–}C_2H_5\text{)} \xrightarrow{\text{IB}} \text{(ketoketenimine, } C{=}N{-}C_2H_5\text{)} \xrightarrow{\text{RCOOH}} \text{(enol ester, } O{-}\overset{O}{\overset{\|}{C}}{-}R\text{, } \overset{}{C}(=O){-}NH{-}C_2H_5\text{)}$$

$$\xrightarrow{+NH_2-R'} R-CO-NH-R' + {}^{\ominus}O_3S-C_6H_4-COCH_2CONH-C_2H_5$$

The acylacetamide derivative which is formed as a by-product is soluble in water, and therefore can be separated easily. Studies aimed at the development of improved isooxazolium salts have been performed by OLOFSON and MARINO[39].

2.2.3.5. *Activated Amide Method*

The application of a series of reactive *N*-acyl derivatives for the formation of peptide bonds has also been described. According to investigations by Th. WIELAND and G. SCHNEIDER *N-aminoacylimidazoles* are suitable for peptide syntheses:

$$R-\overset{O}{\overset{\|}{C}}-N(\text{imidazole}) + R'-NH_2 \longrightarrow R-\overset{O}{\overset{\|}{C}}-NH-R' + \text{imidazole (N–H)}$$

Although the coupling reaction is catalysed by the imidazole liberated, the yields are too small for practical purposes. Better results were obtained using *carbonyldiimidazole*. This reagent reacts with *N*-acylamino acids at room temperature to give the activated amide and carbon dioxide. The activated amide is then coupled with the amino component without intermediate isolation.

$$\text{Im-CO-Im} + \text{R-COOH} \xrightarrow{-CO_2} \text{R-CO-Im} + \text{Im-H}$$

$$\text{R-CO-Im} + NH_2\text{-R}' \longrightarrow \text{R-CO-NH-R}' + \text{Im-H}$$

Thionyldiimidazole and *phosphoryldiimidazole* (phenylphosphoric acid diimidazole) react similarly. Substituted *N-acylpyrazoles*, which can be prepared by the anhydride method from 3,5-dimethylpyrazole and acylamino acids or from acylamino acid hydrazides and acetylacetone, have also been recommended for this purpose.

2.2.3.6. Isocyanate Method

Starting from amino acid ester hydrochlorides and phosgene S. GOLDSCHMIDT and M. WICK synthesised α-isocyanato fatty acid esters (*N*-carbonylamino acid esters), which react with acylamino acids to yield the corresponding peptide derivatives with liberation of CO_2. Together with the phosphazo method described below, this procedure was for a long time supposed to be a prototype of so-called amino activation. As already mentioned (p. 78) such a classification is incorrect, since in fact the reaction proceeds with carboxyl activation through a mixed anhydride of the acylamino acid and carbamic acid:

$$\left[H_3\overset{\oplus}{N}\text{-CH(R)-COOR}'\right]^{\oplus} Cl^{\ominus} \xrightarrow[-3HCl]{+COCl_2} O{=}C{=}N\text{-CH(R)-COOR}'$$

$$O{=}C{=}N\text{-CH(R)-COOR}' \xrightarrow{+\ \text{R-COOH}} \text{R}'\text{OOC-CH(R)-NH-C(=O)-O-C(=O)-R}$$

$$\text{R}'\text{OOC-CH(R)-NH-C(=O)-O-C(=O)-R} \xrightarrow{-CO_2} \text{R-C(=O)-NH-CH(R)-COOR}'$$

The extensive racemisation of phenylalanine in the synthesis of the ANDERSON peptide (see p. 169) from benzyloxycarbonylglycylphenylalanine and *N*-carbonylglycyl ethyl ester[40] provides additional support for the suggested reaction mechanism.

2.2.3.7. *Phosphazo Method*

The reaction of amino acid esters with phosphorus trichloride in pyridine results in the so-called 'phosphazo compounds' (St. GOLDSCHMIDT, 1953), which, without isolation, couple with N-protected amino acids or peptides to give peptide derivatives:

$$PCl_3 + 2\,H_2N\text{-}\underset{|}{\overset{R}{C}}H\text{-}COOR' \longrightarrow P\begin{cases} =N\text{-}CH(R)\text{-}COOR' \\ \text{-}NH\text{-}CH(R)\text{-}COOR' \end{cases}$$

$$\xrightarrow{+2\,R''\text{-}COOH} 2\,R''\text{-}CO\text{-}NH\text{-}CH(R)\text{-}COOR' + HPO_2$$

That this reaction proceeds via a mixed anhydride can be shown unequivocally by racemisation tests. Thus, for reasons to be explained later, if N-protected peptides or benzoylamino acids are used as carboxyl components, in this method the C-terminal amino acid of the carboxyl component is always racemised.

2.2.3.8. *Ynamine Method*

Nearly quantitative yields were obtained in the coupling of benzyloxycarbonylamino acids and amino acid esters with ynamines such as dimethylamino-*tert*-butylacetylene, diethylamino-*tert*-butylacetylene, diethylaminoethylacetylene (R. BUYLE and H. G. VIEHE, 1964), in which an activated ester is formed as an intermediate (J. F. ARENS and co-workers, 1965):

$$C_6H_5\text{-}CH_2\text{-}O\text{-}CO\text{-}NH\text{-}CH(R)\text{-}COOH + R_1\text{-}C\equiv C\text{-}N(R_2)_2 + H_2N\text{-}CH(R)\text{-}COOR'$$

$$\longrightarrow C_6H_5\text{-}CH_2\text{-}O\text{-}CO\text{-}NH\text{-}CH(R)\text{-}CO\text{-}NH\text{-}CH(R)\text{-}COOR' + R_1\text{-}CH_2\text{-}C(=O)\text{-}N(R_2)_2$$

(R_1 = *tert*-butyl or ethyl; R_2 = methyl or ethyl)

During the reaction between optically active N-protected peptides and amino acid esters or peptide esters under these conditions, both a high racemisation rate and addition of the ynamine to the azlactone produced as intermediate (with the formation of enamines) were observed. These side reactions of the ynamine procedure were investigated by F. WEYGAND and his co-workers in 1968.

2.2.3.9. Acid Chloride Method

Acid chlorides of carbethoxyamino acids were used for peptide synthesis by E. FISCHER:

$$H_5C_2-O-CO-NH-\underset{|}{\overset{R}{CH}}-C\begin{matrix}\nearrow O\\ \searrow Cl\end{matrix} + H_2N-\overset{R}{\overset{|}{CH}}-COONa + NaOH \longrightarrow$$

$$H_5C_2-O-CO-NH-\underset{R}{\underset{|}{CH}}-CO-NH-\underset{R}{\underset{|}{CH}}-COONa + NaCl + H_2O$$

However, when the synthesis was carried out in non-aqueous solvents, one equivalent of the amino component was used as HCl-acceptor. Moreover, there was no method of removing the carbethoxy residue under gentle conditions.

The classical investigations of α-halogo carboxylic acid chlorides and amino acid chloride-hydrochlorides by FISCHER's school have already been referred to on p. 80.

After the introduction of the benzyloxycarbonyl-protecting group by M. BERGMANN and co-workers more attention was paid to the acid chloride method. But hopes of combining acid chloride activation with N-blocking by the benzyloxycarbonyl method were frustrated, since benzyloxycarbonyl-amino acid chlorides readily eliminate benzyl chloride to form *N*-carboxy anhydrides on short heating or prolonged storage at room temperature:

$$C_6H_5-CH_2-O-CO-NH-\overset{R}{\overset{|}{CH}}-C\begin{matrix}\nearrow O\\ \searrow Cl\end{matrix} \longrightarrow C_6H_5-CH_2-Cl + \begin{matrix}R-CH-CO\\ |\quad\quad\ \ \ \ O\\ NH-CO\end{matrix}$$

The acid chlorides of tosyl- and phthaloylamino acids, however, are still in occasional current use for peptide synthesis. According to recent studies by M. ZAORAL and Z. ARNOLD, besides phosphorus pentachloride and thionyl chloride, dimethylchlorodimethyleneammonium chloride (which is readily prepared from phosgene and dimethylformamide: see scheme) is also a suitable chlorinating agent.

$$(H_3C)_2N{-}CH{=}O \quad + \quad COCl_2 \quad \longrightarrow \quad \left[(CH_3)_2\overset{\oplus}{N}{=}CH{-}Cl\right]^{\oplus} \quad Cl^{\ominus}$$

In some modern variants of FISCHER's acid chloride method the acid chlorides are produced *in situ* by reactive halogen derivatives and then coupled with the amino component immediately. For this purpose L. HESLINGA and J. F. ARENS proposed α-chlorovinyl ether and α,α-dichlorodiethyl ether, while A. RIECHE and H. GROSS suggested the more accessible dichloromethyl ether.

2.2.3.10. *Aminoacyl Insertion*

The so-called aminoacyl insertion is a very interesting principle of peptide formation. M. BRENNER found that *O*-(α-aminoacyl)-salicyloylamino acids undergo base-catalysed rearrangement to give salicyloylamino acids. Thus, the preparation of salicyloyl tripeptide esters from *O*-glycylsalicyloyl dipeptide esters was successful:

$$C_6H_4(O{-}CO{-}CH_2{-}NH_2){-}CO{-}NH{-}CH(R){-}CO{-}NH{-}CH(R'){-}COOR'' \xrightarrow{\text{(base)}}$$

$$C_6H_4(OH){-}CO{-}NH{-}CH_2{-}CO{-}NH{-}CH(R){-}CO{-}NH{-}CH(R'){-}COOR''$$

However, this procedure is of only limited practical importance, since after completion of chain elongation it is very difficult to remove the salicyloyl residue. In addition, the introduction of the *O*-aminoacyl residue requires considerable preparative effort, and the method could not be successfully applied to either serine or threonine derivatives.

The extension of the principle of aminoacyl insertion to monoacylated N,N′-diaminoacyl hydrazines is significant, however, since these are easily accessible from N-protected amino acids or peptide hydrazides by reaction with *N*-carboxy anhydrides in acetic acid:

$$\text{Y-NH-}\underset{\displaystyle\text{R}}{\text{CH}}\text{-CO-NH-NH}_2 + \text{NCA (RCH-C(=O)-O-C(=O)-NH)} \xrightarrow{-CO_2} \text{Y-NH-}\underset{\displaystyle\text{R}}{\text{CH}}\text{-CO-NH-NH-CO-}\underset{\displaystyle\text{R}}{\text{CH}}\text{-NH}_2$$

The rearrangement occurs smoothly in the presence of weak organic acids (pivalic or propionic acids):

$$\begin{array}{l} H_2N-CH(R')-C(=O)-NH \\ Y-NH-CH(R)-C(=O)-NH \end{array} \longrightarrow Y-NH-CH(R)-C(=O)-NH-CH(R')-CO-NH-NH_2$$

The N-protected peptide hydrazides produced in this manner can be used for the azide coupling or oxidised to peptide derivatives with a free carboxyl group.

2.2.3.11. *Peptide Syntheses via N-carboxy Anhydrides (NCA)*

The susceptibility of *N*-carboxy anhydrides (LEUCHS anhydrides) towards nucleophilic agents opens up further possibilities for the formation of peptide bonds.

These compounds were first prepared by LEUCHS, a pupil of E. FISCHER, in 1903, by thermal elimination of ethyl chloride from *N*-carbethoxyamino acid chlorides. More convenient starting materials for this reaction are methoxycarbonyl- and benzyloxycarbonylamino acid chlorides. A very simple synthesis is the reaction of free amino acids with phosgene:

$$H_2N-\underset{\displaystyle R}{CH}-COOH + COCl_2 \xrightarrow{-2HCl} \text{RCH-C(=O)-O-C(=O)-NH (ring)}$$

In this type of anhydride the carboxyl group is activated and, at the same time, the amino function is blocked. Under coupling conditions spontaneous liberation of the amino group occurs. Traces of moisture can initiate the polymerisation of *N*-carboxylic acid anhydrides. First, opening of the anhydride ring takes place, forming the carbamic acid from which the amino acid is regenerated by CO_2 removal. The amino acid then opens a second molecule of anhydride by aminolysis, to form a dipeptide carbamic acid, which again after CO_2 elimination reacts with a further *N*-carboxy anhydride and so on to give polymer:

$$\text{RHC(NH-CO-O-CO)} + H_2O \rightarrow \text{R-CH(NH-COOH)-COOH} \xrightarrow{-CO_2} \text{R-CH(NH}_2\text{)-COOH}$$

$$\xrightarrow{\text{RHC(NH-CO-O-CO)}} \text{R-CH(NH-COOH)-CO-NH-CH(R)-COOH} \xrightarrow{-CO_2} \text{R-CH(NH}_2\text{)-CO-NH-CH(R)-COOH}$$

In 1949, by excluding the competing reaction which leads to polymers, J. L. Bailey succeeded in preparing the carbamic-acidic salts of peptide esters by treating amino acid *N*-carboxy anhydrides simultaneously with amino acid esters and triethylamine. On cautious heating (30–40°) the carbamic acid salts decomposed to triethylamine, CO_2, and the desired peptide ester:

$$\text{RHC(NH-CO-O-CO)} + H_2N\text{-CH(R')-COOR''} \xrightarrow{N(C_2H_5)_3} [\text{R''OOC-CH(R')-NH-CO-CH(R)-NH-COO}^{\ominus}]^- [\text{H}\overset{\oplus}{\text{N}}(C_2H_5)_3]^+$$

$$\xrightarrow[30\text{-}40°]{-CO_2;\ -N(C_2H_5)_3} H_2N\text{-CH(R)-CO-NH-CH(R')-COOR''}$$

W. Langenbeck has suggested using tribenzylamine as tertiary base, since the corresponding carbamic acid salts are of lower solubility.

Owing partly to difficulties in preparation, storage and handling of the Leuchs anhydrides, the procedure has not found wide application to the synthesis of oligopeptides.

However, an elegant modification has revived the interest of peptide chemists in this method[41]. In this *controlled peptide synthesis in aqueous media* amino acids and peptides are rapidly acylated by crystalline *N*-carboxy anhydrides in the cold at a pH of 10.2. The elimination of CO_2 from the peptide carbamate formed is then accomplished by lowering the pH value of the solution to 3–5. The pH is then raised to 10.2 again and addition of the next LEUCHS anhydride begins a new cycle. After three steps, in one example, the yield amounted to more than 50 per cent. The following scheme gives the course of reaction:

$$R_1\text{-CH(NH}_2\text{)-COO}^\ominus \xrightarrow[\text{pH 10}]{+\ R_2\text{-CH-CO-O-CO-NH (cyclic)}} R_2\text{-CH(NH-COO}^\ominus\text{)-CO-NH-CH(R}_1\text{)-COO}^\ominus \xrightarrow[-CO_2]{\text{pH 3-5}}$$

$$R_2\text{-CH(NH}_3^\oplus\text{)-CO-NH-CH(R}_1\text{)-COO}^\ominus \xrightarrow[\text{pH 10}]{R_3\text{-CH-CO-O-CO-NH (cyclic)}} R_3\text{-CH(NH-COO}^\ominus\text{)-CO-NH-CH(R}_2\text{)-CO-NH-CH(R}_1\text{)-COO}^\ominus$$

$$\xrightarrow[-CO_2]{\text{pH 3-5}} R_3\text{-CH(NH}_3^\oplus\text{)-CO-NH-CH(R}_2\text{)-CO-NH-CH(R}_1\text{)-COO}^\ominus \quad \text{etc.}$$

In order to eliminate the danger of an undesired carbamate exchange, extremely fast stirring is necessary. It is essential to the smooth course of reaction of acylation to have precise control of the pH (10.1–10.5 for amino acids; 10.2 for peptides), because at pH above 10.5 hydantoic acids are formed as by-products by hydrolysis. No racemisation occurs. The *NCA method* is excellent for the synthesis of fragments without the isolation of intermediates, since polyfunctional amino acids with the exception of lysine and cysteine do not need special protection of their side chain functions (*minimum side chain protection*).

Since the thiocarbamate salts are more stable than the corresponding carbamates, the sulphur analogues of NCAs, *2,5-thiazolidinediones* (*N-thiocarboxy anhydrides, NTAs*), have also been applied to peptide syntheses. The acylation can be performed at a lower pH (9–9.5), so that hydrolytic conversion to hydantoic acid is excluded[42,43].

The *NCA–NTA technique* was proved successful in the total synthesis of the S-protein of ribonuclease (see Section 3.6.2.5.)[44]. For the formation of this polypeptide chain containing 104 amino acid residues, 17 fragments were synthesised; 40 per cent of the acylation reactions were accomplished by means of NCAs or NTAs, and the remainder through *N*-hydroxysuccinimide esters. Combination of the fragments (each consisting of between 6 and 17 amino acid esters) to give the S-protein was performed by azide couplings. In presence of the S-peptide the S-protein displayed RNase-S activity.

2.2.3.12. Redox Reaction Coupling

According to investigations in the laboratories of Yu. W. MITIN and T. MUKAIYAMA (1968), the formation of peptide bonds can be accomplished by means of derivatives of trivalent phosphorus such as triethyl phosphite ($R'' = OC_2H_5$) or triphenylphosphine ($R'' = C_6H_5$) in combination with 'soft oxidising agents' of the type ArSX (X = Cl, SCN, S-aryl, NHR‴, etc.). In a redox reaction an intermediate *acyloxyphosphonium salt* (I) is formed from the corresponding organic phosphorus compound, the oxidising agent ($A^{\oplus}$) and the carboxylate grouping of the carboxyl component. In the presence of a tertiary base (B) this reacts with the amino component to give the peptide derivative:

$$R\text{-}COO^{\ominus} + P(R'')_3 + A^{\oplus} \longrightarrow R\text{-}COOP^{\oplus}(R'')_3A^{\ominus} \quad (I)$$

$$I + NH_2\text{-}R' + B \longrightarrow R\text{-}CO\text{-}NH\text{-}R' + BH^{\oplus}A^{\ominus} + O{=}P(R'')_3$$

(R = residue of the carboxyl component, R′ = residue of the amino component)

Th. WIELAND and co-workers (1971) and S. YAMADA and Y. TAKEUCHI (1971) used $Cl^{\oplus}$ (from CCl_4) as oxidising agent, whereas $Br^{\oplus}$ (from $CBrCl_3$) has been recommended by L. E. BARSTOW and V. J. HRUBY. Since the use of several phosphines has been shown to produce considerable racemisation, S. YAMADA and co-workers proposed amides of phosphorous acid ($R'' = N(CH_3)_2$, $N(C_2H_5)_2$, and *N*-methylpiperazine, etc.), which, together with their oxidation products, are easily removed by dilute acids.

References

1 KLAUSNER, Y. S. and BODANSZKY, M. (1972). *Synthesis*, 453

2 FLETCHER, G. A. and JONES, J. H. (1972). *Internat. J. Protein Res.*, **4**, 347

3 SHEEHAN, J. C. and HESS, G. P. (1955). *J. Amer. Chem. Soc.*, **77**, 1067

4 WUENSCH, E. and DRESS, F. (1966). *Chem. Ber.* **99**, 110; WEYGAND, F., HOFFMANN, D. and WUENSCH, E. (1966). *Z. Naturforsch.*, **21b**, 426

5 KEMP, D. S. *et al.* (1970). *J. Amer. Chem. Soc.*, **92**, 1043

6 KEMP, D. S. (1972). *Proc. 11th Europ. Peptide Symp.*, Vienna, 1971, p. 1, North-Holland, Amsterdam

7 GROSS, H. and BILK, L. (1968). *Tetrahedron*, **24**, 6935

8 JESCHKEIT, H. (1969). *Z. Chem.*, **9**, 111

9 WEYGAND, F., STEGLICH, W. and CHYTIL, N. (1968). *Z. Naturforsch.*, **23b**, 1391

10 Jeschkeit, H. (1969). *Z. Chem.*, **9**, 266
11 Koenig, W. and Geiger, R. (1970). *Chem. Ber.*, **103**, 788, 2024, 2034
12 Sheehan, J. C., Presten, J. and P. A. Cruickshank (1965). *J. Amer. Chem. Soc.*, **87**, 2492; Sheehan, J. C., Cruickshank, P. A. and Boshart, G. L. (1961). *J. Org. Chem.*, **26**, 2525; Losse, G. and Sirch, H. J. (1967). *Z. Chem.*, **7**, 234
13 Wolman, Y., Kivity, S. and Frankel, M. (1967). *Chem. Comm.*, 629
14 Curtius, Th. (1902). *Ber. Dtsch. Chem. Ges.*, **35**, 3226
15 Kemp, D. S., Bernstein, Z. and Rebek, J., Jr. (1970). *J. Amer. Chem. Soc.*, **92**, 4756
16 Sieber, P. and B. Riniker (1972). *Proc. 11th Europ. Peptide Symp.*, Vienna, 1971, p. 49, North-Holland, Amsterdam
17 Schnabel, E. (1962). *Annalen*, **659**, 168
18 Anderson, G. W., Zimmermann, J. E. and Callahan, F. M. (1967). *J. Amer. Chem. Soc.*, **89**, 5012
19 Albertson, N. F. In *Organic Reactions*, Vol. 12, p. 157, Wiley, New York
20 Tilak, M. A. (1970). *Tetrahedron Letters*, 849
21 Belleau, B. and Malek, G. (1968). *J. Amer. Chem. Soc.*, **90**, 1651
22 Weygand, F., Huber, P. and Weiss, K. (1967). *Z. Naturforsch.*, **22b**, 1034; (1969). ibid., **24b**, 314
23 Wieland, Th. *et al.* (1971). *Angew. Chem.*, **83**, 333
24 Wieland, Th., Schaffer, W. and Bokelmann, E. (1951). *Annalen*, **573**, 99
25 Jakubke, H. D. (1966). *Z. Chem.*, **6**, 52
26 Garg, H. G. (1970). *J. Sci. Ind. Res. India*, **29**, 236
27 Kisfaludy, L., Kovacs, J. *et al.* (1970). *J. Org. Chem.* **35**, 3563
28 Kovacs, J. *et al.* (1967). *J. Amer. Chem. Soc.*, **89**, 183
29 Johnson, B. J. *et al.* (1968). *Chem. Comm.*, 73; (1968). *J. Org. Chem.* **33**, 4521; (1970). *ibid.*, **35**, 255
30 Beaumont, S. M., Handford, B. O. and Young, G. T. (1965). *Acta Chim. Acad. Sci. Hung.*, **44**, 37
31 Jakubke, H. D. (1965). *Z. Naturforsch.*, **20b**, 273; Jakubke, H. D. and Voigt, A. (1966). *Chem. Ber.*, **99**, 2419
32 Jakubke, H. D., Voigt, A. and Burkhardt, S. (1967). *Chem. Ber.*, **100**, 2367
33 Felton, St. M. and Bruice, Th. C. (1969). *J. Amer. Chem. Soc.*, **91**, 6721
34 Kemp, D. S. and Chien, S. W. (1967). *J. Amer. Chem. Soc.*, **89**, 2743
35 Taschner, E. *et al.* (1965). *Acta Chem. Acad. Sci. Hung.*, **44**, 67
36 Ichimura, K. (1970). *Bull. Chem. Soc. Japan*, **43**, 1572
37 Woodward, R. B. and Olofson, R. A. (1961). *J. Amer. Chem. Soc.*, **83**, 1007
38 Woodward, R. B. *et al.* (1961). *J. Amer. Chem. Soc.*, **83**, 1010
39 Olofson, R. A. and Marino, Y. L. (1970). *Tetrahedron*, **26**, 1779
40 Losse, G. and Goedicke, W. (1967). *Chem. Ber.*, **100**, 3314
41 Denkewalter, R. G., Hirschmann, R. *et al.* (1966). *J. Amer. Chem. Soc.*, **88**, 3163
42 Denkewalter, R. G., Hirschmann, R. *et al.* (1968). *J. Amer. Chem. Soc.*, **90**, 3254
43 Denkewalter, R. G., Hirschmann, R. *et al.* (1971). *J. Org. Chem.*, **36**, 49
44 Hirschmann, R. and Denkewalter, R. G. (1970). *Naturwissenschaften*, **57**, 145

2.2.4. Peptide Synthesis on Polymeric Supports

Although classical peptide synthesis has had great success which has been crowned by total syntheses of insulin, ACTH, glucagon, secretin and calcitonin, the limits of conventional procedures may have been reached. For the synthesis of polypeptides and higher proteins, simplification and automation of the synthetic operations becomes absolutely necessary.

In 1962 R. B. MERRIFIELD of the Rockefeller Institute in New York put forward an entirely new concept which simplified the classical approach in a revolutionary way. This novel method is based on the idea of attaching an amino acid through its carboxyl group to an insoluble, easily filterable polymer and the assembly of the peptide chain in a stepwise manner starting from the C-terminal end. For this purpose an N-protected amino acid is made to react with a reactive grouping of a synthetic resin. Then the protecting group of the amino acid covalently anchored to the supporting particle is removed, the resulting aminoacyl polymer is purified and coupled with a second N-protected amino acid: that is, the peptide chain is built up by stepwise chain elongation in the interior of the resin matrix. The final stage of a MERRIFIELD synthesis is the cleavage of the covalent bond between the C-terminal amino acid of the polypeptide chain and the polymeric support. Several variants of this synthetic technique have been described, which together with the MERRIFIELD technique are discussed in the following sections.

2.2.4.1. The Use of Insoluble High-molecular Supports

In contrast to the procedure developed by R. B. MERRIFIELD, it is of course possible to join an amino acid to a solid support through the amino function so as to enable stepwise synthesis starting from the N-terminal end. Such an approach has been advocated by R. L. LETSINGER. The application of polymeric activated esters to the synthesis of linear and cyclic peptides will be described in this section even though in this method the growing peptide chain is not assembled on a solid support.

2.2.4.1.1. Stepwise Chain Elongation from the C-terminal End (Solid-Phase Peptide Synthesis, MERRIFIELD synthesis[1–8])

The principle of MERRIFIELD synthesis is shown schematically in Figure 16. A copolymer of styrene and divinylbenzene (DVB) usually serves as the polymeric support. Polystyrene cross-linked with 2 per cent DVB is appropriate for peptides with up to about 20 amino acids[9], whereas for the syn-

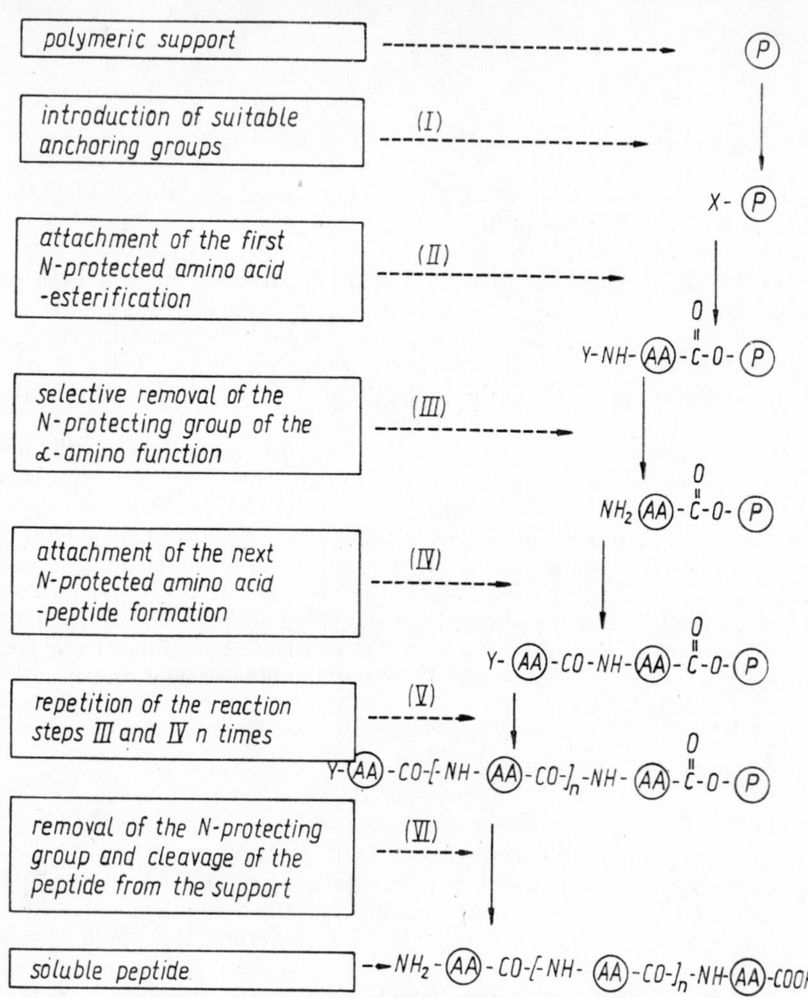

Figure 16. Schematic course of MERRIFIELD synthesis

thesis of longer peptide chains a polystyrene cross-linked with only 1 per cent DVB is more suitable owing to its increased swelling ability and, therefore, enhanced permeability to reagents[10]. Using these less cross-linked resins, however, there is the danger of partial solubility in the solvents used in the solid-phase synthesis. In addition, phenol-formaldehyde resins have been proposed instead of cross-linked polystyrene[11,12]. In order to overcome the dependence on diffusion, BAYER and co-workers[13] employed so-called 'pellicular resins'[14] and 'brush resins'[15], respectively. The former are obtained by polymerisation of a thin layer of a polystyrene resin onto glass beads with a diameter of about 100 μm. Their very poor mechanical stability, however, allows their use only in columns, but 'brush resins' have excellent chemical and mechanical stability. By grafting 1,4-dihydroxymethylbenzene onto appropriate inorganic supports, such as silica gel or glass powder, functional groups which stand out from the surface like bristles of a brush are fixed covalently. Both these types of support, however, have only low capacities.

For some years TREGEAR and co-workers have been working on the solid phase synthesis of peptides using graft-copolymers of styrene and Kel-F or Teflon as a carrier. The advantage of this technique over the normal MERRIFIELD-type resins (low cross-linked polystyrene) might be that the grafted styrene chains practically behave like dissolved styrene in the reaction mixture. Consequently the rigid-matrix problem with its steric hindrance might be absent in this approach.

According to SHEPPARD[5], several of the deficiencies of the MERRIFIELD synthesis arise from the difference in polarity between the hydrocarbon backbone of the cross-linked polystyrene matrix and the peptide chain attached to it. Consequently, it might be desirable for the polymeric support and the growing peptide chain to have structural similarities so that insoluble synthetic polyamides and even polypeptides with appropriate physicochemical properties might be suitable as MERRIFIELD supports. In fact, polar polymeric supports, such as Sephadex LH-20 and modified G-25 types[2,16], had been used previously in peptide synthesis. Investigations on the usefulness of modified polyacryl amides as supports are being carried out[5].

The *introduction of suitable anchoring groups* (*I*) onto the polymeric support is necessary for the covalent attachment of the first residue—that is, the C-terminal amino acid of the peptide to be constructed. The classical anchoring group is the chloromethyl group ($X = Cl{-}CH_2{-}$), which is relatively easily introduced by treatment with chloromethyl ether in the presence of stannic chloride. Under normal conditions about 10–20 per cent of the aromatic rings in the copolymer are substituted. The degree of substitution can be varied, as it depends on the amount of catalyst and the reaction temperature. Bromomethyl or iodomethyl groups ($X = Br{-}CH_2{-}$ or $I{-}CH_2{-}$)[17] as well as chloracetyl and bromoacetyl groups ($X = Cl{-}CH_2{-}CO{-}$ or $Br{-}CH_2{-}CO{-}$) are reactive enough[18] to enable the attachment of N-protected amino acids and peptides[18] under very mild conditions. Hydroxymethyl ($X = HO{-}CH_2{-}$)[19], hydroxybutyryl ($X = HO{-}(CH_2)_3{-}$), $HO{-}(CH_2)_n{-}N(CO{-}CH_3){-}CH_2{-}$[20] and $HO{-}CH_2{-}CH_2{-}SO_2{-}CH_2{-}$ groups[21] are all possible alternative attaching groups, and sulphonium ($X = Cl^{\ominus}(CH_3)_2{-}\overset{\oplus}{S}{-}CH_2{-}$)[22] and benzhydryl halide residues

($X = Br(Cl){-}CH(C_6H_5){-}$)[23] have also been suggested. The anchoring groups $X = HO{-}C_6H_4{-}S{-}CH_2{-}$[24] and $X = H_2N{-}NH{-}CO{-}O{-}C(CH_3)_2{-}CH_2{-}CH_2{-}$[25] are especially ingenious and suitable for the solid-phase synthesis of partially protected peptide fragments. After N-methylation with diazomethane acylsulphonamides are readily cleaved by basic reagents[26], and this suggests a further possible mode of attachment.

Dehydroalanine acts as vehicle for the attachment of peptides to insoluble support (E. Gross and co-workers, 1973):

$$R{-}NH{-}CHR'{-}CO{-}NH{-}\underset{}{\overset{\overset{\displaystyle CH_2}{\|}}{C}}{-}COO{-}CH_2{-}\text{Ⓟ}$$

$$\Big\downarrow \begin{array}{l} N\ HCl\ /\ HOAc \\ 1\ \text{equ. } H_2O\ (50°;\ 30\ \text{min}) \end{array}$$

$$R{-}NH{-}CHR'{-}\overset{\overset{\displaystyle O}{\|}}{C}{-}NH_2 \quad + \quad CH_3{-}\overset{\overset{\displaystyle O}{\|}}{C}{-}\overset{\overset{\displaystyle O}{\|}}{C}{-}O{-}CH_2{-}\text{Ⓟ}$$

This procedure can be used in synthesis of peptides with C-terminal amide groupings.

The method of *attachment of the first amino acid* (*II*) to the polymeric support depends on the the anchoring group introduced. In most cases ester bonds are formed. The numerous anchoring groups mentioned above indicate that there is no ideal method of attaching the peptide to the resin bond yet available. Such a bond must be obtainable simply, it must be absolutely stable during repeated de-protecting and coupling reactions and it must be removable under mild conditions after the synthesis is finished.

The esterification of the standard chloromethyl–Merrifield resin is usually performed in ethanol, ethyl acetate, benzene or tetrahydrofuran under reflux for 24–48 hours:

$$\text{Ⓟ}{-}CH_2{-}Cl + [(C_2H_5)_3\overset{\oplus}{N}H]^{\oplus}\ {}^{\ominus}OOC{-}\overset{\overset{\displaystyle R}{|}}{C}H{-}NH{-}Y \xrightarrow[-[(C_2H_5)_3\overset{\oplus}{N}H]^{\oplus}Cl^{\ominus}]{} \text{Ⓟ}{-}CH_2{-}O{-}\overset{\overset{\displaystyle O}{\|}}{C}{-}\overset{\overset{\displaystyle R}{|}}{C}H{-}NH{-}Y$$

According to A. Loffet, the use of tetramethylammonium hydroxide as base not only results in higher yields but also avoids the formation of quaternary ammonium groups on the support.

Using the more reactive bromomethylated or iodomethylated support, the esterification is effected at room temperature within a few hours[17]. Cross-linking of the resin, which usually occurs during chloromethylation of the support, can also occur later through reaction between free amino groups of amino acids or peptides attached to the support and the residual chloromethyl residues. For this reason excess chloromethyl groups should be blocked by heating the support with sodium acetate in dimethylformamide. Some authors also report observation of reaction between triethylamine and chloromethyl residues leading to quaternary ammonium groups. Whether the ion exchange capacity of such supports gives rise to problems in MERRIFIELD synthesis is not yet completely clarified.

Some amino acids cannot be esterified onto chloromethylated resin without side reactions. For instance, Boc-methionine is attached as an ester in a yield of only about 50 per cent, whereas the remaining portion is attached to the support through the sulphur atom[27]. A better method is the use of Boc-Met(O)-P, followed by reduction of Boc-Met(O)-P to Boc-Met-P (K. BRUNFELDT and co-workers, 1971). DORMAN and LOVE[22] treated chloromethyl resin with dimethyl sulphide and replaced the chloride of the resulting dimethyl-arylmethylene-sulphonium support by carbonate or hydrogen carbonate. After neutralising the basic resin with the N-protected amino acid, drying and heating to 80–85°, the ester bond is formed with liberation of dimethyl sulphide. This reaction proceeds unambiguously, leads to higher yields and achieves more complete conversion of the chloromethyl residues available.

The ester bond between the N-protected starting amino acid and hydroxymethyl[19] and hydroxyalkylamino resins[20], respectively, can be formed either by carbonyldiimidazole or *N,N*-dimethylformamide dineopentylacetal as shown:

$$\text{Ⓟ}-CH_2-OH \quad + \quad HOOC-R \quad + \quad (CH_3)_2NCH\left[-O-CH_2-C(CH_3)_3\right]_2 \longrightarrow$$

$$\text{Ⓟ}-CH_2-O-\overset{\overset{\displaystyle O}{\|}}{C}-R \quad + \quad (CH_3)_2N-C(=O)H \quad + \quad 2\,(CH_3)_3C-CH_2-OH$$

TESSER[21] treated a chloromethyl resin with thioglycol and oxidised the resulting thio ether with perbenzoic acid to give the sulphone. The N-protected amino acid could then be attached by means of DCC:

$$\text{Ⓟ}-CH_2-SO_2-CH_2-CH_2-OH \quad + \quad HOOC-\underset{}{\overset{\overset{\displaystyle R}{|}}{C}}H-NH-Y \xrightarrow{DCC}$$

$$\text{Ⓟ}-CH_2-SO_2-CH_2-CH_2-O-\overset{\overset{\displaystyle O}{\|}}{C}-\overset{\overset{\displaystyle R}{|}}{C}H-NH-Y$$

WEYGAND[18] showed that the chloroacetyl and bromoacetyl resins which may be simply prepared via FRIEDEL–CRAFTS reactions, can be linked with *N*-acyl peptides at room temperature in the presence of ethyl-diisopropylamine. After the synthesis is finished, the phenacyl ester bond can be cleaved by thiophenolate which liberates the peptide acid or by ammonia and hydrazine, which give the corresponding amide and hydrazide, respectively.

The requirement that the resin must show total stability during the coupling and mild de-protection steps of solid-phase synthesis are extremely hard to achieve in practice. However, by applying the so-called 'safety catch' principle it is possible, in principle, to overcome these difficulties[5,28]. In this approach one uses an anchoring bond which is absolutely stable under the conditions of all the intermediate stages but which can be removed under mild conditions at the conclusion after labilisation by a specific chemical reaction.

After the first amino acid has been attached, an amino acid analysis must be performed. For this purpose a sample is hydrolysed totally and the amino acid content is ascertained by colorimetry or by means of an analyser (see p. 46). A knowledge of the amino acid loading of the resin is needed to gauge the dosage of reagents in the subsequent reaction steps and as a standard for the calculation of the total yield of the peptide synthesised. A loading from 0.1 to 0.6 mmol of amino acid per gram of resin has proved to be convenient.

The *selective removal of the N-protecting group from the α-amino function* (III) is the next step in the Merrifield synthesis. The choice of a suitable α-amino-protecting group depends on the stability of the anchoring bond between the starting amino acid and the polymeric support as well as on the protecting groups used for the side chain functions.

Merrifield[9] initially used the benzyloxycarbonyl group (see p. 82) for the temporary masking of the α-amino function. This, however, did not turn out to be a good choice, since the support had to be nitrated, in order to enable a selective removal of the Z-group by HBr/glacial acetic acid. This is possible because the nitrobenzyl ester bond is resistant to this removing agent. After the synthesis was complete, liberation of the peptide from the support was performed by alkaline hydrolysis. On the other hand, if a support substituted by hydroxylalkylamino groups[20] is used, Z-amino acids can be employed in the synthesis owing to the stability of this anchoring bond towards HBr/glacial acetic acid.

The Boc-group is much more suitable as an α-amino protecting group for the solid phase peptide synthesis (see p. 84). As an N-protecting group with urethane structure it (like the Z-group) is effective in preventing the occurrence of racemisation during stepwise synthesis, and it can be removed easily by acidolysis. 1 N HCl/glacial acetic acid, 4 N HCl/dioxane, trifluoroacetic acid and trifluoroacetic acid/methylene chloride mixtures are mostly used as removing agents. The relative advantages and disadvantages of these variations on the method of removal cannot be discussed here. The preferred reagent for N(α) de-protection of Boc-glutaminylpeptide–resin conjugates is trifluoroacetic acid, because the use of other acidolytic removal methods can, in some cases, lead to the formation of pyroglutamic acid[30]. Addition of 1 per cent mercaptoethanol protects the sensitive tryptophan during acidolytic deblocking, especially if 4 N HCl/dioxane is used. In the synthesis of long peptides some of the benzyl ester bonds to the resin do not survive the relatively frequent treatment with HCl/glacial acetic acid, and diminished yields result[31].

The *tert*-amyloxycarbonyl[32], furfuryloxycarbonyl[33] and *p*-methoxybenzyloxycarbonyl groups[34] have not yet found wide application.

Merrifield[35] also investigated the application of the Nps-group (see p. 90) in solid-phase synthesis. The mild removal conditions offer greater tactical flexibility, but separation of the by-products formed in the deblocking step is difficult[36,37].

The Bpoc-group (see p. 25) can be removed by 75 per cent acetic acid and 0.5 per cent trifluoroacetic acid, respectively, and so permits the protection of side chain functions by *tert*-butyloxycarbonyl, *tert*-butyl ether and *tert*-butyl ester residues[38]. Other valuable combinations include the Bpoc-group with an alkyloxycarbonyl hydrazide support[24] and the *N*-(2-benzoyl-1-methylvinyl) group with a benzhydryl resin[23].

The α,α-dimethyl-3,5-dimethoxybenzyloxycarbonyl group has been recommended as a very labile protecting group for Merrifield synthesis[29]. It can be cleaved by photolysis as well as by 5 per cent trifluoroacetic acid in methylene chloride within 15 minutes. Boc- and trityl residues are not attacked under these conditions.

The planning of Merrifield synthesis with regard to the tactics to be used is not without problems. Once a particular anchoring bond has been chosen, the choice of group to be used for the α-amino protection is very limited. Starting from the combination that has been most often used so far (Boc-protecting group for the α-amino function and benzyl ester-like anchoring bond), some possibilities for side chain protection will be illustrated.

For arginine, ω-nitro and ω-tosyl derivatives are very suitable, but protective protonation of the guanidino group has not proved practicable. Catalytic hydrogenation and treatment with HF/anisole are convenient for the removal of the nitro group; the tosyl group can be cleaved by sodium/liquid ammonia or with HF-anisole.

The ε-amino function of lysine can be masked by the trifluoroacetyl- (removal by 1 M piperidine), the tosyl- (deprotection by Na/liq. NH_3), the 2,4-dimethyl-3-pentyloxycarbonyl-[39] or the Z-group. When the Z-group is used, however, some cleavage is observed during N(α) de-protection[40]: this may be serious in the synthesis of longer peptide chains where repeated treatment with 1 N HCl/acetic acid may be involved. The selectivity of cleavage can be enhanced either by use of more acid-labile, substituted Z-groups, such as Z(p-NO_2)-group (cleavage by catalytic hydrogenation), or by using milder conditions for removing the α-Boc-protecting group[40]. Boron trifluoride etherate in glacial acetic acid and mercaptoethane sulphonic acid ($HS-CH_2-CH_2-SO_3H$)—suggested by LOFFET[41]— both cleave the Boc-group very quickly: the ε-Z-residue is stable under these conditions.

The N^{im}-benzyl protection has frequently been applied to histidine (cleavage by Na/liq. NH_3 or by catalytic hydrogenation, although this often causes difficulties). The 1-benzyloxycarbonylamino-2,2,2-trifluoroethyl group—removable by anhydrous HF or HBr/glacial acetic acid—and the N^{im}-dinitrophenyl residue[42]—removable by mercaptoethanol—are alternatives. The tosyl group, which can be removed by hydrogen fluoride, has been recommended by SAKAKIBARA[43]. In some cases histidine has been used without protection.

The ω-carboxyl functions of aspartic and glutamic acids are best protected as benzyl esters. HBr/trifluoroacetic acid simultaneously removes the peptide from the support and removes the side chain protection. Nitrobenzyl esters are acid-resistant (final removal is by catalytic hydrogenation) and are therefore suitable, but the *tert*-butyl ester group is cleaved by the conditions normally used for N(α) de-protection, and is therefore not useful.

Although tyrosine has often been used without protection, conversion to the *O*-benzyl ether group, which can be removed with HBr/trifluoroacetic acid or HF (D. YAMASHIRO and C. H. LI, 1973), has been recommended, although there is a danger of forming 3-benzyltyrosine on acidolysis.

The hydroxyl functions of serine and threonine should be blocked by *O*-benzylation. Synthesis with unprotected threonine has, in some cases, resulted in chain branching, but the situation has been eased by the fact that synthetic *O*-benzyl threonine[44] has become more easily available.

For a long time the SH function of cysteine was protected almost exclusively by the benzyl group which is removable by sodium/liquid ammonia or by anhydrous HF. ZAHN and co-workers[45] have investigated the usefulness of several other S-protecting groups, since, especially for the synthesis of insulin chains, *S*-benzyl protection is not the most convenient because of the drastic conditions needed for its removal. The *S*-ethylmercapto group has been successfully applied to the synthesis of the insulin A-chain[46]. Trityl and tetrahydropyranyl residues are partially cleaved even by 1N HCl/glacial acetic acid. *S*-Acyl groups (Z-, Ac-, Bz-, etc.) are liable to migrate to nitrogen[45]. The problem of cysteine protection in MERRIFIELD synthesis has not yet been solved satisfactorily.

If the removal of the Boc-group is effected by 1N HCl/glacial acetic acid in presence of 1 per cent mercaptoethanol and if the cleavage of the peptide from the polymeric support is performed by HF/mercaptoethanol/anisol, it is possible to incorporate tryptophan in a solid phase synthesis.

After the acidolytic elimination of the Boc-protecting group by HCl/glacial acetic acid, the α-amino function is protonated. By treatment with triethylamine in chloroform, dimethylformamide or methylene chloride the amino groups are liberated, and they may be estimated by titration of the chloride eluted by the triethylamine.

For the *attachment of the second N-protected amino acid* (IV) nearly all the important conventional methods for forming peptide bonds (see Section 2.2.3.) have been used. So far, the best results have been achieved using the carbodiimide method (see p. 108):

$$(CH_3)_3C\text{-}O\text{-}CO\text{-}NH\text{-}CH(R')\text{-}COOH + H_2N\text{-}CH(R)\text{-}CO\text{-}O\text{-}CH_2\text{-}Ⓟ \xrightarrow{DCC} (CH_3)_3C\text{-}O\text{-}CO\text{-}NH\text{-}CH(R')\text{-}CO\text{-}NH\text{-}CH(R)\text{-}CO\text{-}O\text{-}CH_2\text{-}Ⓟ$$

In a typical synthesis, in order to obtain as near quantitative a yield as possible in each coupling stage, 3 to 4 equivalents of acylating agent are used and a reaction time between 2 and 4 hours is allowed for each peptide linkage, but, for example, in the MERRIFIELD synthesis of a cytochrome c-analogue consisting of 104 amino acids[47], the excess of acylating agent for attaching the last 45 amino acids was increased to 30–70 equivalents and the time allowed per coupling was prolonged to 24 hours in the last phase. In contrast, ANFINSEN and co-workers have synthesised bradykinin in less than 5 hours using a modified technique[48]. Their procedure involves adding a tenfold excess of dicyclohexylcarbodiimide to a similar excess of the carboxyl component prior to each coupling step; coupling is allowed to proceed for 3 minutes only and the coupling cycle is then repeated.

The DDC coupling is carried out at room temperature and it is not surprising that some formation of acyl urea via intramolecular rearrangement occurs in the activated carboxyl component (p. 109). Dimethylformamide promotes this side reaction. For the coupling reaction, therefore, only methylene chloride seems to be suitable, because, for example, in benzene, dioxane, ethanol and ethyl acetate the resin swells insufficiently. The dicyclohexylurea formed is relatively insoluble. The separation of this coproduct is effected by washing with ethanol, glacial acetic acid and also by the treatment with 1N HCl/glacial acetic acid which follows each coupling step for the cleavage of the Boc-group. HAGENMEIER[49] was able to improve the coupling yields by mixing the Boc-amino acids with DCC in 2 : 1 ratio, filtering off the dicyclohexylurea and then adding the activated amino acid in a ratio of 3 : 1.

From the economic point of view the use of a large excess of the expensive Boc-amino acids and DCC is undesirable especially as there are difficulties in recovering the unreacted amino acids. For these reasons activated esters—for example, *p*-nitrophenyl, *N*-hydroxyphthalimide, *o*-nitrophenyl, *N*-hydroxysuccinimide, thiophenyl, 5-chloroquinolyl-(8), quinolyl-(8), 2,4,5-trichlorophenyl and pentachlorophenyl esters—have been used, since they can in principle be recovered. Insulin chains have in fact been built up using Boc-amino acid *p*-nitrophenyl esters exclusively[50] and Boc-glutamine and Boc-asparagine are usually introduced via activated esters; these necessitate much longer reaction times than couplings by means of DCC. A direct comparison of several coupling methods in syntheses of a model peptide has confirmed that the DCC method at present represents the most effective coupling procedure in solid phase synthesis[51]. The yields obtained by the DCC procedure—ascertained by the DORMAN method[52]—were not attained either by the anhydride procedure or with WOODWARD's reagent. These investigations showed that *p*-nitrophenyl esters in methylene chloride had a distinctly lower aminolysis rate than, for example, 5-chloroquinolyl-(8) esters. According to MORLEY[53], the reaction rate of 2-pyridyl esters in methylene chloride is also greater than that of *p*-nitrophenyl esters, whereas the reverse is true in dimethylformamide. These solvent effects are in line with those observed by KEMP for *p*-nitrophenyl ester aminolysis[54] and confirm that the mechanism of quinolyl-(8) ester aminolysis[55] is analogous to that of the 2-pyridyl esters. Besides the activated esters mentioned above, selenophenyl, *N*-hydroxyurethanes[51], several activated esters (*p*-nitrophenyl, *N*-hydroxysuccinimide, pentachlorophenyl and 2,4,5-trichlorophenyl esters) together with bifunctional catalysers, mixed anhydrides with isobutyl chloroformate[20,51,56,57], symmetrical anhydrides[34,58], azides[2,59,60], isoxazolium salts[51,59,61], etc., have been applied to MERRIFIELD syntheses with varying success.

At present comparison of the solid phase coupling methods is difficult because of the large number of variables. The continuous change of the support-bound amino component, the dependence on the sequence concerned, the influence of the amino acid residues in neighbouring peptide chains, the effect of chain length and the general complexity of the situation all combine to frustrate attempts to make generalised statements about the kinetics of peptide formation on polymeric supports.

By *repeating the reaction steps III and IV* (V) (see Figure 16) the desired peptide sequence can be built up on the polymeric support. Owing to the attachment of the growing peptide chain to the insoluble polymer, the purification operations necessary in conventional stepwise syntheses can be replaced by simple washing and filtration procedures. After attaching the first N-protected amino acid to the support, all the subsequent operations are performed in a special glass reactor with a fritted disc (Figure 17).

It is of paramount importance for the purity of the peptide being synthesised that all de-protection and coupling steps proceed quantitatively. The chloride titration developed by VOLHARD indicates the amount of free amino groups present after each de-blocking step. Theoretically this should remain constant even after *n* repeated de-blocking and coupling reactions. The completeness of coupling however, cannot be determined by the chloride titration method. WEYGAND and OBERMEIER[62] have developed a combination of the Edman degradation (see p. 276) and mass spectrometric dilution analysis for determination of the unreacted portion of polymer-bound amino component after each coupling reaction, while Anfinsen[59] used the

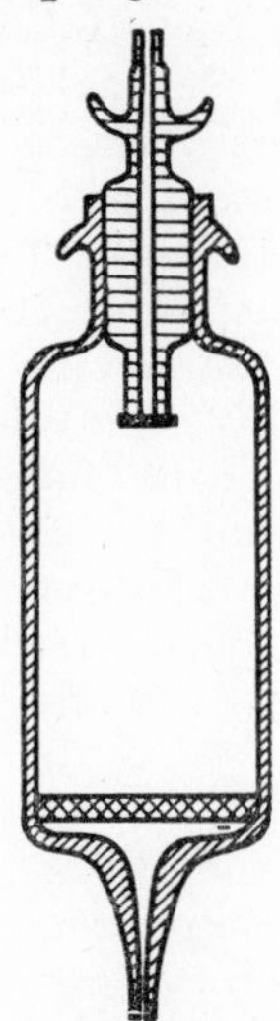

Figure 17. Reaction vessel for Merrifield synthesis

dansyl method (see p. 274). According to Esko and co-workers[63], residual free amino groups react with 2-hydroxy-1-naphthaldehyde: if the resulting Schiff base is treated with benzylamine, the *N*-(2-hydroxy-1-naphthylidene)-benzylamine formed by exchange can be estimated by colorimetry. The semiquantitative ninhydrin assay devised by Bayer and co-workers[64] also uses the colorimetric technique. Kaiser *et al.*[65] have described a colour test based on ninhydrin which permits quantitative determination of the completeness of coupling. A fluorescamine test for rapid detection of trace amounts of unreacted residues has been introduced by Felix *et al.*[66]. Geising and Hoernle[67] have estimated free amino groups by means of ^{14}C-labelled phenylisothiocyanate. Two very interesting analytical methods were devised by Beyermann and co-workers[68]. In the first, the extent of the coupling reaction is determined using *N*-(2[^{14}C]Boc)-amino acids[56]. At fixed intervals samples are taken from the reaction vessel, and the extent of coupling is calculated from the decrease in radioactivity. By introducing a scintillation flow-through type of cell, it is planned to apply this technique to 'feed-back' control of an automatic synthesizer. The second method described by the same group permits rapid determination of the extent of amino acid or peptide bound to the support using [^{35}S]sulphuric acid.

All the methods described above and also procedures which use chromatographic or mass spectrometric control of purity of partial sequences to determine the extent of coupling[69,70] require repeated sampling from the reaction vessel. In the synthesis of large peptides this analytical control of each step leads, of course, to a considerable loss of yield[71]. For this reason methods which allow a continuous control of synthesis are more attractive. Using special activated esters, such as *p*-nitrophenyl esters[19,49,72] or 5-chloro-8-hydroxyquinoline esters[73], the activating component liberated during the aminolysis can be determined by colorimetry and the progress of coupling can be followed in this way. The disappearance of Nps-amino acids during coupling can be determined similarly[74].

A relatively simple analytical method has been worked out by DORMAN[52]. Any free amino groups remaining after coupling are protonated with pyridine hydrochloride, and subsequent elution of chloride ions by triethylamine and VOLHARD titration gives a measure of the number of unreacted amino groups. A second version of the method makes use of the direct titration of unreacted amino groups with 0.1 N perchloric acid in glacial acetic acid/methylene chloride (1 : 1) after coupling as well as after deprotection[75].

Even with a yield of 99 per cent at every coupling step truncated and failure sequences must be excepted[76,77], and there have therefore been efforts directed at blocking unreactive amino functions[78,80]. Although these procedures should not to be underrated, it is doubtful whether the formation of failure sequences can be suppressed completely in this way[81].

In practice 100 per cent yield in each coupling step is not attained, and cleavage from the support yields a peptide mixture which poses great purification problems. These are somewhat simplified if dipeptides or even larger fragments are used instead of monomer units, because of the greater molecular weight differences of the constituents of the final mixture. For fragment coupling only methods free of racemisation may be employed. The WUENSCH–WEYGAND procedure[34,59], the azide method and WOODWARD's reagent[59] have been used for the addition of di-, tri- and tetrapeptide without racemisation. Among other examples[82,83] an elegant application by SAKAKIBARA *et al.*[84] for the synthesis of a collagen model peptide deserves special mention. After condensation of Boc–Pro–Pro to a glycyl resin the attachment of the tripeptide Boc–Pro–Pro–Gly was repeated until the molecular weight was about 5000. VISSER and KERLING[85] have showed that a pentapeptide built up on a support in a stepwise manner could be coupled to an octapeptide while still attached to the resin, and YAJIMA and co-workers[86] transferred the classical principle of fragment condensation to solid phase synthesis by esterification of a tetrapeptide with a bromomethylated support and then coupled it to a hexapeptide. There are some special limitations to the coupling methods applicable to the fragment condensation. For example, with the WUENSCH–WEYGAND procedure, irreversible blocking of terminal amino functions has been observed[65]: this can be attributed to known side reactions (p. 110).

In spite of all the disadvantages described, the solid phase synthesis has opened up new possibilities with regard to automation.

The MERRIFIELD *Peptide Synthesizer* machine consists of two main parts. In Figure 18—on the left—is shown the programmer. Beside it are the reaction vessel and the mechanical mixing arrangement and the dosing pump. On the right-hand side are the waste flask and the connection to the vacuum pump. The storage vessels for the amino acid derivatives, solvents and reagents are arranged above them. In the middle of the picture two special valve systems driven by two motors are visible. These are for the transfer of amino acid derivatives, reagents and solvents, programmed as regards time and amount, to the glass reactor.

Starting from this construction improvements, especially in programming of such 'Peptide synthesisers' have been described[87–90].

Figure 18. Peptide Synthesizer according to MERRIFIELD[135]

The first commercially available peptide synthesising machine (the SCHWARZ Bio Research Automated Peptide Synthesizer) is an instrument developed at the Danish Institute for Protein Chemistry[88].

The reaction vessels have been improved and, in addition to the shaking principle[91,92], modifications have been described in which the resin is agitated either by recirculating the reaction solution in a thermostated vessel[73] or by bubbling nitrogen[87] or by mechanical stirring[93]. In both the last two methods of agitation and also in two other rather complicated devices[94, 95], there is the danger that some resin particles will adhere to the wall of the vessel and no longer be wetted by the reaction solution. The centrifugal reactor for MERRIFIELD syntheses designed by BIRR *et al.*[29] is

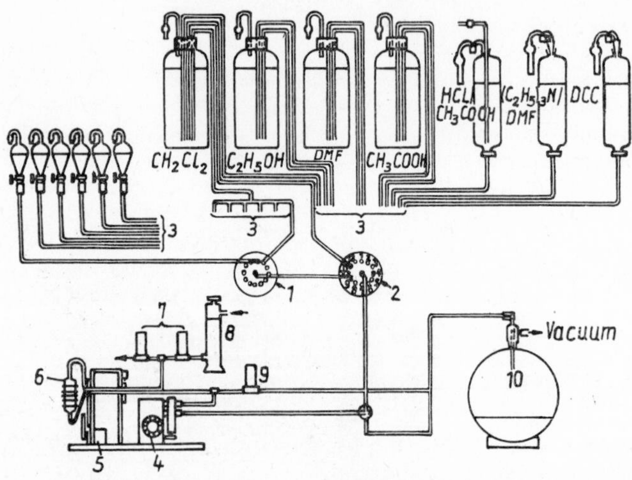

Figure 19. Scheme of the MERRIFIELD Peptide Synthesizer[135]: 1, amino acid selector valve; 2, solvent selector valve; 3, lines to the corresponding valve ports; 4, metering pump; 5, shaker; 6, reaction vessel; 7, solenoid valves for air outlet and air inlet, respectively; 8, drying tower; 9, solvent outlet solenoid; 10, waste flask

a very interesting innovation. The *cleavage of the synthesised peptide from the support* (VI) and the subsequent purification of the product is the last step of a MERRIFIELD synthesis. The removal of the peptide from the support is effected by reagents that cleave the anchoring bond between the terminal amino acid and the resin. The benzyl ester bond is usually cleaved by acidolysis. Hydrogen bromide in trifluoroacetic acid appears to be the most convenient reagent for this purpose. The use of glacial acetic acid as solvent carries the potential danger of acylating unprotected hydroxyamino acids[96]. Although a removal time of 90 minutes is often allowed, according to investigations by LOSSE and NEUBERT[97] 45 minutes is sufficient. Other laboratories have used even shorter removal times[93, 95, 98]. With anhydrous hydrogen fluoride in the presence of anisole nearly all the side-chain-protecting groups are cleaved simultaneously with the benzyl ester bond[32, 99, 101]. Sodium ethyoxide has also been used[102]. Base-catalysed transesterification with ethanol or methanol in the presence of *N*-methylpiperidine, ethylamine or a strong anion exchanger have been suggested as mild procedures[103,104]. Hydrazinolytic cleavage from the support[36,38,105,106] yields peptide hydrazides which may be used in fragment condensations both in solution and on polymeric supports. Of course, this procedure must not be used with peptides with side-chain-protecting groups labile to hydrazinolysis[106]. An alternative is the variant by WANG and MERRIFIELD[25]. In the synthesis of peptides with C-terminal amide groupings, removal from the support with ammonia is the method of choice[37,107–110]. In presence of side chain carboxyl functions difficulties can occur during ammonolysis and in some cases transesterification has been noted[19,103]. However, the development of *p*-methoxybenzylamine and modified benzhydrylamine supports by PIETTA and MARSHALL[111] may provide a promising new method for syntheses of peptides with C-terminal amide groups. Of particular interest is the synthesis of [8-lysine] vasopressin—which has a C-terminal amide grouping—published by MEIENHOFER[112]. In this synthesis Boc–Lys–Gly–NH_2 was anchored to polymeric benzyl chloroformate (see Section 2.2.4.1.2). The utilisation of amino acid side chain functions for resin-attachment had been advocated earlier by L. Y. SKYLAROV *et al.* in 1969. By linking N-protected histidine derivatives to a support via dinitrophenyl groups which are labile to thiolysis (starting from *N*-(5-fluoro-2,4-dinitro-)glycyl resin) J. GLASS (1972) opened up the possibility of chain elongation towards either the N-terminal or the C-terminal end.

At present, the most difficult problem in the MERRIFIELD synthesis of longer peptide chains is the purification of the finished product. Because of the disadvantages already pointed out, such as the formation of truncated and incomplete sequences, even relatively short peptides have to be subjected to intensive purification by means of countercurrent distribution, chromatography or electrophoresis. If the contaminating peptides are similar in molecular weight and structure to the desired synthetic product, their separation is very complicated if not impossible by the techniques currently available. According to calculations performed by the groups of BAYER[13] and BEYERMANN[113], there is an exponential increase of the number of failure sequences with increasing chain length, whereas the proportion of correct sequence varies linearly. Therefore routine MERRIFIELD synthesis is effectively limited to peptides with 10 to 20 amino acid residues.

In spite of this there have been some notable solid phase syntheses of large peptides, such as those of a polypeptide chain with acyl carrier-protein activity having 74 amino acid units[114], an analogue of cytochrome c with 104 amino acids[47], ribonuclease A with 124 residues[31] and a sequence of 188 amino acids, which was purported to be the structure of the human growth hormone. Although none of these syntheses gave products that fulfilled the criteria of purity of ordinary normal preparative chemistry, biological activity was detectable in these mixtures of polypeptides and proteins: in the case of the final purified product of the ribonuclease synthesis this amounted to 78 per cent of the activity of the native enzyme. However, the biological activity of a synthetic product is not necessarily evidence that it has the correct sequence. This was illustrated in the synthesis by LI and YAMASHIRO[71] of a protein with a sequence of 188 amino acids assumed to be that of the human growth hormone. The final product of the solid phase synthesis was separated from the resin by means of hydrogen fluoride and had *ca.* 10 per cent growth-promoting activity, whereas the natural hormone which had been similarly abused by HF treatment had *ca.* 35 per cent growth-promoting activity. Shortly after this work, however, NIALL[115] showed that the sequence which had been the object of the synthesis was not in fact that of human growth hormone. The present level of development of the MERRIFIELD synthesis is still inadequate for the synthesis of larger peptides and proteins. The difficulties encountered at each stage require systematic investigation. It is clear—for example, from the synthesis of hexaleucine by FANKHAUSER and BRENNER[116] using several different analytical methods in parallel—that analytically more reliable monitoring of the reaction on supports is the prerequisite for further improvement and optimisation of the MERRIFIELD synthesis. The main aims must be the development of more convenient polymeric supports, of more effective coupling methods, of protecting groups with higher selectivity, and careful kinetic studies on all the individual reactions as well as the improvement of purification methods. Only then will the MERRIFIELD synthesis become the method of choice for peptide synthesis.

2.2.4.1.2. *Stepwise Chain Elongation from the N-Terminal End* (LETSINGER *Synthesis*)

In 1963 LETSINGER and KORNET[117, 118] published a version of peptide synthesis on an insoluble polymeric support, in which contrary to R. B. MERRIFIELD's conception, the amino function of the N-terminal amino acid is covalently attached to the support.

As the supporting material, a popcorn copolymer consisting of 99.5 per cent styrene and 0.5 per cent divinylbenzene was used. Later the proportion of divinylbenzene was reduced right down to 0.1 per cent. Such polymers are completely insoluble and possess only slight swelling ability, but because they have relatively few chemical cross-links, popcorn polymers permit free diffusion.

To introduce a convenient functional group the polymer was suspended in nitrobenzene, and diphenylcarbamoyl chloride and aluminium chloride were added. The resulting amide was hydrolysed by acid. Subsequent reduction of the carboxyl group by lithium aluminium hydride resulted in a polymeric benzyl alcohol which reacted with phosgene in benzene to give the polymeric benzyl chloroformate:

$$\text{(P)-COOH} \xrightarrow{LiAlH_4} \text{(P)-}CH_2\text{-OH} \xrightarrow{COCl_2} \text{(P)-}CH_2\text{-O-C}(=O)\text{Cl}$$

The latter was then treated with, for example, leucine methyl ester in presence of triethylamine and saponified:

$$\text{(P)-}CH_2\text{-O-C}(=O)\text{Cl} + H_2N\text{-}CH(CH_2\text{-}CH(CH_3)_2)\text{-COOCH}_3 \xrightarrow[OH^{\ominus}]{N(C_2H_5)_3} \text{(P)-}CH_2\text{-O-}\overset{O}{\overset{\|}{C}}\text{-NH-}CH(CH_2\text{-}CH(CH_3)_2)\text{-COOH}$$

The mixed anhydride method, with isobutyl chloroformate, was used for coupling with glycine benzyl ester:

$$\text{(P)-Leu} + \text{GlyOBzl} \xrightarrow{\text{anhydride method}} \text{(P)-Leu-GlyOBzl}$$

The dipeptide was cleaved from the support with 15 per cent hydrogen bromide in glacial acetic acid with simultaneous removal of the benzyl ester group:

$$\text{(P)-Leu-GlyOBzl} \longrightarrow \text{Leu-Gly} + \text{(P)}$$

The disadvantage of this approach is that, after the dipeptide formation, only activation methods free of racemisation can be used for further attachment of C-protected amino acids. Although C-terminal glycine and proline residues can be activated in any way and although several racemisation-free coupling methods are available, to date, the LETSINGER procedure has not been widely used. The lack of suitable carboxyl protecting groups has proved to be another limiting factor. FELIX and MERRIFIELD[60] have applied the approach by using the azide method. After attaching an amino acid *tert*-butyloxycarbonyl hydrazide to the polymeric benzyl chloroformate, the unreacted functional groups of the resin were blocked with dimethylamine. The hydrazide-protecting group was then removed with 4N HCl/dioxane and the azide was formed by treatment with n-butyl nitrite in tetrahydrofuran[122]. The support-bound amino acid azide was then coupled to the next amino acid *tert*-butyloxycarbonyl hydrazide. Similarly, using a chlorosulphonated support, DAHLMANS (1972) anchored the first amino acid as its *tert*-butyl ester, then removed the C-protecting group with CF_3COOH/CH_2Cl_2 (1 : 1) and attached the next amino acid *tert*-butyl ester by means of DCC/*N*-hydroxybenztriazole. Phosphonium iodide in trifluoroacetic acid was used to remove the peptide chain from the resin.

The LETSINGER approach has also been applied to the stepwise synthesis of oligonucleotides[119–121].

2.2.4.1.3. Peptide Synthesis Using Insoluble Polymeric Reagents

These procedures are not true solid phase peptide syntheses, since it is not the growing peptide chain which is on the polymeric support. Insoluble polymeric amino acid or peptide active esters are treated with low-molecular amino compounds. The N-protected amino acid residue which is attached to the polymeric support by an active ester link can be used in excess and separated by filtration or centrifugation after the reaction is finished.

Starting from poly-(4-hydroxy-3-nitrostyrene) cross-linked by 4 per cent divinylbenzene, Katchalski and co-workers[123] prepared activated polymeric esters of N-protected amino acids by DCC, azide, or anhydride procedures, etc.:

$$\underset{\text{(ring: } NO_2,\ OH)}{-(CH_2-CH-)_n-} + Y-NH-\underset{}{\overset{R}{\overset{|}{CH}}}-COOH \longrightarrow \underset{\text{(ring: } NO_2,\ O-\underset{\underset{O}{\|}}{C}-\underset{\underset{R}{|}}{CH}-NH-Y)}{-(CH_2-CH-)_n-}$$

The best results were obtained by the mixed anhydride method, with which between 1 and 2 mmoles of acylamino acid could be bound per gram of resin.

The polymeric activated esters are stable, can be stored for long periods and are similar to the polymeric starting materials in their physical properties.

The usefulness of such reagents in the preparation of linear peptides has been demonstrated convincingly by syntheses of glutathione and bradykinin; the latter in 60 per cent yield.

Wieland and his colleagues[124] have used *p,p'*-dihydroxydiphenylsulphone cross-linked with formaldehyde as a polymeric support for the preparation of N-protected activated esters and have recently investigated coloured activated esters of the phenylazophenyl type which are bound to Dowex 1×2 via a *p*-sulphonic acid grouping. By running a solution of a C-protected amino acid or peptide derivative through a column filled with activated ester resin, or by shaking a suspension of the reactants in tetrahydrofuran at room temperature, the acyl peptide esters are obtainable in nearly quantitative yield. In the same laboratory interesting experiments have been carried out to try to automate the process (see opposite).

At first the *tert*-butyloxycarbonyl group was used for temporary blocking of the amino function, but the continuous increase of the salt concentration created problems, and thermolabile and photolabile N-protecting groups were investigated. Groups removable by photolysis, such as the 3,5-dimethoxybenzyloxycarbonyl residue, seem to be the most suitable for this purpose.

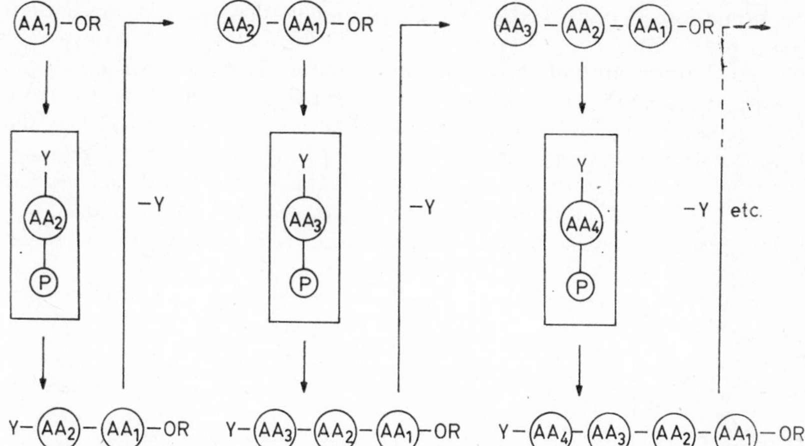

Further work has produced the α,α-dimethyl-3,5-dimethoxybenzyloxycarbonyl residue[29], a new photolabile protecting group, which should meet the requirements of this approach.

Other polymeric reagents include the insoluble poly-(hexamethylene carbodiimide)[125] and other polymeric active ester components, such as poly-(5-vinyl-8-hydroxyquinaldine)[126], poly-(ethylene-*N*-hydroxysuccinimide)[127] and branched copolymers from DL-lysine and 3-nitro-L-tyrosine[123]. PANSE and LAUFER[128] have described yet another variation on the theme of polymeric reagents. By esterification of the triethylammonium salt of 3-hydroxy-4-nitrobenzoic acid with the classical chloromethylated MERRIFIELD support they obtained a polymer which was applied to the preparation of polymeric *o*-nitrophenyl esters in the usual way.

In 1971, polymer-linked *N-ethoxycarbonyl-2-ethoxy-1,2-dihydroquinoline* was suggested as a regenerable coupling reagent by J. BROWN and R. E. WILLIAMS.

The principle of activating the carboxyl group on a polymeric support has been applied to the synthesis of cyclic peptides by KATCHALSKI and his collaborators[129]. The N-protected peptide sequence to be cyclised is attached to poly-(4-hydroxy-3-nitrostyrene) using DCC, and then the N-protecting group is removed. On adding triethylamine, the cyclic peptide is formed in high yield by intramolecular aminolysis:

$$\text{Z–(AA)}_n\text{–OH} + \text{(P)} \xrightarrow{\text{DCC}} \text{Z–(AA)}_n\text{–(P)}$$

$$\xrightarrow[\text{–Z}]{\text{HBr/CH}_3\text{COOH}} \text{HBr} \cdot \text{(AA)}_n\text{–(P)} \xrightarrow[\text{–(P)}]{\text{tert base}} \underbrace{\text{HN–(AA)}_n\text{–C=O}}$$

The formation of unwanted linear oligo- and polypeptides is largely suppressed by use of this procedure. By contrast, FLANIGAN and MARSHALL[130] built up the sequence to be cyclised on a polymeric 4-(methylthio)-phenyl support in a stepwise manner. Oxidation to the sulphone was carried out with *m*-chloroperbenzoic acid. After acidolytic cleavage of the terminal Boc-protecting group the desired cyclic peptide was obtained by intramolecular aminolysis of the activated C-terminal ester bond in 2 per cent triethylamine/DMF solution. This very elegant method can, of course, only be used in the absence of amino acids sensitive to oxidation.

2.2.4.2. The Use of Soluble High-molecular-weight Supports

SHEMYAKIN *et al.*[131] have devised a polymeric support that is soluble in organic solvents as an alternative to the MERRIFIELD technique.

Polystyrene with a molecular weight of 200 000, obtained by emulsion polymerisation, served as the soluble support. The resin was chloromethylated in the usual way and esterified with the C-terminal *tert*-butyloxycarbonyl amino acid. The stepwise chain elongation from the C-terminal end was achieved with the corresponding *tert*-butyloxycarbonyl amino acid *N*-hydroxysuccinimide esters. In order to separate excess reagents and by-products after each coupling reaction, the reaction mixture was diluted, —for example, with water—and the peptide bound to the precipitated water-insoluble support was filtered off. Unreacted *tert*-butyloxycarbonyl amino acid *N*-hydroxysuccinimide esters and by-products remained in solution. The tetrapeptide Gly–Gly–Leu–Gly was obtained in 65 per cent yield based on the glycyl polymer used, after cleavage from the support with hydrogen bromide in trifluoroacetic acid and subsequent ion exchange chromatography. A further example is in the synthesis of [5-glycine, 10-glycine] gramicidin S[132]: the linear pentapeptide precursor was obtained by stepwise synthesis on the soluble polymeric support in 70 per cent yield.

Of course, this procedure has some disadvantages. On the one hand, loss in yield results from inhomogeneity of the aminoacyl polystyrene, and this is increased by the precipitation operations with water. Parts of the support loaded with peptide remain in solution. Furthermore, as the length of the peptide chain increases, the solubility in organic solvents diminishes. In spite of these limitations and the fact that, by comparison with the solid phase technique, the experimental operations are more time-consuming, and probably more difficult to automate, interest in this method ought to be maintained. The incomplete reactions in the heterogeneous phase that plague the MERRIFIELD syntheses should suffice to ensure this. GREEN and GARSON[133] have investigated soluble polymeric supports with a view to extending their application in peptide synthesis. For example, the attachment of the Boc-amino acid to the chloromethylated support was carried out with sodium carbonate in dimethylformamide. This cannot lead to the quaternisation of chloromethyl groups and has already been employed in MERRIFIELD synthesis[68]. As a method of coupling, the mixed anhydride procedure, with isobutyl chloroformate, proved particularly successful, and there is, in general, a greater flexibility in the choice of the activating groups than there is in solid phase synthesis.

The problems associated with the precipitation reactions can be circumvented by a novel procedure suggested by BAYER and co-workers[134]. The advantage of this procedure is that all reactions are performed in homogeneous solution and the low-molecular reagents are separated by ultrafiltration (membrane filtration).

Polyethyleneglycol (molecular weight 20000) has proved to be an excellent polymeric soluble support for syntheses in aqueous solution, but other polymeric supports, such as polyvinylalcohol or a copolymer from polyvinylamine and polyvinylpyrrolidone, can also be used.

Compared with the solid phase peptide synthesis, the homogeneous phase approach gives better results in the de-protection and coupling reactions and there is scope for greater flexibility. Although all the variations of conventional peptide synthesis can be transferred to the so-called *liquid phase technique,* the time-consuming purification and separation processes are here restricted to ultracentrifugation.

A definitive evaluation of this very elegant approach to peptide synthesis must await its wider practical use.

References

1 MERRIFIELD, R. B. (1969). *Advan. Enzymol.*, **32**, 221

2 STEWART, J. N. and YOUNG, J. D. (1969). In *Solid Phase Peptide Synthesis,* Freeman, San Francisco

3 LOSSE, G. and NEUBERT, K. (1970). *Z. Chem.*, **10**, 48

4 OKUDA, T. (1968). *Naturwissenschaften,* **55**, 209

5 SHEPPARD, R. C. (1972). *Proc. 11th Europ. Peptide Symp.*, Vienna, **1971**, p. **111**, North-Holland, Amsterdam

6 *Biochemical Aspects of Reactions on Solid Supports,* ed. G. R. Stark, Academic Press, New York (1971)

7 LOSSE, G. and ULBRICH, R. (1973). *Wissenschaftliche Zeitschrift der Technischen Universität Dresden,* **22**, 263

8 MARSHALL, C. R. and MERRIFIELD, R. B. (1970). *Handbook of Biochemistry,* 2nd edn, ed. H. A. Sober, p. C-145, Chemical Rubber Co., Cleveland

9 MERRIFIELD, R. B. (1965). *Endeavour,* **24**, 3; (1963). *J. Amer. Chem. Soc.*, **85**, 2149

10 MERRIFIELD, R. B. (1966). *Hypotensive Peptides,* p. 1, Springer, New York

11 LOSSE, G., MADLUNG, C. and LORENZ, P. (1968). *Chem. Ber.*, **101**, 1257

12 INUKAI, N., NAKANO, K. and MURAKAMI, M. (1968). *Bull. Chem. Soc. Japan,* **41**, 182

13 Bayer, E. *et al.* (1971). *Proc. 10th Europ. Peptide Symp.*, Padova, 1969, p. 65, North-Holland, Amsterdam; *J. Amer. Chem. Soc.*, **92**, 1735

14 Horvath, C. G. *et al.* (1967). *Anal. Chem.*, **39**, 1422

15 Halasz, I. (1969). 5th Internat. Symp. Advan. Chromatogr., Las Vegas

16 Vlasov, G. P. and Bilibin, Yu. (1969). *Izv. Akad. Nauk (U.S.S.R.), Ser. Khim.*, 1400

17 Tilak, M. A. (1968). *Tetrahedron Letters*, 6323

18 Weygand, F. (1968). *Proc. 9th Europ. Peptide Symp.*, Orsay, 1968, p. 183, North-Holland, Amsterdam

19 Bodanszky, M. and Sheehan, J. T. (1966). *Chem. Ind.*, 1567

20 Tilak, M. A. and Hollinden, C. S. (1968). *Tetrahedron Letters*, 1297

21 Tesser, G. I. and Ellenbrock, B. W. J. (1967). *Proc. 8th Europ. Peptide Symp.*, Noordwijk, 1966, p. 144, North-Holland, Amsterdam

22 Dorman, L. C. and Love, J. (1968). *J. Org. Chem.*, **34**, 1273

23 Southard, S. L. *et al.* (1969). *Tetrahedron Letters*, 3505

24 Marshall, D. L. and Liener, I. E. (1970). *J. Org. Chem.*, **35**, 867

25 Wang, S. and Merrifield, R. B. (1969). *J. Amer. Chem. Soc.*, **91**, 6488

26 Kenner, G. W., McDermott, J. R. and Sheppard, R. C. (1971). *Chem. Comm.*, 636

27 Sieber, P. and Iselin, B. (1968). *Helv. Chim. Acta*, **51**, 622

28 Rudinger, J. (1963). *Pure Appl. Chem.*, **7**, 335

29 Birr, Ch., Lockinger, W. and Wieland, Th. (1972). *Proc. 11th Europ. Peptide Symp.*, Vienna, 1971, p. 175, North-Holland, Amsterdam

30 Takashima, H., du Vigneaud, V. and Merrifield, R. B. (1968). *J. Amer. Chem. Soc.*, **90**, 1323

31 Gutte, B. and Merrifield, R. B. (1969). *J. Amer. Chem. Soc.*, **91**, 501

32 Sakakibara, S. *et al.* (1968). *Bull. Chem. Soc. Japan*, **41**, 1273

33 Losse, G. and Neubert, R. (1970). *Tetrahedron Letters*, 1267

34 Weygand, F. and Ragnarsson, U. (1966). *Z. Naturforsch.*, **21b**, 1141

35 Najjar, V. A. and Merrifield, R. B. (1965). *Biochemistry*, **5**, 3765

36 Kessler, W. and Iselin, B. (1966). *Helv. Chim. Acta*, **49**, 1330

37 Ives, D. A. (1968). *Can. J. Chem.*, **46**, 2318

38 Sieber, P. and Iselin, B. (1968). *Helv. Chim. Acta*, **51**, 622

39 Sakakibara, S. *et al.* (1970). *Bull. Chem. Soc. Japan*, **43**, 3322

40 Schnabel, E., Klostermeyer, H. and Berndt, H. (1972). *Proc. 11th Europ. Peptide Symp.*, Vienna, 1971, p. 69 North-Holland, Amsterdam

41 Loffet, A. (1972). *Proc. 11th Europ. Peptide Symp.*, Vienna, 1971, p. 249, North-Holland, Amsterdam

42 Chillemi, F. and Merrifield, R. B. (1969). *Biochemistry*, **8**, 4344

43 Fujii, T. and Sakakibara, S. (1970). *Bull. Chem. Soc. Japan*, **43**, 3954

44 Mizoguchi, T. *et al.* (1968). *J. Org. Chem.*, **33**, 907

45 Hammerstrom, K., Lunkerheimer, W. and Zahn, H. (1970). *Makromolek. Chem.*, **133**, 41

46 Weber, U. (1969). *Z. Physiol. Chem.*, **350**, 1421

47 SANO, S. and KURIHARA, M. (1969). *Z. Physiol. Chem.*, **350**, 1183
48 CORLEY, L., SACHS, D. H. and ANFINSEN, C. B. (1972). *Biochem. Biophys. Res. Comm.*, **47**, 1353
49 HAGENMAIER, H. (1972). *Z. Physiol. Chem.*, **353**, 1973
50 HOERNLE, S. (1967). *Z. Physiol. Chem.*, **348**, 1355; WEBER, U. *et al.* (1967). *ibid.*, **348**, 1715
51 JAKUBKE, H. D. and BAUMERT, A. (1972). *Proc. 11th Europ. Peptide Symp.*, Vienna, 1971, p. 132, North-Holland, Amsterdam
52 DORMAN, L. C. (1969). *Tetrahedron Letters*, 2319
53 DUTTA, A. S. and MORLEY, J. S. (1972). *Proc. 11 th Europ. Peptide Symp.*, Vienna, 1971, p. 21, North-Holland, Amsterdam
54 KEMP, D. S. (1972). *Proc. 11 th Europ. Peptide Symp.*, Vienna, 1971, p. 1, North-Holland, Amsterdam
55 JAKUBKE, H. D., VOIGT, A. and BURKHARDT (1967). *Chem. Ber.*, **100**, 2367
56 KRUMDIECK, C. L. and BAUGH, C. M. (1969). *Biochemistry*, **8**, 1568
57 SEMKIN, E. P., GAFUROVA, N. D. and SHCHUKINA, L. A. (1967). *Khim. Pirodn. Soedin. Akad. Nauk (U.S.S.R.)*, **3**, 220
58 WIELAND, Th., BIRR, Ch. and FLOR, F. (1971). *Angew. Chem.*, **83**, 333
59 OMEN, G. S. and ANFINSEN, C. B. (1968). *J. Amer. Chem. Soc.*, **90**, 6571
60 FELIX, A. M. and MERRIFIELD, R. B. (1970). *J. Amer. Chem. Soc.*, **92**, 1385
61 KHOSLA, M. C. *et al.* (1968). *Biochemistry*, **7**, 3417
62 WEYGAND, F. and OBERMEIER, R. (1968). *Z. Naturforsch.*, **23b**, 1390
63 ESKO, K., KARLSON, S. and PORATH, J. (1968). *Acta Chem. Scand.*, **22**, 3342
64 BAYER, E. *et al.* (1971). *Proc. 10 th Europ. Peptide Symp.*, Padova, 1969, p. 65, North-Holland, Amsterdam
65 KAISER, E. *et al.* (1970). *Anal. Biochem.*, **34**, 595
66 FELIX, A. M. and JIMENEZ, M. H. (1973). *Anal. Biochem.*, **52**, 377
67 GEISING, W. and HOERNLE, S. (1972). *Proc. 11 th Europ. Peptide Symp.*, Vienna, 1971, p. 146, North-Holland, Amsterdam
68 BEYERMANN, H. C. *et al.* (1972). *Proc. 11 th Europ. Peptide Symp.*, p. 138, Vienna, 1971, North-Holland, Amsterdam
69 NEUBERT, K. (1970). Dissertation, Technical University Dresden
70 BAYER, E. *et al.* (1968). *Proc. 9 th Europ. Peptide Symp.*, Orsay, 1968, p. 162, North-Holland, Amsterdam
71 LI, C. H. and YAMASHIRO, D. (1970). *J. Amer. Chem. Soc.*, **92**, 7608
72 BODANSZKY, M. and BATH, R. J. (1969). *Chem. Comm.*, 1259
73 RUDINGER, J. and GUT, V. (1967). *Proc. 8 th Europ. Peptide Symp.*, Noordwijk, 1966, p. 89, North-Holland, Amsterdam
74 GUT, V. and RUDINGER, J. (1968). *Proc. 9 th Europ. Peptide Symp.*, Orsay, 1968, p. 185, North-Holland, Amsterdam
75 BRUNFELDT, K., ROEPSTORFF, P. and THOMSEN, J. (1969). *Acta Chem. Scand.*, **23**, 2906
76 BAYER, E. (1969). *Chem. Labor Betrieb*, **20**, 193
77 BAYER, E. *et al.* (1970). *J. Amer. Chem. Soc.*, **92**, 1735

78 Wieland, Th., Birr, Ch. and Wissenbach, H. (1969). *Angew. Chem.*, **81**, 782
79 Markley, L. D. and Dorman, L. C. (1970). *Tetrahedron Letters*, 1787
80 Wissmann, H. and Geiger, R. (1970). *Angew. Chem.* **82**, 937
81 Chou, F. C. H., Chawla, R. K., Kibler, R. F. and Shapiro, R. (1971). *J. Amer. Chem. Soc.*, **93**, 937
82 Izumija, N. (1969). *Tanpakushitu, Kakusan, Koso*, **14**, 641
83 Gisin, B. F., Merrifield, R. B. and Tosteson, D. C. (1969). *J. Amer. Chem. Soc.*, **91**, 2691
84 Sakakibara, S. *et al.* (1968). *Bull. Chem. Soc. Japan*, **41**, 1273
85 Visser, S. and Kerling, K. E. T. (1970). *Rec. Trav. Chim.*, **89**, 880
86 Yamija, H., Kawatani, H. and Watanabe, H. (1970). *Chem. Pharm. Bull. (Japan)*, **18**, 1333
87 Loffet, A. and Close, I. (1968). *Proc. 9 th Europ. Peptide Symp.*, Orsay, 1968, p. 189, North-Holland, Amsterdam
88 Brunfeldt, K., Halstroem, J. and Roepsdorff, P. (1968). *Proc. 9 th Europ. Peptide Symp.*, Orsay, 1968, p. 194, North-Holland, Amsterdam
89 Mansveld, G. W. H. A., Hindriks, H. and Beyerman, H. C. (1968). *Proc. 9 th Europ. Peptide Symp.*, Orsay, 1968, p. 197, North-Holland, Amsterdam
90 Sugiyama, H., Miura, Y. and Seto, Sh. (1968). *Yuki Gosei Kagaku Kyokai Shi.*, **26**, 1010; (1969). *C.A.*, **70**, 58264
91 Merrifield, R. B. and Corigliano, M. A. (1968). *Biochem. Prep.*, **12**, 98
92 Kusch, P. (1966). *Kolloid-Z.*, **208**, 138; (1966). *Angew. Chem.*, **78**, 611
93 Grahl-Nielson, O. and Tritsch, G. T. (1969). *Biochemistry*, **8**, 187
94 Khosla, M. C., Smeby, R. R. and Bumpus, F. M. (1967). *Science*, **156**, 253
95 Ney, K. H. and Polzhofer, K. P. (1968). *Tetrahedron*, **26**, 6619
96 Merrifield, R. B. (1964). *Biochemistry*, **3**, 1385
97 Losse, G. and Neubert, K. (1968). *Z. Chem.*, **8**, 228
98 Khosla, M. C. (1967). *Biochemistry*, **6**, 754
99 Merrifield, R. B. and Gutte, B. (1969). *J. Amer. Chem. Soc.*, **91**, 501
100 Marshall, G. R. (1967). *Advan. Exp. Med. Biol.*, **2**, 48
101 Lenard, J. and Robinson, A. B. (1967). *J. Amer. Chem. Soc.*, **89**, 181
102 Loffet, A. (1967). *Experientia*, **23**, 406
103 Beyerman, H. C. *et al.* (1968). *Chem. Comm.*, 1668
104 Halpern, B. *et al.* (1968). *Tetrahedron Letters*, 1563
105 Visser, S. *et al.* (1968). *Rec. Trav. Chim.*, **87**, 559
106 Ohno, M. and Anfinsen, C. B. (1967). *J. Amer. Chem. Soc.*, **89**, 5994
107 Manning, M. (1968). *J. Amer. Chem. Soc.*, **90**, 1348
108 Bayer, E. and Hagenmeier, H. (1968). *Tetrahedron Letters*, 2037
109 Beyerman, H. C. *et al.* (1968). *Rec. Trav. Chim.*, **87**, 257
110 Takashima, H., du Vigneaud, V. and Merrifield, R. E. (1968). *J. Amer. Chem. Soc.*, **90**, 1323
111 Pietta, P. G. and Marshall, G. R. (1970). *Chem. Comm.*, 650
112 Meienhofer, J. (1972). *Proc. 11 th Europ. Peptide Symp.*, Vienna, 1971, p. 199, North-Holland, Amsterdam

113 BAAS, J. M. A., BEYERMAN, H. C., VAN DEN GRAAF, B. and DE LEER, E. W. B. (1971). *Proc. 10 th Europ. Peptide Symp.*, Padova, 1969, p. 173, North-Holland, Amsterdam
114 MARSHALL, G. R. *et al.* (1971). *J. Amer. Chem. Soc.*, **93**, 1799
115 NIALL, H. D. (1971). *Nature New Biology*, **230**, 90
116 FANKHAUSER, P., BRENNER, M. *et al.* (1972). *Proc. 11 th Europ. Peptide Symp.*, Vienna, 1971, p. 153, North-Holland, Amsterdam
117 LETSINGER, R. L. and KORNET, M. J. (1963). *J. Amer. Chem. Soc.*, **85**, 3045
118 LETSINGER, R. L. *et al.* (1964). *J. Amer. Chem. Soc.*, **86**, 5163
119 LETSINGER, R. L. and MAHADEVAN, V. (1966). *J. Amer. Chem. Soc.*, **88**, 5319
120 SHIMIDZU, T. and LETSINGER, R. L. (1968). *J. Org. Chem.*, **33**, 708
121 LETSINGER, R. L. (1965). *J. Amer. Chem. Soc.*, **87**, 3526
122 HONZL, J. and RUDINGER, J. (1961). *Coll. Czech. Chem. Comm.*, **26**, 2333
123 FRIDKIN, M., PATCHORNIK, A. and KATCHALSKI (1966). *J. Amer. Chem. Soc.*, **88**, 3164; (1968). *ibid.*, **90**, 2953; (1967). *Proc. 8 th Europ. Peptide Symp.*, Noordwijk, 1966, p. 91, North-Holland, Amsterdam
124 WIELAND, Th. and BIRR, Ch. (1966). *Angew. Chem.* (*Int. Ed.*), **5**, 310; (1967). *Chimia*, **21**, 581
125 WOLMAN, Y., KIVITY, S. and FRANKEL, M. (1967). *Chem. Comm.*, 629
126 MANECKE, G. and HAAKE, E. (1968). *Naturwissenschaften*, **55**, 343
127 BLOUT, E. R. *et al.* (1968). *J. Amer. Chem. Soc.*, **90**, 2696
128 PANSE, G. T. and LAUFER, D. A. (1970). *Tetrahedron Letters*, 4181
129 FRIDKIN, M., PATCHORNIK, A. and KATCHALSKI, E. (1965). *J. Amer. Chem. Soc.*, **87**, 4646
130 FLANIGAN, E. and MARSHALL, G. R. (1970). *Tetrahedron Letters*, 2403
131 SHEMYAKIN, M. M. *et al.* (1965). *Tetrahedron Letters*, 2323
132 SHEMYAKIN, M. M. *et al.* (1967). *Proc. 8 th Europ. Peptide Symp.*, Noordwijk, 1966, p. 100, North-Holland, Amsterdam
133 GREEN, B. and GARSON, L. R. (1969). *J. Chem. Soc.*, 401
134 MUTTER, M., HAGENMAIER, H. and BAYER, E. (1971). *Angew. Chem.* **83**, 883
135 MERRIFIELD, R. B. *et al.* (1966). *Anal. Chem.*, **38**, 1905

2.2.5. Synthesis of Cyclic Peptides

Only the synthesis of homomeric cyclic peptides will be considered in this section.

It is essential to choose reaction conditions which will exclude intermolecular polycondensation of the activated linear starting peptide sequence in favour of the desired intramolecular cyclisation. From the kinetic point of view the intramolecular formation of a peptide bond, since it is a unimolecular reaction, is independent of concentration, in contrast to the bimolecular linear polycondensation. It is therefore necessary to carry out the cyclisation under high dilution conditions (RUGGLI–ZIEGLER dilution principle).

2.2.5.1. Synthesis of Homodetic Cyclic Peptides

For the synthesis of cyclic peptides it is first necessary to synthesise the appropriate linear peptides as starting materials. The linear peptide is synthesised by standard methods and then either is activated at the C-terminus in the presence of a free N-terminal amino function or the N-protecting group may be removed from an N-protected, carboxyl-activated peptide. The latter method using activated esters is mostly applied to cyclisation. Several tactical variations, such as a combination of the cyanomethyl ester with the *N*-trityl group, or a combination of the *p*-nitrophenyl ester with the trityl and especially the *N*-benzyloxycarbonyl group, have been successful. After the acidolytic cleavage of the N-protecting group, intermediate blocking of the amino function by salt formation occurs. The free amine is generated when required by adding base. Ring-closure can also be performed by the azide method or, starting from the free peptide, by DCC or water-soluble carbodiimides, ethoxyacetylene, *N*-ethyl-5-phenylisoxazolium-3′-sulphonate, etc. The so-called single-stage cyclisation (Th. Wieland and collaborators, 1967) with ethyl chloroformate in presence of pyridine hydrochloride must also be mentioned.

A priori, each linear peptide unit would be expected to yield a single cyclopeptide with the same sequence. However, R. Schwyzer and collaborators found that cyclic hexapeptides are usually obtained upon attempted cyclisation of tripeptide derivatives. Evidently, dimerisation of the linear monomer occurs prior to intramolecular peptide formation. The various factors involved in this phenomenon have been discussed by the same authors[1].

Kopple and Savrda[10] have investigated the cyclodimerisation of Gly–Leu–Gly in water using *N*-ethyl-*N*-γ-(dimethylaminopropyl)-carbodiimide: they detected not only the cyclic hexapeptide but also the cyclic nonapeptide *cyclo*-(–Gly–Leu–Gly–)$_3$, together with higher polymers and glycylleucine-dioxopiperazine. The structure of the cyclic nonapeptide was elucidated by mass spectroscopy after permethylation.

As a rule dimerisation reactions have been reported in cyclisations of linear tri- and pentapeptides—that is, in cyclisations of units having an odd number of amino acid residues. Nevertheless there are some exceptions. For instance, in the synthesis of gramicidin S (see Section 2.3.3.1) from the linear activated pentapeptide derivative, N. Izumiya and co-workers obtained the cyclic pentapeptide in 32 per cent yield in addition to the desired decapeptide.

The sterically unfavourable *cis*-configuration of three peptide bonds which is required in the transition state makes the cyclisation of tripeptides difficult. In 1965, however, Rothe and co-workers[2] synthesised *cyclo*-(Pro–Pro–Pro). This was the first synthesis of a cyclic peptide with 9-membered ring.

Since cyclisation attempts with linear glycyl-prolyl-glycine resulted in dimerisation to give the cyclohexapeptide (which has a strain-free 18-membered ring) association through H-bonds of two anti-parallel peptide chains cannot be decisive for the dimerisation reaction. The influence on the conformation, brought about by introducing a proline residue, is obviously not sufficient to force the amino and carboxyl groups into an intramolecular reaction. The conformation required for the intramolecular reaction is provided only in prolyl-prolyl-proline, since the cyclisation of prolyl-prolyl-glycine produces only the cyclohexapeptide.

However, conditions and synthetic procedures (aminoacyl insertion, etc.) have been found[3] which in tripeptides containing proline, sarcosine, glycine and β-alanine promote the formation of cyclic tripeptides at the expense of cyclodimerisation.

2,5-Dioxopiperazines, whose 6-membered ring structure requires the two constituent amino acids to be *cis*, behave differently. The natural tendency to form 6-membered rings and the additional mesomeric stabilisation of both the peptide bonds is regarded as the reason for the ease of preparation and stability of these simple cyclic peptides[4].

The use of polymeric supports in the synthesis of cyclic peptides has been reported (see Section 2.2.4.1.3).

2.2.5.2. Synthesis of Heterodetic Cyclic Peptides

Some biologically important heterodetic cyclic peptides owe their cyclic structure to the formation of disulphide bonds. For example, oxytocin and vasopressin contain two cysteine residues connected to each other by an intramolecular disulphide bridge. During the synthesis of the linear starting

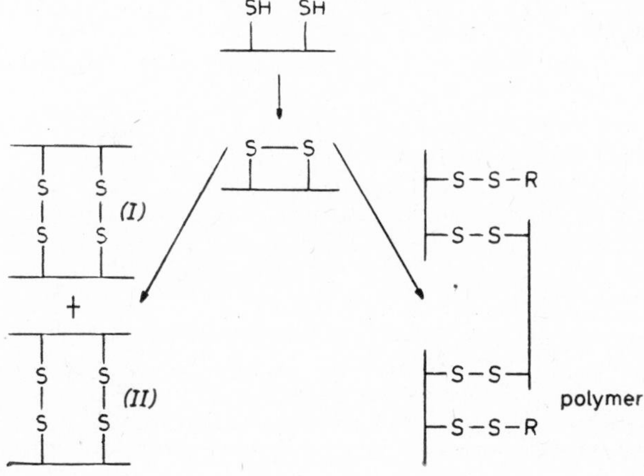

peptides both thiol functions are protected appropriately by the same S-protecting group, and then they are de-protected separately or jointly with the remaining protecting groups. The SH functions regenerated in this manner are oxidised, often with air oxygen in dilute solution, to give the correct disulphide bridge, generally, in 20–40 per cent yield. In addition, two cyclic dimers with parallel (I) and anti-parallel (II) structures and also polymeric products are formed as unwanted by-products. This approach was employed by the Nobel Prize winner V. du Vigneaud[5] for the first synthesis of oxytocin (see p. 196).

A much more complicated set of problems arises in syntheses of unsymmetrical peptides with several disulphide bridges. Notable in this respect is insulin, whose structure was successfully determined by F. Sanger in 1955. In 1963 H. Zahn and collaborators described a total synthesis of insulin, and shortly afterwards two other groups announced independent syntheses. In Section 2.3.2.7 the synthesis of insulin will be described in detail.

Using various amino, carboxyl and side chain protecting groups which could be removed with high selectivity, L. Zervas, I. Photaki *et al.* investigated alternative approaches for the synthesis of unsymmetrical cystine peptides, and were able to prepare a heptapeptide fragment of sheep insulin with the intrachain disulphide bridge A_6–A_{11}, whose third SH function was protected by the easily removable trityl or benzhydryl residue[6].

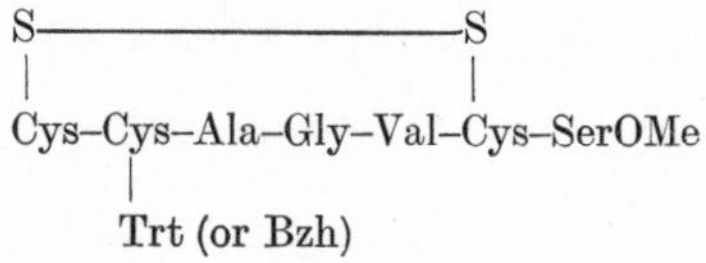

```
S-------------------------S
|                         |
Cys–Cys–Ala–Gly–Val–Cys–SerOMe
     |
     Trt (or Bzh)
```

R. G. Hiskey and his collaborators, using the *sulphenylthiocyanate method*, developed a procedure which enables the formation of selected disulphide linkages in open-chain cystine peptides and also in cyclic unsymmetrical ones[7]. Using S-protecting groups of different lability (*S*-trityl or *S*-benzhydryl residue), the disulphide bridges can be joined in the desired way by controlled treatment with thiocyanogen. The power of this approach was demonstrated impressively in the synthesis of a model peptide with one intrachain and two interchain disulphide bonds[8]

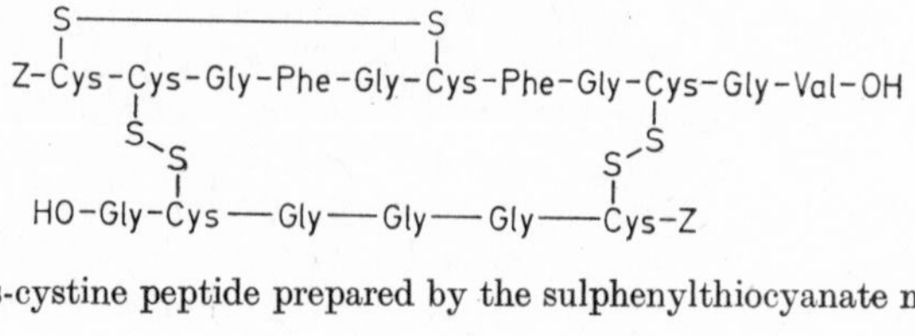

```
  S-------------------S
  |                   |
Z-Cys-Cys-Gly-Phe-Gly-Cys-Phe-Gly-Cys-Gly-Val-OH
      |                           |
      S                           S
       \S                       S/
        |                       |
HO-Gly-Cys---Gly---Gly---Gly---Cys-Z
```

Figure 20. Tris-cystine peptide prepared by the sulphenylthiocyanate method[8]

The tris-cystine peptide synthesised by R. G. HISKEY and collaborators (Figure 20) demonstrates that by this approach insulin could be built up via directed formation of the disulphide linkages, and an insulin A-chain completely protected with the correct intrachain disulphide bridge has been synthesised[9]. Since the cysteine residues in positions 7 and 20 are differentially protected by a benzhydryl group and a trityl group, respectively, the requisites for a rational insulin synthesis are met in principle by this structure.

References

1 SCHWYZER, R., CARRION, J. P., GORUP, B., NOLTING, H. and TUNG-KYI, A. (1964). *Helv. Chim. Acta*, **47**, 441

2 ROTHE, M., STEFFEN, K. D. and ROTHE, I. (1965). *Angew. Chem.*, **77**, 347

3 ROTHE, M. *et al.* (1972). *Proc. 11 th Europ. Peptide Symp.*, Vienna, 1971, p. 388, North-Holland, Amsterdam

4 AUGUSTIN, M. (1966). *Wiss. Z. Univ. Halle*, **XV**, 553

5 DU VIGNEAUD, V. *et al.* (1953). *J. Amer. Chem. Soc.*, **75**, 4879

6 ZERVAS, L., PHOTAKI, I. *et al.* (1963). *Proc. 5 th Europ. Peptide Symp.*, Oxford, 1962, p. 27, Pergamon Press, Oxford; (1965). *J. Amer. Chem. Soc.*, **87**, 4922

7 HISKEY, R. G. *et al.* (1965). *J. Amer. Chem. Soc.*, **87**, 3965; (1967). *Tetrahedron*, **23**, 3923; (1967). *J. Org. Chem.*, **32**, 97, 2772; (1968). *J. Amer. Chem. Soc.*, **90**, 2677

8 HISKEY, R. G. (1968). *Proc. 9 th Europ. Peptide Symp.*, Orsay, 1968, p. 209, North-Holland, Amsterdam

9 HISKEY, R. G. (1972). *Proc. 11 th Europ. Peptide Symp.*, Vienna, 1971, p. 107, North-Holland, Amsterdam

10 KOPPLE, K. D. and SAVRDA, J. (1972). *Proc. 11 th Europ. Peptide Symp.*, Vienna, 1971, p. 400, North-Holland, Amsterdam

2.2.6. Racemisation during the Formation of the Peptide Bond

All peptide synthesis operations which are performed on a functional group attached to an asymmetric centre are attended by a risk of racemisation. For the synthesis of large optically active peptides coupling free of racemisation is of decisive importance, since racemic contaminants cannot be separated by crystallisation as is sometimes possible in the case of smaller peptides. Because of the sheer number of reaction steps, even a slight loss of optical purity at each stage of the synthesis would play havoc with the steric integrity of the final product. The paramount significance of the racemisation problem is demonstrated by the fact that for the synthesis of a polypeptide involving 100 amino acid residues, even if a degree of racemisation of only 1 per cent during each coupling reaction is assumed, it can be calculated that after 100 cycles a final product with only 60.9 per cent of the desired stereoisomer will result. Since peptide synthesis is aimed at the preparation of peptides and polypeptides whose biological activity largely depends on their optical purity, great care has to be devoted to the question of racemisation.

2.2.6.1. Mechanisms of Racemisation

Optically active amino acids and peptides are racemised by reversible ionisation of the C(α)—H bond:

$$\text{R-NH-}\underset{\underset{\text{X}}{|}}{\underset{\text{C=O}}{|}}{\overset{\text{R}'}{\overset{|}{\text{C}}}}\text{-H} \rightleftharpoons \text{R-NH-}\underset{\underset{\text{X}}{|}}{\underset{\text{C=O}}{|}}{\overset{\text{R}'}{\overset{|}{\text{C}}}}|^{\ominus} + \text{H}^{\oplus}$$

This exchange is catalysed by bases or acids, especially at high temperature in polar solvents. Since racemisation catalysed by acids usually occurs only at very low pH and high temperatures, and therefore never under coupling conditions, this kind of racemisation will not be dealt with in more detail. The racemisation catalysed by bases, however, is very important, and two different mechanisms will be discussed. The reversible cleavage of the proton can take place directly or via an intermediate azlactone.

2.2.6.1.1. Racemisation by Direct Proton Abstraction from the Asymmetric C Atom

The stability of the C—H bond on the asymmetric centre depends mainly on the substituents R, R′ and X. Free amino acids (R = H, X = OH) are very resistant to racemisation. Only in the presence of strong bases and at elevated temperatures does racemisation occur. Although the carboxylate grouping discourages further proton abstraction, there is mesomeric stabilisation of the di-anion:

$$\text{H}_2\text{N-}\underset{\underset{|\underline{\text{O}}|^{-}}{|}}{\underset{\text{C=O}}{|}}{\overset{\text{R}}{\overset{|}{\text{C}}}}\text{-H} \underset{+\text{H}^{\oplus}}{\overset{-\text{H}^{\oplus}}{\rightleftharpoons}} \left[\text{H}_2\text{N-}\underset{\underset{|\underline{\text{O}}|^{\ominus}}{|}}{\underset{\text{C=O}}{|}}{\overset{\text{R}}{\overset{|}{\text{C}}}}|^{\ominus} \leftrightarrow \text{H}_2\text{N-}\underset{\underset{|\underline{\text{O}}|^{\ominus}}{|}}{\underset{\text{C-}\overline{\underline{\text{O}}}|^{\ominus}}{\|}}{\overset{\text{R}}{\overset{|}{\text{C}}}} \right]$$

Electron-withdrawing groups R in the β-position, such as are present in the amino acids cysteine, serine, threonine, tryptophan, histidine, tyrosine, phenylalanine, aspartic acid, etc., should theoretically favour racemisation by direct proton cleavage from the α-C atom. Benzyloxycarbonylamino acid esters, especially those of cysteine, β-cyanoalanine and phenylalanine, are slowly racemised by direct proton abstraction in the presence of bases. This tendency is particularly evident in activated esters of such N-protected amino acids, because the ionisation is further promoted by the electron-withdrawing effect of the activating component.

B. LIBEREK has shown that several activated esters of phthaloyl amino acids with electron-attracting substituents in the β-position are racemised by direct proton abstraction:

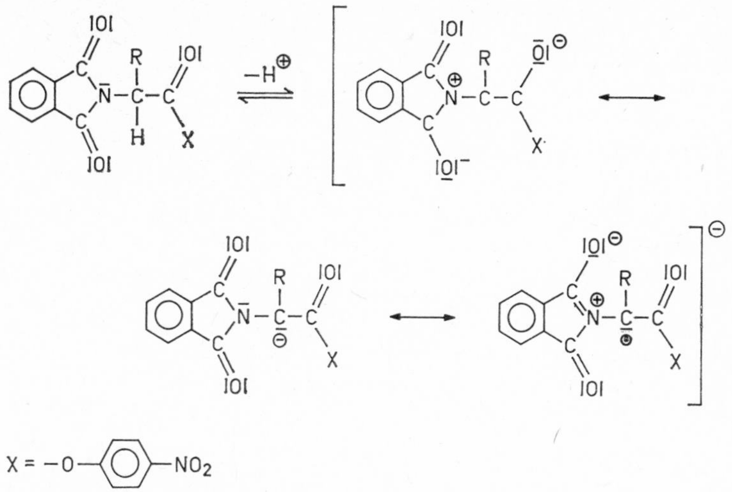

X = —O—C_6H_4—NO_2

Proton abstraction from the α-C atom in mixed anhydrides may be promoted by intramolecular hydrogen bonding[1], but the racemisation which can result from their use in coupling probably arises from the azlactone intermediate[2,3].

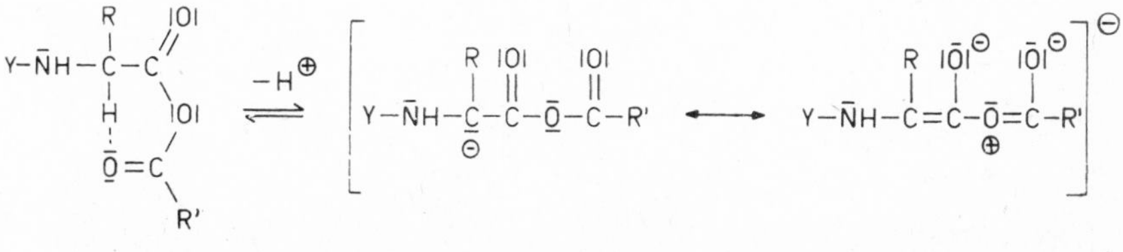

According to GOODMAN and GLASER[3], racemisation during azide syntheses, such as in the reaction of benzoyl-L-alanine azide with L-phenylalanine benzyl ester[4] or of benzoyl-L-leucine azide with glycine ethyl ester[2] cannot be interpreted by this mechanism but is due to direct proton abstraction from the asymmetric C-atom.

To explain the racemisation of the *N*-benzyloxycarbonyl-*S*-benzyl-L-cysteine *p*-nitrophenyl ester in presence of bases a '*β-elimination-readdi-*

tion' mechanism was first proposed[5,6]. After elimination of benzyl mercaptan the corresponding dehydroalanine derivative is formed; readdition of the benzyl mercaptan gives the racemic product. Further investigations, however, by KOVACS *et al.*[7] showed that the racemisation of activated esters of Z-S-Bzl-L-cysteine does not proceed according to this mechanism but takes place by direct exchange. In their studies of the base-catalysed racemisation and deuterium exchange of Z-S-Bzl-L-cysteine pentachlorophenyl ester in chloroform in the presence of triethylamine and monodeuterated methanol, they discovered the first instance of *isoracemisation* of an amino acid derivative. This term, introduced by CRAM[8], is applied to a racemisation process which proceeds without proton exchange.

2.2.6.1.2. *Azlactone Mechanism*

Although there are examples, mentioned above, of racemisation by direct proton abstraction from the asymmetric C atom, the vast majority of the cases of racemisation during coupling involve an optically labile azlactone intermediate. In the activation stage of amino acids acylated on nitrogen (I), elimination of HX may take place to form azlactones, which are known to racemise readily[3]. The rate of formation of the azlactone (II) depends on several factors. Powerful electron-attracting substituents X increase electrophilicity of the carboxyl C atom, and promote the intramolecular nucleophilic attack of the acyl O atom. *N*-acyl residues (R′), such as benzoyl, chloroacetyl, trifluoroacetyl, etc., also facilitate the ring-closing reaction, because they increase the nucleophilicity of the carbonyl O atom. Therefore, benzoyl- and acetylamino acid derivatives are rather easily racemised[9]. The azlactone mechanism is depicted as follows:

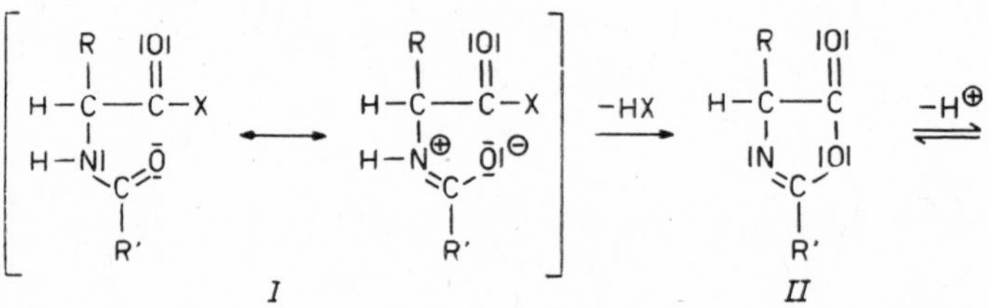

$$\text{I} \xrightarrow{-HX} \text{II} \underset{}{\overset{-H^{\oplus}}{\rightleftharpoons}} \text{III}$$

I

II

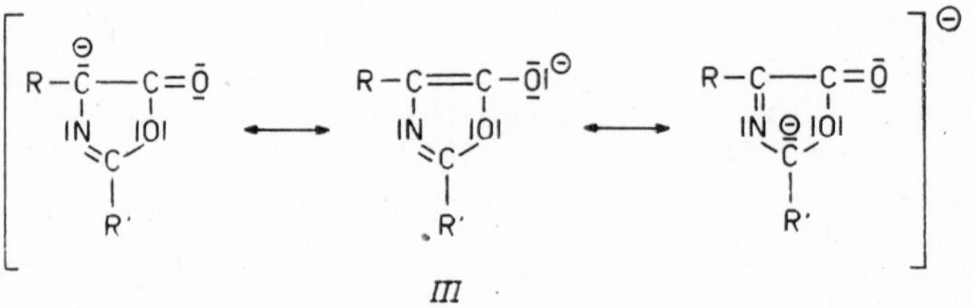

III

The azlactones formed (II) readily undergo reversible exchange of the α-proton in the presence of base, since the negative charge on the anions is delocalised (III). In addition to azlactone formation during carboxyl activation, azlactone formation is also possible in the coupling reaction of acylamino acids or N-protected peptides. Details of the evidence and further investigations have been lucidly reviewed by YOUNG[10] and by GOODMAN and GLASER[3]. Removal by base of the amide proton should favour ring closure to the azlactone. KEMP has postulated two mechanisms for base-catalysed azlactone formation[11,12]. In the case of the specific base-catalysis the base liberates the amide proton in a fast step, while the ring closure is the rate-determining step:

$$B| + -C(=O)-N(H)-C-C(=O)-X \underset{\text{fast}}{\rightleftharpoons} BH^{\oplus} + -C(-\bar{O}|^{\ominus})=N-C-C(=O)-X \xrightarrow{\text{slow}} \text{(azlactone)} + X^{\ominus}$$

In polar solvents the formation of the amide anion is facilitated so that, under these conditions, the specific base-catalysis mechanism predominates. This is true, for example, for various activated phenyl esters in dimethylformamide. In non-polar solvents the general base-catalysed azlactone formation predominates.

Under synthetic conditions the situation is extremely complicated. Thus in the aminolysis of activated acylamino acid esters—as well as of other activated N-protected peptides—the amino component acts as both nucleophile and base. The equilibria are summarised in the following simplified scheme[10,13]:

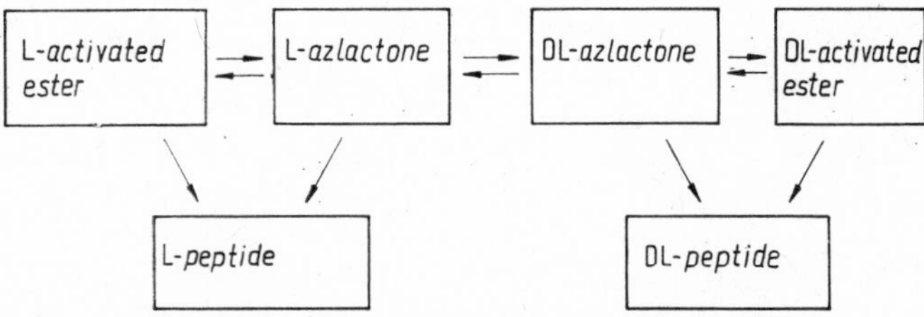

Figure 21. Racemisation routes possible in peptide syntheses with activated esters

Although not indicated in Figure 21, the amino component functions as nucleophile or base in each reaction step. The relative rates depend on the nucleophilicity/basicity ratio of the particular amino component used, on the solvent, the temperature, and the structure of the azlactone being formed.

After the development of methods for the preparation of optically active crystalline azlactones by GOODMAN *et al.*[14], studies on the rates of racemisation and ring opening of azlactones became possible. For the racemisation pseudo-first or -second order kinetics were encountered, depending on the amino component and solvent, whereas the ring opening was invariably a second-order reaction. The rate of azlactone cleavage depends on the nucleophilicity of the amine component, but that of racemisation depends on its basicity.

The increase of racemisation by adding special salts—termed 'the chloride ion effect' by G. T. YOUNG—is attributed to the increase in ionic strength of the solution promoting the charge separation necessary for racemisation[3]. KEMP[11] has confirmed this by studies of the coupling between Z–Gly–PheONp and GlyOEt in DMF with added triethylammonium chloride or triethylammonium fluoroborate. However, the addition of glycine ethyl ester hydrochloride resulted in a ninefold lowering of the proportion of racemate. Also, in DMF without added salts, the proportion of racemate increases as the concentration of reactants decreases. Both of these observations are extremely difficult to explain convincingly.

Some 1,2-dinucleophilic compounds with a high nucleophilicity/basicity ratio, such as hydrazine and hydroxylamine, open the azlactone ring very rapidly without detectable racemisation[3,15]. For this reason hydroxylamine derivatives, such as *N*-hydroxypiperidine, *N*-hydroxyurethane, *N*-hydroxysuccinimide, *N*-hydroxyphthalimide, *N*-hydroxyglutarimide, 1-hydroxybenztriazole, hydroxamic acids, etc. (Table 18), are particularly interesting both as activating components and as additives for the DCC method (see Section 2.2.3.1). Although DCC is an excellent reagent for preparing azlactones and azlactones are racemised even by the very weak base DCC, the properties of the 1,2-dinucleophilic compounds mentioned above enable the corresponding activated esters to be obtained in high optical purity.

If for some reason the formation of an azlactone intermediate is not possible, racemisation can take place only by the direct exchange mechanism discussed in Section 2.2.6.1.1. The risk of racemisation occuring is therefore very greatly reduced.

Thus the freedom from racemisation of the azide method and of the coupling reactions of activated esters whose aminolysis relies on intramolecular general base-catalysis (*N*-hydroxypiperidyl, 8-quinolyl, *o*-hydroxyphenyl esters and 3-acyl-oxy-2-hydroxy-*N*-ethylbenzamides, etc.) is based on the impossibility of azlactone formation. In addition, because *N*-acylated proline derivatives are not capable of forming an azlactone ring, C-terminal proline residues can always be activated without danger of racemisation.

Benzoyl, acetyl and formyl amino acids easily form azlactones, and only N-protecting groups of the urethane type offer an effective racemisation protection for amino acids (but not for peptides). In the case of the urethane-type protecting groups mesomeric stabilisation of the polarised carbonyl groups prevents azlactone formation:

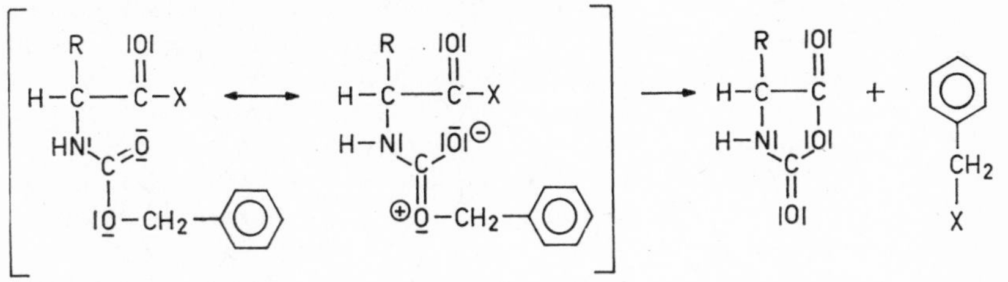

$$\left[\text{R–CH(NH–CO–O–CH}_2\text{–C}_6\text{H}_5\text{)–CO–X} \longleftrightarrow \text{R–CH(NH–C(–O}^{\ominus}\text{)=O}^{\oplus}\text{–CH}_2\text{–C}_6\text{H}_5\text{)–CO–X} \right] \longrightarrow \text{R–CH–C(=O)–O–C(=O)–NH (ring)} + \text{C}_6\text{H}_5\text{–CH}_2\text{–X}$$

Benzyloxycarbonyl amino acid chlorides (X = Cl) can form *N*-carboxylic acid anhydrides under appropriate conditions by elimination of benzyl chloride but they never form azlactones. Phthaloyl amino acids are also incapable of forming azlactone derivatives.

2.2.6.2. Methods for the Detection of Racemisation

The degree of racemisation that occurs during a coupling reaction and the optical purity of the final peptide product need not necessarily be related to each other. Racemic product that is actually formed, especially in the case of short peptide derivatives, may be separated at the crystallisation stage of the work-up, and thus racemisation-free peptide synthesis may be inferred erroneously. For the detection of racemisation numerous physical, biological and enzymatic procedures have been described: only a few examples will be discussed in the following section.

2.2.6.2.1. ANDERSON–CALLAHAN *Test*[16]

After coupling benzyloxycarbonylglycyl-L-phenylalanine to glycine ethyl ester, the optically pure compound can be separated from the racemate by fractional crystallisation. From a 2 per cent solution in ethanol the less soluble racemate crystallises first, and it can be separated and weighed easily (limit of sensitivity *ca.* 1–2 per cent).

This method gives the amount of racemisation occurring in the activation of the C-terminal phenylalanine:

$$\text{Z–Gly–Phe} + \text{GlyOEt} \rightarrow \text{Z–Gly–Phe–GlyOEt}$$

It is essential that the N-protected dipeptide used be optically pure.

Application of the isotopic dilution method devised by D. S. KEMP *et al.* to the ANDERSON test has given a highly sensitive test system suitable for determining amounts of racemate between 1.0 and 0.001 per cent[17].

2.2.6.2.2. YOUNG *Test*[9]

When first proposed, acetyl-L-leucyl-glycine ethyl ester was used as the model peptide for this polarimetric method. Because of the poor crystallinity of the model peptide this procedure permitted only a rough estimation of the amount of racemisation in a peptide coupling. The N-protected dipeptide ester formed by coupling of benzoyl-L-leucine with glycine ethyl ester is more suitable:

$$\text{Bz–Leu} + \text{GlyOEt} \rightarrow \text{Bz–Leu–GlyOEt}$$

The extent of racemisation can be determined by the optical rotation of the crude product. In addition, after saponification of the ester, the amount of the racemic benzoyl dipeptide can be estimated by fractional crystallisation.

The application of the isotope dilution method to this test[17] using 7-[^{14}C]-benzoyl-L-leucine has resulted in an increase of sensitivity from *ca.* 1–2 per cent to 0.001–0.01 per cent.

2.2.6.2.3. WEYGAND *Test*[18]

WEYGAND *et al.*[18] developed a racemisation test based on the gas chromatographic separation of diastereoisomeric *N*-trifluoroacetyl dipeptide methyl esters. A very suitable model peptide is the tripeptide ester formed by coupling an activated benzyloxycarbonyl-L-leucyl-L-phenylalanine derivative to L-valine *tert*-butyl ester:

$$\text{Z–Leu–Phe} + \text{ValOBu}^t \rightarrow \text{Z–Leu–Phe–ValOBu}^t$$

After removing the protecting groups from the tripeptide with trifluoroacetic acid/anisole, the dipeptide phenylalanyl-valine is obtained by partial hydrolysis without racemisation. This dipeptide is then esterified with methanolic HCl and trifluoroacetylated with trifluoroacetic acid methyl ester in the presence of triethylamine. The ratio of Tfa–Phe–ValOMe to Tfa–D-Phe–ValOMe is determined by gas chromatography. Since the L-valine *tert*-butyl ester used as amino component cannot suffer any loss of optical purity, this ratio gives the degree of racemisation in the coupling reaction. The reliability and accuracy of this method depend on the fact that, in all the operations made, no purification steps are necessary and the gas chromatographic analysis of the trifluoroacetyl dipeptide methyl ester can be accomplished without its isolation even in presence of solvents, by-products and contaminants. The limit of detection is 0.1–1.0 per cent. A simplified gas chromatographic racemisation test has also been described by the same authors[19].

2.2.6.2.4. Further Methods of Detection

Bodanszky and Conklin[20] have developed a procedure to detect racemisation in which the amount of Ac–D-aIle–GlyOEt formed in any coupling of Ac–Ile with Gly–OEt is determined. The amount of racemisation in the coupling reaction is calculated from the proportion of D-aIle determined by amino acid analysis.

An NMR racemisation test based on the different signals of the alanine methyl doublets in the spectra of diastereoisomeric acyl–Ala–PheOMe[21] has also been introduced, and its sensitivity has been increased tenfold by use of the [^{13}C]H-satellite signal of the L–L compound as an internal standard[22].

Manning and Moore[23] hydrolysed the synthetic peptide and treated the hydrolysate with the NCA of an L-amino acid. The DL-diastereoisomers present can then be determined in the amino acid analyser with an accuracy of better than 0.1 per cent.

The quantitative separation of Gly–Ala–Leu and Gly–D-Ala–Leu on ion exchange chromatography is the basis of the test system developed by Izumiya and Muraoka[24]: after coupling Z–Gly–Ala to LeuOBzl both the protecting groups are removed by hydrogenolysis, and the DL diastereoisomer is determined on the analyser.

All the methods for the estimation of racemisation described so far determine racemisation for a particular method in a model system only. It is also important to be able to determine the optical purity of the final product of an actual synthesis, because a satisfactory synthesis can only be claimed when the peptide synthesised fulfils the normal criteria of purity of preparative organic chemistry. Therefore, great importance has to be attributed to the operations of purification. Only in exceptional cases is it possible to purify peptides by repeated crystallisation. In most cases one is dependent on countercurrent distribution, on a variety of chromatographic methods (such as gel chromatography, affinity chromatography, ion exchange chromatography, preparative paper and thin layer chromatography) and on the different electrophoretic separation methods. The special difficulties involved in the purification of peptides built up on polymeric supports have already been pointed out.

The most suitable method of investigating the steric uniformity of the final product is the degradation by stereospecific enzymes, such as leucine aminopeptidase, or one of several carboxypeptidases, which attack rapidly only L-peptide bonds, although there are substrates containing D-amino acid residues which are slowly cleaved. The application of enzymes to the monitoring of purification and structural elucidation of peptides and proteins has been reviewed by ZUBER[25].

2.2.6.3. Evaluation of the most Important Coupling Methods with respect to Racemisation

The value of any judgements made in this difficult area is limited by the sensitivity of the racemisation test used, and the difficulties are exacerbated by the fact that not all the activating methods have been evaluated using the same racemisation test. Detailed investigations were made by WEYGAND *et al.*[26] in 1966, using the racemisation test based on gas chromatographic separation of the diastereomeric *N*-trifluoroacetyl dipeptide methyl esters (p. 170). In Table 19 the coupling methods are listed in order of increasing racemisation. However, the extent of racemisation depends to a great extent on the precise coupling conditions and, furthermore, since these investigations were made more sensitive detection methods have been developed: some of the values given in the table are therefore misleading, as we shall show in the following discussion.

The *azide method* is generally accepted as the method most free of racemisation (0.01–0.04 per cent racemate)[27]. During the azide coupling any contact with tertiary bases (especially triethylamine) must be prevented, however, because under these conditions acyl azides are extensively racemised[27].

N-hydroxypiperidyl esters can be synthesised in an optically pure state and guarantee racemisation-free coupling, but their usefulness is somewhat diminished by their low aminolysis rate.

8-Quinolyl esters are proof against racemisation (KEMP's test: 0.16 per cent racemate) and sufficiently reactive. However, optically pure N-protected peptide quinolyl-(8) esters cannot be prepared directly unless the inverse GOODMAN method[28] is used. For stepwise synthesis of peptides from the C-terminal end, especially in the MERRIFIELD synthesis[29], Boc-amino acid 5-chloroquinolyl-(8) esters have proved successful.

p-Nitrophenyl esters of N-protected peptides, like the corresponding 8-quinolyl esters, cannot be synthesised directly without some risk of racemisation. In ethyl acetate, tetrahydrofuran, chloroform and even pyridine as solvent the aminolysis of *p*-nitrophenyl esters appears to proceed virtually without racemisation[11,17], but in dimethylformamide, dimethylsulphoxide (solvents needed to obtain a high aminolysis rate of the *p*-nitrophenyl esters)[11] D. S. KEMP found 0.53 and 1–2 per cent of racemate, respectively[11]. Using methylene chloride as solvent for the coupling reaction, 0.63 per cent racemate was detected.

The *mixed anhydride method* using isobutyl chloroformate (see Sec-

Table 19. Studies on racemisation in the synthesis of Z–Leu–Phe–ValOBu^t from Z–Leu–Phe and ValOBu^t (modified from ref. 26)

No.	Method	Solvent	Temp. (°C)	Time (h)	D-Phe–Val (%)
1	Azide	Ethyl acetate	0	48	0
2	*N*-Hydroxypiperidyl ester	THF	40	44	0
3	8-Quinolyl ester	THF	40	44	0
4	Vinyl ester	Malonic ester	30	86	0 (0.6)*
5	Thiophenyl ester	THF	40	22	0
6	*p*-Nitrophenyl ester	Acetonitrile	30	21	0 (2.2)*
7	Ethoxyacetylene	Ethyl acetate	20	24	2.4
8	Carbonyldiimidazole	THF	0	20	2.8
9	WOODWARD	Acetonitrile	0	22	3.2
10	Hydrazide oxidation	THF	0	0.16	4.45
11	Mixed anhydride	THF	−35	16	4.9
12	DCC	THF	−10	48	6.0 (12.5)†
13	Ynamine	THF	0	23.5	26.5
14	Phosphorazo	Pyridine	100	3	28.3

* Control value in the preparation of activated esters.
† Coupling at 20°.

tion 2.2.3.3) is, if the conditions described by G. W. ANDERSON are employed, essentially free of racemisation[17].

3-Acyloxy-2-hydroxy-N-ethylbenzamide derivatives couple with only a slight tendency towards racemisation in DMF (0.13 per cent) and can also be prepared directly by means of 2-ethyl-7-hydroxybenzisooxazolium fluoroborate[12]. The drawback with these derivatives is their great sensitivity to hydrolysis.

The *DCC method* (see Section 2.2.3.1) is a coupling procedure which normally results in some racemisation. Thus in tetrahydrofuran 6.1 per cent of racemate and in dimethylformamide (a solvent in which DCC coupling is slow) 7.1 per cent of racemic product was detected[11]. However, additives such as *N*-hydroxysuccinimide, *N*-hydroxycarbamates, *N*-hydroxybenztriazole and similar compounds can help to achieve a racemisation-free coupling. Studies on the racemisation which results when *N*-hydroxysuccinimide is used as additive indicated higher proportions of racemate in dimethylformamide (at 4° 0.20 per cent; at 24° 0.24 per cent) than in tetrahydrofuran (at 4° 0.24 per cent; at 22° 0.12 per cent)[11]. It was also shown that in DMF the proportion of racemate increased with decreasing concentration (at 4° 0.62 per cent).

N-hydroxysuccinimide esters of N-protected peptides are very labile even under mild conditions[3,30]. If *N*-hydroxysuccinimide esters are synthesised in dimethylformamide by means of DCC and coupled in the same solvent, the amount of racemisation is considerable. Much better results have been obtained with tetrahydrofuran[11].

According to Kovacs *et al.*[31], *pentachloro- and pentafluorophenyl esters* can be produced in an optically pure form by causing N-protected peptides to react with a complex of 1 mol of DCC and 3 mol of pentachlorophenol or pentafluorophenol, respectively. Alternatively the mixed anhydride method may be used: this method, which is usually free of racemisation, can also be used to prepare other optically pure esters, such as *p*-nitrophenyl and 8-quinolyl esters.

No racemisation has been detected in the *controlled peptide synthesis in aqueous medium by means of NCAs* (see Section 2.2.3.11.).

References

1 Neuberger, A. (1948). *Advan. Protein Chem.*, **4**, 297

2 Kemp, D. S. and Rebek, J. (1970). *J. Amer. Chem. Soc.*, **92**, 5792

3 Goodman, M. and Glaser, C. (1970). *Peptides: Chemistry and Biology*, ed. B. Weinstein and S. Lande, p. 267, Dekker, New York

4 Determann, H. *et al.* (1966). *Annalen*, **694**, 190

5 Young, G. T. (1959). *Coll. Czech. Chem. Comm., Special Issue*, **24**, 114; Determann, H. and Wieland, Th. (1963). *Annalen*, **670**, 136

6 Bodanszky, M. and Bodanszky, A. (1967). *Chem. Comm.*, 591

7 Kovacs, J. *et al.* (1968). *Chem. Comm.*, 1066; (1970). *ibid.*, 53; (1971). *J. Amer. Chem. Soc.*, **93**, 1541

8 Almy, J. and Cram, D. J. (1969). *J. Amer. Chem. Soc.*, **91**, 4459; Cram, D. J. and Gosser, L. (1964). *J. Amer. Chem. Soc.*, **86**, 5457; Cram, D. J. (1965). In *Fundamentals of Carbanion Chemistry*, Academic Press, New York

9 Williams, M. W. and Young, G. T. (1963). *J. Chem. Soc.*, 881

10 Young, G. T. (1967). *Proc. 8th Europ. Peptide Symp.*, Noordwijk, 1966, p. 55, North-Holland, Amsterdam

11 Kemp, D. S. (1972). *Proc. 11th Europ. Peptide Symp.*, Vienna, 1971, p. 1, North-Holland, Amsterdam

12 Kemp, D. S. and Chien, S. W. (1967). *J. Amer. Chem. Soc.*, **89**, 2745

13 Goodman, M. *et al.* (1967). *Tetrahedron*, **23**, 2031

14 Goodman, M. *et al.* (1964). *J. Amer. Chem. Soc.*, **86**, 2918; (1966). *ibid.*, **88**, 3887

15 Goodman, M. *et al.* (1968). *Chem. Engng. News*, **46**, 40

16 ANDERSON, G. W. and CALLAHAN, F. M. (1958). *J. Amer. Chem. Soc.*, **80**, 2902
17 KEMP, D. S. *et al.* (1970). *J. Amer. Chem. Soc.*, **92**, 1043
18 WEYGAND, F., PROX, A., SCHMIDHAMMER, L. and KOENIG, W. (1963). *Angew. Chem.*, **75**, 282
19 WEYGAND, F., HOFFMANN, D. and PROX, A. (1968). *Z. Naturforsch.*, **23**b, 279
20 BODANSZKY, M. and CONKLIN, I. E. (1967). *Chem. Comm.*, 733
21 HALPERN, B. *et al.* (1967). *J. Amer. Chem. Soc.*, **89**, 5051
22 HIRSCHMANN, R. *et al.* (1968). *J. Amer. Chem. Soc.*, **90**, 3254
23 MANNING, J. M. and MOORE, S. (1968). *J. Biol. Chem.*, **243**, 5591
24 IZUMIYA, N. and MURAOKA, M. (1969). *J. Amer. Chem. Soc.*, **91**, 2391
25 ZUBER, H. (1960), *Chimia*, **14**, 405
26 WEYGAND, F., PROX, A. and KOENIG, W. (1966). *Chem. Ber.*, **99**, 1451
27 KEMP, D. S. *et al.* (1970). *J. Amer. Chem. Soc.*, **92**, 4756
28 GOODMAN, M. and STEUBEN, K. G. (1962). *J. Org. Chem.*, **27**, 3409; (1959). *J. Amer. Chem. Soc.*, **81**, 3980
29 JAKUBKE, H. D. and BAUMERT, A. (1972). *Proc. 11th Europ. Peptide Symp.*, Vienna, 1971, p. 132, North-Holland, Amsterdam
30 ZIMMERMANN, J. E. and ANDERSON, G. W. (1967). *J. Amer. Chem. Soc.*, **89**, 715
31 KOVACS, J. *et al.* (1967). *J. Amer. Chem. Soc.*, **89**, 183; (1970). *J. Org. Chem.*, **35**, 3563

2.2.7. The Combination of Protecting Groups and Coupling Methods

As a rule, in the synthesis of di- and tripeptides no significant difficulties occur with regard to the choice of protecting groups and coupling methods. However, for the synthesis of higher peptide sequences careful planning is essential. The array of methods for protection and coupling may seem to be wide enough for all the operations of peptide synthesis to be performed without complications, but few of these methods are universally applicable and it is for this reason that the success of a peptide synthesis is very much a question of strategy and tactics.

2.2.7.1. Strategy of Peptide Synthesis

In principle, a peptide chain can be built up by stepwise chain elongation from either the N-terminal end or the C-terminal end. Although stepwise synthesis of polypeptide chains is involved in their biosynthesis and might therefore be thought to be the method of choice, the preparative difficulties of this approach make fragment condensation a better proposition.

In this connection it is interesting to mention synthetic operations with peptides of biological origin. Partial synthesis[1] involves the use of naturally derived fragments for both the stepwise and fragment condensation approaches to peptide syntheses. In this approach one starts with the cleavage of a natural peptide or protein to obtain ready-made intermediates with natural sequences. For a stepwise partial synthesis it is necessary to remove one or more residues sequentially from the end of a peptide chain—for example, by means of the phenylisothiocyanate reaction. For fragment condensation partial synthesis, peptide bonds may be broken enzymatically (trypsin) or chemically (cyanogen bromide cleavage at methionine peptide bonds). Brief details of a procedure for the conversion of porcine into human insulin by a partial synthesis approach have been reported by RUTTENBERG[2].

2.2.7.1.1. Stepwise Chain Elongation

In general, stepwise chain elongation in conventional syntheses is useful only for peptides containing up to 10–15 amino acids. As the chain length increases, the starting and final products differ less and less in their physical properties, and the isolation of the desired synthetic product becomes more and more difficult. After the development of the *solid phase* peptide *synthesis* by R. B. MERRIFIELD, this approach appeared to become the method of choice for stepwise syntheses.

Stepwise chain elongation from the N-terminal end is at present of minor practical importance, although, as indicated by R. L. LETSINGER, it can be accomplished on a polymeric support. The main reason for this lies in the danger of racemisation in the activation of acyl peptides. At present, only peptides with C-terminal glycine or proline can be activated by any of the known methods without risk of racemisation. Because of this, and the fact that the classical azide coupling is suitable for routine stepwise chain elongation, no practical significance attaches to this synthetic approach. It is interesting that this situation was not changed by the development of additional racemisation-free coupling methods, such as the WUENSCH–WEYGAND procedure or the variant of KOENIG and GEIGER. Even the modification of the *Letsinger synthesis* by R. B. MERRIFIELD and his collaborators (see Section 2.2.4.1.2) has not so far resulted in wider application of the stepwise chain elongation from the N-terminal end. The preparative advantage of using unprotected amino acids as the amino component is offset by the low yields in the coupling reaction, and in such cases both the starting material and final product have free carboxyl groups and this makes their separation more difficult.

Stepwise chain elongation from the C-terminal end, using N-protecting groups of the urethane type (Z-, Boc-, Bpoc-, etc.), has the advantage that the peptide bond is formed without racemisation, whatever the activating method chosen. This principle of strategy was introduced into conventional synthesis by M. BODANSZKY and V. DU VIGNEAUD and applied by them and by other workers to syntheses of biologically active peptides with great success. In principle it is better to use the C-terminal starting amino acid as a carboxyl-protected derivative. Otherwise, after each coupling step and subsequent removal of the N-protecting group the intermediate is present as zwitterion, and the next acylation reaction may be rendered more difficult by solubility limitation. However, it must be admitted that the removal of the N-protecting group, which is required after every coupling step, constitutes a methodological disadvantage. On the one hand, the deprotection may be accompanied by undesired side reactions, and on the other hand, difficult tactical problems can arise with regard to the choice of convenient protecting groups for trifunctional amino acids. However, these criticisms do not detract from the immense importance of this synthetic approach. M. BODANSZKY's total synthesis of the heptacosapeptide amide secretin demonstrates that the upper limit of 10–15 amino acid units, which is suggested by most practical experience, can be exceeded in particular cases. In general, the conventional stepwise chain elongation from the C-terminal end is convenient for building up smaller peptides or partial sequences which are then linked to each other by means of the fragment condensation.

The *stepwise peptide synthesis without isolation of intermediates*, as developed by the groups of KNORRE[3], SHEEHAN[4], PODUŠKA[5], TILAK[6] and HIRSCHMANN[7], has considerable potentiality. Even if the technique cannot be applied to the synthesis of longer chains, there is the prospect of avoiding the wasteful purification operations necessary after every coupling stage in the conventional stepwise synthesis.

The logical extension of this approach was MERRIFIELD's ingenious idea of anchoring the starting amino acid covalently to a polymeric support and to build up the peptide chain on the support in a stepwise manner. Seven years before MERRIFIELD's initial papers, at the 6th Canadian Forum of High Polymers in 1955, NICHOLLS[8] reported a new method of peptide synthesis in which an amino acid was fixed to an ion-exchanger containing quaternary ammonium hydroxide groups and, after changing the solvent, was treated with a carboxyl activated second amino acid in benzene to give peptide. However, it is due to MERRIFIELD that the *solid phase peptide synthesis* has attained its present significance in synthetic protein chemistry. The simplicity of the operation and the possibility of automation are the most characteristic advantages of this strategy. The laborious precipitation and purification steps needed in conventional synthesis are replaced by simple filtration and washing stages in solid phase synthesis. However, this leads to homogeneous products only if each coupling step proceeds in completely quantitative yield. In spite of the use of large excesses of acylating agents, which carries the danger of *N*-acylation of peptide bonds, a 100 per cent yield at every coupling step is not attained in practice and the separation of truncated and failure sequences from the desired final product is virtually impossible with available methods, especially with large biologically active peptides or proteins. The presence of biological activity cannot be used as a criterion for a successful synthesis. On the other hand, these criticisms should not be overdone, because dogmatic persistence in exclusively conventional methods may obstruct the further progress of synthetic protein chemistry and may discourage the work necessary for improvement of peptide synthesis on polymeric supports. The present level of refinement of the MERRIFIELD technique, in general, justifies its use for the synthesis of biologically active peptides and fragments of up to 10–20 amino acid units.

Peptides can also be built up in a stepwise manner from the C-terminal end by means of *insoluble polymeric agents*. Polymeric activated esters or polymeric reagents are treated with a solution of the amino component. The growing peptide chain remains in solution, whereas excess reagents and by-products can be separated by filtration.

Using esters with a basic centre, such as the 4-picolyl esters developed by G. T. YOUNG, peptides can be synthesised from the C-terminal end in a stepwise manner. Coupling takes place in solution and after each step the growing peptide chain is reversibly bound to an ion exchange resin, in order to separate by-products. This technique has already proved successful in the synthesis of several bradykinin analogues.

The evaluation of the general applicability and limitations of these latter methods is not yet possible.

2.2.7.1.2. Fragment Condensation

It must be repeatedly emphasised that the formation of peptide bonds in the synthesis of biologically active peptides must be performed without racemisation. Splitting up the sequence to be synthesised at convenient prolyl and glycyl bonds gives fragments which can be activated without danger of racemisation and are, therefore, ideal for the fragment condensation approach. Of course, in natural polypeptides a favourable distribution of such glycine and proline residues in the chain is not always present in the structure. In general, even with the most careful planning it cannot be predicted with any certainty which division of the chain will prove the most convenient in a fragment condensation. Thus it is always an advantage to plan the approach to be as flexible as possible. In this way, should a particular route prove impracticable, loss of time and material can be kept to a minimum.

For linking together the fragments to form larger peptides, the activating methods of choice are the azide method and the WUENSCH–WEYGAND procedure. In some cases side reactions have been observed using DCC/*N*-hydroxysuccinimide, and in critical situations additives whose use is not accompanied by side reactions (*N*-hydroxybenztriazole, or other *N*-hydroxycompounds) must be chosen. As a further method for coupling fragments the mixed anhydride method using isobutyl chloroformate may be used. This guarantees a racemisation-free coupling if special conditions are oberved. The activating method convenient for a particular fragment condensation depends largely on the over-all tactical approach to the synthesis, and this will be discussed in detail later. For a conventional synthetic plan using a minimum protection of side chain functions—as was the case in the total synthesis of ribonuclease S-protein by R. HIRSCHMANN and collaborators—only the azide method is applicable, because the WUENSCH–WEYGAND method and the GEIGER–KOENIG procedure require the full protection of side chain functions.

Owing to the greater number of coupling methods free from racemisation which are now available, there is much greater freedom of choice for the partition of a long peptide chain into convenient fragments. It is now possible to give more attention to the choice of the amino component. This is important since it is an advantage that the N-terminal residue of a fragment should have as high a nucleophilicity as possible.

It is sometimes possible to accomplish *fragment condensation on polymeric supports* by protecting the carboxyl function of a peptide fragment which is to act as an amino component with an insoluble polymeric benzyl ester grouping. For example, in the synthesis of the decapeptide corresponding to the tryptophan-containing ACTH sequence 5–14, YAJIMA *et al.*[9] esterified the tetrapeptide Boc–Lys(Z)–Pro–Val–Gly with a MERRIFIELD support, and, after removing the Boc-protecting group, coupled it with the fragment Z–Glu(OBzl)–His–Phe–Arg(NO_2)–Trp–Gly by the DCC method.

2.2.7.2. Tactics of Peptide Synthesis

The suitable selection of protecting groups and coupling methods is of decisive importance for the planning of a peptide synthesis. For the reversible blocking of the terminal groups and the basic, acidic, alcoholic and phenolic side chain functions a variety of protecting groups is available (see Section 2.2.2). The choice of side chain protection is dictated by the selectivity required in de-protection reactions and also by the over-all strategy of the particular synthesis concerned.

The partial sequences needed for fragment condensations are most conveniently synthesised by stepwise chain lengthening from the C-terminal end, either by conventional methods or by the MERRIFIELD technique. When the conventional method is used, one has a choice between a procedure with *full protection of side chains* and a procedure with *minimum protection of side chains,* and this choice is determined by the requirements of the subsequent fragment coupling. The advantage of having all side chains masked is that there is a maximum protection against undesired side reactions during the numerous purification and work-up steps which may be involved in the course of a long synthesis. Inevitably, serious problems of solubility occur and this at present sets a limit of 30–50 amino acids with side chain functions to the units that can be built up from fragments. For this reason the further development of side-chain-protecting groups which possess solubility-promoting properties in addition to high selectivity would be extremely valuable.

In their synthesis of fragments required for the total synthesis of ribonuclease T_1 HOFFMANN and his collaborators[10] used activated esters so that tyrosine, serine, threonine and histidine derivatives could be used without side-chain-protection, and the use of the benzyloxycarbonyl hydrazide group for protecting the C-terminal carboxyl function enabled fragment condensations by azide coupling.

The procedure with a minimum protection of side chains was even more dramatically demonstrated by R. HIRSCHMANN, R. G. DENKEWALTER *et al.* in their total synthesis of ribonuclease S-protein. Apart from tryptophan, the sequence contains all the usual trifunctional amino acids, but only the ε-amino and SH groups were blocked. This meant that, for the synthesis of the fragments in the range of tri- to nonapeptides, only the NCA/NTA method (see Section 2.2.3.11) and the *N-hydroxysuccinimide ester procedure* could be used. Using minimum protection of side chain functions inevitably means that the danger of side reactions involving the side chains is considerable, so that in many cases the individual fragments must be purified very carefully. Linking the fragments to form larger units could only be accomplished by the azide method when this tactical approach had been chosen. However, the use of only a few side-chain-protecting groups meant that there were no major problems of solubility. Only in the azide coupling of the two largest fragments containing 44 and 60 amino acid residues, respectively, was there an obvious decrease in yield.

As an example, two approaches to the synthesis of a decapeptide amide will now be discussed. All the different possibilities cannot be included but the approaches described will illustrate many of the principles involved.

In the case of a *stepwise chain elongation from the C-terminal end* by means of activated esters, the blocking of the imidazole-nitrogen of histidine is necessary. Since the SH function of cysteine is to be protected as its benzyl derivative, a sodium-in-ammonia cleavage will be necessary as the last stage of the synthesis. Therefore a benzyl derivative may as well be used to protect the histidine residue and basic side chains of lysine and arginine will be protected by tosyl residues since these can also be removed with sodium in liquid ammonia. The hydroxyl groups of serine and tyrosine are most conveniently protected by conversion into the corresponding benzyl ethers, while the β-carboxyl group of aspartic acid is protected as its benzyl ester. For the temporary protection of the α-amino function throughout the synthesis, *tert*-butyloxycarbonyl groups are used, since they can be removed selectively with trifluoroacetic acid without affecting either tosyl or benzyl

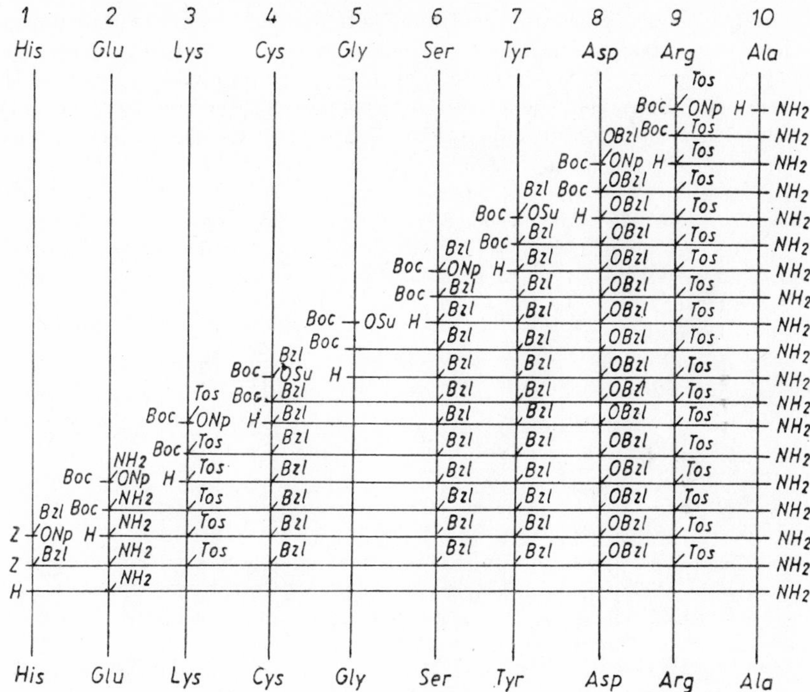

Figure 22. Reaction scheme for the synthesis of a decapeptide amide by stepwise chain elongation

groupings. The synthesis is shown in schematic form in Figure 22. The N-terminal histidine is best attached as its benzyloxycarbonyl derivative instead of its *tert*-butyloxycarbonyl derivative, because the Z-group can be removed together with all the other protecting groups by treatment with sodium in liquid ammonia in the final step.

For *fragment condensation* the most suitable approach is to synthesise the N-terminal pentapeptide 1–5 with glycine at the carboxyl end. For the reversible protection of the carboxyl group esterification is suitable, but a methyl ester will not do, since, after the stepwise synthesis of the peptide is complete, this ester must be saponified by alkali, and this would lead to attack on the carboxamide group of glutamine. However, WEYGAND[11] has described the 2,4-dimethoxybenzyl residue as an easily removable protecting group for acid amide groupings. If an alkyl ester is used as carboxyl-protecting group, only the azide method may be used for the coupling with the fragment 6–10. The C-terminal sequence 7–10 is synthesised by means of activated esters, removing the benzyloxycarbonyl group from the dipeptide by HBr/glacial acetic acid and from the tripeptide by catalytic hydrogenation, respectively. The need for hydrogenation arises because the *tert*-butyl ester group of the aspartic acid is partially removed by HBr/glacial acetic acid. After treatment with *N,O*-bisbenzyloxycarbonyl tyrosine trichlorophenyl ester and subsequent hydrogenolysis, benzyloxycarbonyl

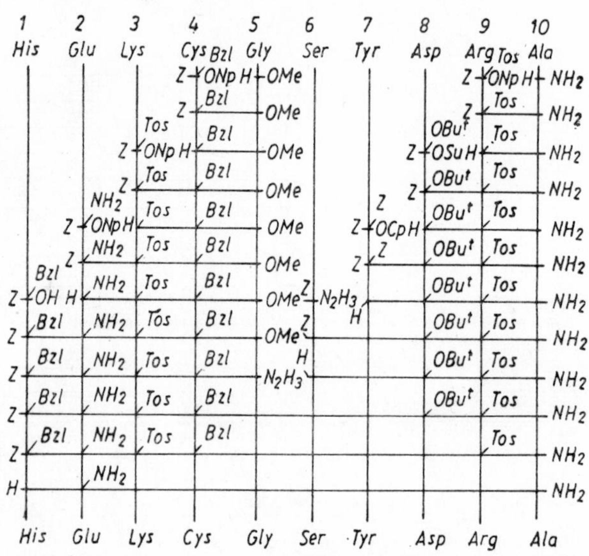

Figure 23. Reaction scheme for the synthesis of a decapeptide amide by fragment condensation

serine is attached via the azide without protection of the side chain function. Catalytic hydrogenation yields the partially protected pentapeptide amide 6–10, which is then used as amino component for the fragment condensation. The synthesis of the N-terminal pentapeptide fragment is summarised in Figure 23. Apart from the N-terminal histidine which is attached by the DCC method, all the other amino acid derivatives are coupled via activated esters. The cleavage of the N-protecting group which is necessary at each stage can be performed by acidolysis with hydrogen bromide in glacial acetic acid. After conversion into the hydrazide and azide, the protected pentapeptide methyl ester is coupled to the C-terminal fragment. The *tert*-butyl ester group is removed by treatment with trifluoroacetic acid, and then the remaining protecting groups are eliminated by sodium in liquid ammonia.

Of course, this model peptide could also be built up on a polymeric support or by other variations of conventional procedures. There is no completely general set of rules for the strategy and tactics of peptide synthesis which enables the correct combination of the various possibilities to be selected for each sequence.

Table 20 gives data on the stability and method of cleavage of some important protecting groups under standard conditions. The value of such summaries is often disputed, because it is not always possible to generalise and in a particular case the stability and lability of protecting groups may depend on the actual conditions involved. The first summary of this kind

Table 20. The cleavage and stability of some protecting groups

Cleavage by	Protection of															
	Amino groups							Carboxyl groups			Side chain groups					
	Z-	Z(OMe)-	Boc-	Nps-	Bpoc-	Trt-	Tfa-	-OMe -OEt	-OBzl	-OBu^t	OBzl-	OBu^t-	SBzl-	Acm-	NO_2 / Arg	Bzl . / His
HBr / CF_3COOH	removable	removable	removable	removable	removable	removable	removable	not removable	removable	removable	removable	removable	not removable	not removable	not removable	not removable
CF_3COOH	not removable	removable	removable	removable	removable	removable	not removable	not removable	not removable	removable	not removable	removable	not removable	not removable	not removable	not removable
dilute CH_3COOH	not removable	not removable	not removable	not removable	removable	removable	not removable	not removable	not removable	not removable	not removable	not removable	not removable	not removable	not removable	not removable
liq. HF	removable	removable	removable	removable	removable	removable	not removable	not removable	removable	removable	removable	removable	removable	not removable	removable	removable
Hydrogenolysis H_2 / Pd	removable	removable	not removable	side reactions	not removable	removable	not removable	not removable	removable	not removable	removable	not removable	side reactions	side reactions	removable	removable
Reduction with Na / liq. NH_3	removable	removable	not removable	?	?	?	side reactions	side reactions	removable	?	removable	not removable	removable	?	side reactions	removable
Hydrolysis NaOH	not removable	not removable	not removable	not removable	not removable	not removable	removable	removable	removable	not removable	not removable	not removable	not removable	not removable	not removable	not removable

removable · not removable · side reactions · ? not elucidated

was given for a number of N-protecting groups by Boissonnas and Preitner[12] in 1953. The rate of cleavage of differently substituted benzyloxycarbonyl glycine derivatives by hydrogen bromide in glacial acetic acid has been studied by Blaha and Rudinger[13], and Losse *et al.*[14] have reported studies of the kinetics of the acidolysis of N- and C-protecting groups in peptide syntheses. However, not all the protecting groups have been subjected to all the usual cleavage conditions and it is therefore difficult to give a complete picture. In making a choice between the possibilities, all the combinations of groups must be considered carefully: wide experience is the greatest asset one can have, and the table must be taken as a rough guide only.

References

1 Offord, R. E. (1972). *Biochem. J.*, **129**, 499
2 Ruttenberg, M. A. (1972). *Science*, **177**, 623

3 Knorre, D. G. and Shubina, T. N. (1965). *Proc. 7 th Europ. Peptide Symp.* Budapest, 1964: *Acta Chim. Acad. Sci. Hung.*, **44**, 77
4 Sheehan, J. C., Preston, J. and Cruickshand, P. A. (1965). *J. Amer. Chem. Soc.*, **87**, 2492
5 Poduška, K. (1968). *Coll. Czech. Chem. Comm.*, **33**, 3779
6 Tilak, M. A. (1970). *Tetrahedron Letters*, 849
7 Hirschmann, R. *et al.* (1966). *J. Amer. Chem. Soc.*, **88**, 3163
8 Nicholls, R. V. V. (1955). *Angew. Chem.*, **67**, 333
9 Yajima, H. *et al.* (1970). *Chem. Pharm. Bull.* (*Japan*), **18**, 1333
10 Hofmann, K. *et al.* (1972). *Proc. 11 th Europ. Peptide Symp.*, Vienna, 1971, p. 227, North-Holland, Amsterdam
11 Weygand, F. *et al.* (1966). *Tetrahedron Letters*, 3483; (1968). *Chem. Ber.*, **101**, 3623, 3642
12 Boissonnas, R. A. and Preitner, G. (1953). *Helv. Chim. Acta*, **36**, 875
13 Blaha, K. and Rudinger, J. (1965). *Coll. Czech. Chem. Comm.*, **30**, 585
14 Losse, G. *et al.* (1968). *Annalen*, **715**, 196

2.3. Important Naturally Occurring Peptides[1–4]

Over the past 20 years the number of peptides which have been discovered in animals, plants and microorganisms has increased enormously. The isolation of these compounds, many of which occur in very low concentrations and often in association with proteins, has only been possible with the use of modern techniques such as countercurrent distribution, electrophoresis, gel filtration, ion exchange chromatography, etc.; the elucidation of the primary structures was achieved by methods which are described more fully in Section 3.6.1.1.

The majority of the naturally occurring peptides consist exclusively of the amino acids which are commonly found in proteins ('proteinogenic' amino acids). However, frequently other simple amino acids are also found as constituents: for example, β-alanine and γ-aminobutyric acid occur in the dipeptides carnosine (β-alanyl-histidine), homocarnosine (γ-aminobutyryl-histidine) and anserine (β-alanyl-*N*-methyl-histidine), which are widespread in muscle tissues. Even *N*-alkyl and D-amino acids are encountered. Unusual links are frequently found: γ-glutamyl peptides are widely distributed in plant material, especially in onions, garlic and the seeds of the leguminosae.

The γ-glutamyl peptides derived from folic acid have the properties of vitamins. Pteroyl triglutamic acid from *Corynebacterium* sp. promotes the growth of certain lactic acid bacteria, and the pteroyl heptaglutamic acid (vitamin B conjugate) which occurs in yeast is important for the growth of chickens.

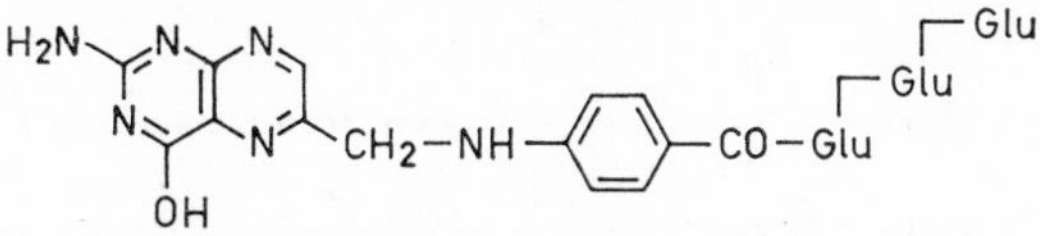

H_2N, N, N, N, N, OH, CH_2–NH–C_6H_4–CO–Glu–Glu–Glu

The importance of the γ-peptide bonds in nitrogen metabolism, especially in transpeptidation reactions, has been widely discussed.

In peptides with N-terminal glutamic acid a pyrrolidone ring system can be generated by reaction of the γ-carboxyl group with the unprotected amino group producing intramolecularly N-blocked pyroglutamyl peptides. Natural representatives of this series have been discovered in algae (eisenin, ⌊Glu–Gln–Ala and pelvetin, ⌊Glu–Gln–Gln), in snake venom (⌊Glu–Asn–Trp and ⌊Glu–Gln–Trp) and in the releasing hormones (cf. p. 198).

Some simple β-aspartyl peptides are common constituents of human urine in addition to γ-glutamyl, prolyl, hydroxyprolyl and other peptides. That the β-peptide bonds are original and not artefacts arising from isomerisation of α-aspartyl peptides has been proved by addition of labelled α-peptides which can be recovered unchanged after the isolation process.

An example of an ω-peptide bond of an amino dicarboxylic acid is in δ-aminoadipyl-cysteinyl-valine, which has been discovered in the mycelium of various penicillium strains and which seems to be important for the biosynthesis of penicillin. ω-Bonds involving diamino carboxylic acids are seldom encountered, but one possible example may be bacitracin, which contains an ε-lysine bond (cf. p. 218).

Most naturally occurring peptides are biologically active. Their natural functions are multiple and in many cases not yet known.

G. Ungar *et al.* have isolated and elucidated the structure of a peptide termed scotophobin (Greek: *skotos* = dark; *phobos* = fear) which, it was claimed, was the first compound isolated of a molecular coding system by which information is processed in the brain. The brains of rodents which had learned a certain behaviour (fear of darkness) were worked up chemically and the active substance isolated was injected into untrained rodents. The receiver animals spontaneously showed behaviour which was similar to that of the donor animals. This first peptide with 'memory transmitting action' possesses the following sequence, which has been confirmed by synthesis (W. Parr and G. Holzer, 1971): Ser–Asp–Asn–Asn–Gln–Gln–Gly–Lys–Ser–Ala–Gln–Gln–Gly–Gly–Tyr–NH_2.

In the following sections the physiologically important members of the glutathione group, the peptide and protein hormones, the peptide antibiotics and the peptide toxins are discussed in detail appropriate to their importance, and the 'strepogenin-active peptides', the peptide alkaloids, the depsipeptides and, finally, the peptoides (phospho-, nucleo-, lipo-, glyco- and chromopeptides) are also discussed.

2.3.1. *Glutathione and its Analogues*[5]

Glutathione (abbreviation GSH) is a tripeptide which occurs widely in both plant and animal tissue. It comprises glutamic acid, cysteine and glycine:

$$H_2N-CH(COOH)-CH_2-CH_2-CO-NH-CH(CH_2SH)-CO-NH-CH_2-COOH$$

When glutathione was isolated from yeast by HOPKINS in 1921 it was the first peptide containing a non-α-peptide bond. Although about 20 ways for its synthesis have been described in the literature, it seems to be more economical to isolate the peptide from natural sources.

The biosynthesis of glutathione is shown in the following equations:

$$\text{Glu} + \text{Cys} + \text{ATP} \xrightarrow{\text{I}} \text{Glu(Cys)} + \text{ADP} + H_3PO_4$$

$$\text{Gly} + \text{Glu(Cys)} + \text{ATP} \xrightarrow{\text{II}} \text{Glu(Cys–Gly)} + \text{ADP} + H_3PO_4$$

Reaction I is catalysed by γ-glutamyl-cysteine synthetase and reaction II by glutathione synthetase.

On exposure to the air, glutathione is oxidised to the disulphide form according to the equation

$$2\,\text{GSH} + {}^1\!/_2\,O_2 \rightarrow \text{GSSG} + H_2O$$

The reaction is catalysed by Cu and Fe salts. It is a reversible reaction; the reduction of the disulphide to produce GSH is catalysed by glutathione reductase. In strongly acidic solutions the mercapto group is stabilised by formation of thiazoline ring system which can be detected by UV spectra:

$$H_3N^{\oplus}-CH(COOH)-CH_2-CH_2-C\begin{matrix} \diagup S\text{———}CH_2 \\ \quad\quad\quad\quad | \\ \diagdown\!\!\overset{\oplus}{N}H-CH-CO-NH-CH_2-COOH \end{matrix}$$

The biochemical functions of glutathione are very varied. It is the coenzyme of glyoxalase, formaldehyde dehydrogenase, indolyl pyruvic acid keto–enol tautomerase, malate acetylacetic acid isomerase and other enzymes. Moreover it is the prosthetic group of glyceraldehyde-3-phosphate dehydrogenase and it plays a part in the action of the DDT-dehydrochlorinase which has been discovered in the DDT-resistant house-fly. Glutathione is also very important in many biological redox reactions because it can stabilise free SH-groups.

Homoglutathione, Glu(Cys–β-Ala), has been isolated from seeds of certain beans (for example, *Phaseolus aureus*). It is similar to the GSH and may substitute for it in some enzymic reactions.

Eye lens contains a glutathione derivative which has a sulphur-linked $-CH(CO_2H)CH_2(CO_2H)$ group. For investigations of the structural specificity of glutathione, tripeptides with analogous structures have been synthesised and studied—for example, with regard to methylglyoxylase activity. This work has shown that isoglutathione Glu–Cys–Gly, asparthione Asp(Cys–Gly) and epiglutathione D-Glu(Cys–Gly)—which is also of interest in connection with tumour chemistry—may substitute the natural peptide although they have lower activity. Not all the actions of glutathione are due to the free SH function. For example, the 'food reflex' which is provoked in *Pysalia gastrozooidens* and *Hydralitoralis* is also produced by ophthalmic acid, which was isolated in 1958 by S. G. WALEY from calf eye lens and differs from glutathione only in the replacement of the SH group by a CH_3 group:

$$H_2N-CH(COOH)-CH_2-CH_2-CO-NH-\underset{\displaystyle CH_2-CH_3}{\underset{|}{CH}}-CO-NH-CH_2-COOH$$

WALEY was able to synthesise ophthalmic acid starting from benzyloxycarbonyl-γ-glutamyl hydrazide which was coupled with aminobutyryl glycine using the azide method. It could also be synthesised enzymatically from the amino acids and a rabbit liver enzyme preparation in the presence of ATP and 3-phosphoglyceric acid. After an incubation time of 6 hours the ophthalmic acid was isolated by ion exchange chromatography. The yield in the preparative enzymic synthesis amounted to 20 per cent.

The lower homologue of ophthalmic acid, norophthalmic acid, was also discovered in calf eye lens. Its primary structure is γ-glutamyl-alanylglycine: it is easily produced by reaction of glutathione with RANEY nickel.

2.3.2. Peptide and Protein Hormones[6–8]

Hormones are chemical substances of the body which are produced by specific organs or tissues and transported via the bloodstream to the site of action. This chemically heterogeneous group of regulator substances includes a number of peptides and proteins. Some important representatives are shown in Table 21. A large number of peptide and protein hormones are produced in the glands of the endocrine system (hypothalamus, pituitary gland, thyroid gland, subthyroid gland, islet cells of pancreas, etc.), whereas others are produced in specialised tissues, so that a distinction is often drawn between gland and tissue hormones.

Many of the known peptide hormones have protein precursors from which they are released by enzymes. Angiotensin, bradykinin, kallidin and insulin are typical examples, and similar metabolic precursors have also been suggested for vasopressin, gastrin, glucagon, parathyroid hormone anp

ACTH. First the high-molecular-weight, biologically inactive, precursors are synthesised on the ribosomes. One or more enzymes then release primary hormones or prohormones. The half-lives of the free hormones are generally less than 30 minutes. They are inactivated by various endo- and exopeptidases. In order to obtain hormones with prolonged actions, great efforts have been made to synthesise analogues which are resistant to the inactivating enzymes; for example, by means of incorporation of D-amino acids, substitution and/or elimination of terminal residues, etc. WALTER[9,10] has made the intriguing suggestion that some peptide hormones might be prohormones for other hormones: for example, the hypothalamus of rabbits and rats contains a membrane-bound exopeptidase which releases the MSH-release-inhibiting hormone Pro–Leu–Gly–NH_2 by cleavage of oxytocin.

The elucidation of the mechanism of the interaction between hormones and their target cells is a central problem and the subject of intensive investigation. To interpret the cell-specific metabolic response evoked by hormones it is now generally postulated[11,12] that the hormones, which were termed 'chemical messengers' by STARLING and BAYLISS as early as 1904, interact primarily with the outer membrane of the target cell. This primary site of action is termed the receptor. The term 'receptor' refers to a molecule or molecular complex of the target cell which recognises and selectively interacts with the hormone and is capable of generating a signal which then initiates the chain of events which leads to the biological response. The particular part of the hormone sequence which contains the information needed to produce the biological effect is called the active site. The recognition function for the specific receptor may be contained in another part of the sequence of the hormone, and is termed the address or binding site. Hormone-receptor studies with hormone analogues have resulted in a differentiation between agonists or partial agonists which can realise a maximum or reduced biological effect and antagonists. Antagonists possess sufficient receptor affinity for binding but are unable to initiate the cell-specific metabolic processes induced by the hormone itself.

Starting from the fact that adenylate cyclase is stimulated by diverse peptide and protein hormones and the knowledge that the enzyme is fixed to the inner membrane surface, it has been postulated that the adenylate cyclase system acts as a part of the receptor complex, possibly with a function as an amplifier. The metabolic processes initiated by hormone interaction produce cyclic adenosine monophosphate (cAMP) in great quantities. For example, a hormone concentration of 10^{-10} can cause within a few seconds an increase of the cAMP concentration from 10^{-8} to 10^{-3}. SUTHERLAND, the discoverer of the cAMP, has described it as a 'second messenger'.

Investigations of the interactions between peptide hormones and potential receptors have been made using analogues with radioactive (^{3}H, ^{125}I) and fluorescent labels.

Table 21. Peptide and protein hormones

Hormone	Site of biosynthesis	Action	Structure
Adreno-corticotropin (ACTH)	Pituitary anterior lobe (adenohypophysis)	Stimulates the suprarenal gland for release of glucocorticoids	Linear peptide, 39 residues
Melanotropin (α- and β-MSH)	Pituitary intermediate lobe	Regulates pigment metabolism; function in man not yet established	Linear peptide, 13 and 18 residues, respectively
Lipotropin, lipotropic hormone (LPH)	Pituitary anterior lobe	Lipolytic	Polypeptide, 91 residues (β-LPH)
Growth hormone (STH, GH)	Pituitary anterior lobe	Responsible for growth	Protein, 190 residues
Prolactin, lactogenic hormone (LTH)	Pituitary anterior lobe	Stimulates milk secretion	Protein, 198 residues
Luteinising hormone (LH)	Pituitary anterior lobe	Formation and secretion of steroid sex hormones; acts on the growth of the gonads	Glycoprotein
Follicle-stimulating hormone (FSH)	Pituitary anterior lobe	Formation and secretion of steroid sex hormones	Glycoprotein, 2 subunits
Thyrotropin, thyrotropic hormone (TSH)	Adenohypophysis	Stimulates the thyroid gland	Glycoprotein, 2 subunits α-TSH = 96 β-TSH = 113 residues
Oxytocin	Hypothalamus	Contracts the uterus (initiating labour); milk ejection	Heterodetic cyclic peptide, 9 residues
Vasopressin	Hypothalamus	Increase in blood pressure; antidiuretic effect	Heterodetic cyclic peptide, 9 residues
Thyrotropin-releasing hormone (TRH)	Hypothalamus	Release of thyrotropin from the pituitary anterior lobe	Linear peptide, 3 residues

Table 21 (cont.)

Hormone	Site of biosynthesis	Action	Structure
Gonadotropin-releasing hormone (GnRH)	Hypothalamus	Stimulates the synthesis and secretion of LH and FSH	Linear peptide, 10 residues
Growth-hormone-releasing hormone (GH–RH)	Hypothalamus	Stimulates the synthesis and secretion of the GH	Linear peptide, 10 residues
Insulin	Pancreas (β-cells)	Lowers the blood sugar level; antilipolytic	Heterodetic cyclic peptide, 51 residues
Glucagon	Pancreas (α-cells)	Glycogenolytic, lipolytic, hyperglycaemic	Linear peptide, 29 residues
Calcitonin	Thyroid gland	Lowers the blood calcium level	Heterodetic cyclic peptide, 32 residues
Parathyroid hormone	Subthyroid gland	Increases the blood calcium level	Polypeptide, 84 residues
Gastrin	Antral mucosa	Stimulates secretion of gastric juice and pepsin	Linear peptide, 17 residues
Secretin	Intestinal mucosa	Stimulates the pancreas for the synthesis and secretion of digestive juice containing $NaHCO_3$	Linear peptide, 27 residues
Gastrin-inhibiting peptide	Antral mucosa	Gastrin antagonist	Polypeptide, 43 residues
Cholecysto-kinin-pancreo-zymin (CCK–PZ)	Intestinal mucosa	Gall-bladder-contracting; stimulates secretion of a enzyme-rich pancreatic juice	Polypeptide, 33 residues
Caerulein	Intestinal mucosa	Shows gastrin and CCK–PZ activity; acts to decrease blood pressure	Peptide, 10 residues

Table 21 (cont.)

Hormone	Site of biosynthesis	Action	Structure
Angiotensin	Blood plasma	Increases the blood pressure, stimulates aldosterone production in the suprarenal gland	Linear peptide, 8 residues
Bradykinin	Blood plasma	Decreases the blood pressure; increases vessel permeability; contracts the smooth muscle	Linear peptide, 9 residues
Substance P	Tissue	Decreases the blood pressure, stimulates the smooth muscle	Decapeptide
Eledoisin	Salivary gland of cuttlefish	Decreases the blood pressure	Linear peptide, 11 residues
Physalaemin	Skin of amphibia (*Physalaemus fuscumaculatus*)	Decreases the blood pressure	Linear peptide, 11 residues
Bombesin	Tissue	Stimulates smooth muscle, kidney and stomach	Linear peptide, 14 residues
Relaxin	Ovary, placenta and/or uterus	Loosening up of the collagenous connective tissue (widening of the pelvic bone ring)	Polypeptide

Naturally this table is incomplete and in the following discussion of selected peptide and protein hormones it will not be possible to consider all aspects of their chemistry and biochemistry. It will not be possible, either, to discuss all the synthetic analogues which have been prepared. In the field of structure–activity relations there is a marked trend towards increasing biological orientation as there is in peptide chemistry as a whole. It is inevitable that yet more biologically active peptides with hormonal properties will be discovered—in the hypothalamus, in intestines, lungs, amphibian skin, lymphocytes etc.—providing structures to be elucidated and correlated with the responses which they induce. For example, during the course of the purification of substance P from bovine hypothalamus extracts in 1973, Carraway and Leeman discovered and isolated a new hypotensive tridecapeptide, which they called neurotensin: it induces hypertension in the rat and can stimulate the contraction of guinea-pig ileum and rat uterus i.e. is a kinin.

2.3.2.1. *Adrenocotricotropin (ACTH, Corticotropin)*[13,14]

The synthesis and secretion of ACTH by the pituitary anterior lobe is stimulated by cotricotropin-releasing hormone. The hormone reaches the target tissue, the suprarenal gland, via the bloodstream and stimulates the synthesis of glucocorticoids via the adenylate cyclase system. The α_p-ACTH has the following sequence, revised by B. RINIKER, P. SIEBER, W. RITTEL and H. ZUBER in 1972:

Ser–	Tyr–	Ser–	Met–	Glu–	His–	Phe–	Arg–	Trp–	Gly–	Lys–	Pro–	Val–	Gly–	Lys–	Lys–
1	2	3	4	5	6	7	8	9	10	11	12	13	14	15	16
Arg–	Arg–	Pro–	Val–	Lys–	Val–	Tyr–	Pro–	Asn–	Gly–	Ala–	Glu–	Asp–	Glu–	Leu–	Ala–
17	18	19	20	21	22	23	24	25	26	27	28	29	30	31	32
Glu–	Ala–	Phe–	Pro–	Leu–	Glu–	Phe									
33	34	35	36	37	38	39									

The corticotropins of other species differ from porcine ACTH in positions 25–33:

Corticotropin	25	26	27	28	29	30	31	32	33
Human	–Asn–	Gly–	Ala–	Glu–	Asp–	Glu–	Ser–	Ala–	Glu–
Bovine	–Asp–	Gly–	Glu–	Ala–	Glu–	Asp–	Ser–	Ala–	Gln–
Sheep	–Ala–	Gly–	Glu–	Asp–	Asp–	Glu–	Ala–	Ser–	Gln–

Since 1956 many different partial sequences have been synthesised, especially by the groups of BOISSONNAS, HOFMANN, LI and SCHWYZER. The total synthesis of ACTH (with Asp and Gln in positions 25 and 30, respectively) was achieved by SCHWYZER and SIEBER[15] in 1963.

As a result of structure–activity studies the ACTH sequence may be divided into three parts. The segment 25–39 is the species-specific part. The partial sequence 1–18 is the shortest biologically active fragment (with an amide residue at the C-terminal end), whereas the N-terminal sequence 1–24 shows the complete biological activity (Synacthen®). It appears (K. HOFMANN *et al.*, 1972) that the N-terminal sequence 1–10, especially the sequence 5–10 (Glu–His–Phe–Arg–Trp–Gly) contains the 'active centre', and the sequence 11–18 is the 'bonding centre': SCHWYZER's conclusion is that the 'message' is contained in the sequence 5–10, whereas the 'address' is present in the sequence 11–24. The (11–24)-tetradecapeptide alone acts as a competitive antagonist, but the sequence 5–24, by comparison, shows a good biological activity.

ACTH, MSH and lipotropin contain the common partial sequence Met–Glu–His–Phe–Arg–Trp–Gly. The other parts of these hormones contain the 'addresses' for forming bonds to the appropriate specific receptor in the different target tissues and are thus responsible for selecting the characteristic adrenocorticotropic, melanotropic or lipolytic actions.

Of special interest is the [β-Ala^{1}, Lys^{17}]-corticotropin-(1–17)-heptadecapeptide-4-amino-*N*-butylamide (synthesised by R. GEIGER in 1971), which shows higher biological activity than the natural hormone despite the fact that it possesses a shortened chain. The β-alanine at the N-terminus prevents enzymic attack by aminopeptidases, and the substituted amide group at the C-terminal ensures stability to carboxypeptidases. Therefore, although this analogue is less strongly bound to the appropriate receptors, it still shows high activity because it is resistant to enzymatic degradation.

2.3.2.2. Melanotropin[16]

The formation of the melanocyte-stimulating hormone (MSH) in the intermediate lobe of the pituitary gland is controlled by the MSH-releasing and MSH-release-inhibiting hormones. MSH stimulates the synthesis of melanin through the operation of the adenylate cyclase system and controls the dispersion of melanin granules in amphibians and the darkening of the skin associated with it. The function of MSH in man is not yet clearly understood.

Two types of melanotropins, α-MSH and β-MSH, are known. The primary structure of α-MSH was elucidated by HARRIS and LERNER in 1957:

Ac–Ser–Tyr–Ser–Met–Glu–His–Phe–Arg–Trp–Gly–Lys–Pro–Val–NH_2
1 2 3 4 5 6 7 8 9 10 11 12 13

The sequence without the N-terminal acetyl and the C-terminal amide groups is identical with α_p-ACTH-(1–13)-peptide.

The β-melanotropin (β-MSH, β-intermedin) of the pig, the structure of which was elucidated by HARRIS and LI, consists of 18 amino acid residues:

Asp–Glu–Gly–Pro–Tyr–Lys–Met–Glu–His–Phe–Arg–Trp–Gly–Ser–Pro–Pro–Lys–Asp
1 2 3 4 5 6 7 8 9 10 11 12 13 14 15 16 17 18

The difference between porcine and bovine β-MSH consists of the replacement of the glutamic acid in position 2 by serine. Results obtained by radioimmunoassay, bioassy and physicochemical studies indicated that the human pituitary does not normally produce α-MSH or β-MSH and it was suggested by SCOTT and LOWRY (1974) that these peptides may be artefacts formed by enzymic degradation of β-lipotropin during extraction.

2.3.2.3. Lipotropin[14]

Lipotropic hormone (LPH) is a polypeptide hormone from the anterior lobe of the pituitary gland which has a lipolytic effect. β -LPH has 91 residues, whereas γ-LPH has 58. The sequence of the γ-LPH is identical with the N-terminal sequence 1–58 of β-LPH; the fragment 47–53 of β-LPH is identical with the sequence 4–10 of ACTH. Lipotropin affects fat mobilisation in intact adipose tissue and in isolated fat cells. The ovine β-lipotropin-(42–91)-pentakontapeptide has been synthesized by YAMASHIRO and LI (1974). This COOH-terminal part of lipotropin exhibited sixfold higher lipolytic activity in isolated fat cells than the native hormone on a weight basis.

2.3.2.4. Growth Hormone[14]

Growth hormone (GH), also termed somatotropic hormone or somatotropin, is produced in the anterior lobe of the pituitary gland. Its production is controlled by the growth-hormone-releasing and the growth-hormone-release-inhibiting hormones. It is responsible for the regulation of growth. It is remarkable for its high species specificty: the bovine GH is without effect in man.

LI *et al.* first isolated the human growth hormone (HGH) in 1956. In 1966 they published a first account of the sequence and in 1969 they published a revised sequence. A year later LI and YAMASHIRO[17] reported the total synthesis of HGH by the MERRIFIELD technique and obtained a peptide which showed a growth-promoting activity of about 10 per cent. This result is surprising, because some time later the proposed sequence which was the objective of the synthesis had to be revised.

In particular, three extra residues were eventually inserted in positions 68, 76 and 77, a tract of 15 residues in positions 78–92, and a glutamic acid residue in position 122 was replaced by glutamine. The significance and physiological role of HGH are still subjects of intensive work. Searches for an active core of HGH are of great importance because chemical synthesis of the entire hormone is prohibitive owing to its size of 191 amino acids residues.

GH from various species retain growth-promoting activity after limited hydrolysis by proteolytic enzymes. In 1974 two active fragments were isolated by LI and GRAF after digestion with plasmin which were derived from sequence regions 1–134 (10–20 per cent activity) and 141–191 (3–6 per cent activity). Detailed studies should eventually prove whether smaller fragments will exhibit similarly high activity.

Synthetic *N*-acetyl-HHG-(95–136), obtained using solid-phase technique, exhibited only lew activity in the rat tibia test.

GH action in some tissues, e. g. skeletal tissue, is mediated by special plasma factors, called somatomedins[18]. The somatomediis are insulin-like peptides of unknown structure (molecular weights ca. 4000–8000).

Clinical use of HGH for hypopituitary children is well established. Further, benefical effects were demonstrated in treatment of muscular dystrophy, bleeding ulcers and osteoporosis.

2.3.2.5. Prolactin[14]

The synthesis of prolactin (luteotropin, lactogenic hormone, LTH) in the anterior lobe of pituitary gland is controlled by the interaction of prolactin-releasing hormone and prolactin-release-inhibiting hormone. Prolactin stimulates milk secretion in mammals and its production increases during pregnancy and lactation. Besides this, prolactin promotes growth, influences pigment metabolism and osmotic regulation, and is responsible for the broody instinct of some animals. In the target organs the biological effect is brought about via the adenylate cyclase system. LI *et al.*[19] have elucidated the primary structure: it is a single-chain protein with 198 residues and 3 disulphide bridges.

2.3.2.6. Oxytocin and Vasopressin[20]

Despite the name 'neurohypophyseal hormones', production of oxytocin and vasopressin does not take place in the neurohypophysis. Both hormones are synthesised in the hypothalamus (oxytocin in the nucleus paraventricularis and vasopressin in the nucleus supraopticus). The carrier proteins neurophysin I and neurophysin II transport oxytocin and vasopressin via nerve fibres through the infundibulum in the hypophyseal stalk and then into the posterior lobe of the pituitary gland. There they are stored, and released on call and delivered via the bloodstream.

Phylogenetic investigations have shown that *vasotocin* is the evolutionary precursor of the neurohypophyseal hormones:

$$\text{Cys–Tyr–Ile–Gln–Asn–Cys–Pro–Arg–Gly–NH}_2 \quad (\text{Cys}^1\text{–Cys}^6 \text{ bridge})$$

It may be that in the transition from the cyclostomata to the fishes a gene duplication occurred resulting in the structurally similar hormones. Vasotocin, which is responsible for the regulation of the water and mineral metabolism of lower vertebrates, is replaced by vasopressin in the mammals: only the replacement of isoleucine in position 3 by a phenylalanine residue is involved in this change. Aspartocin, valitocin, glumitocin, isotocin and mesotocin are also evolutionary precursors of oxytocin; that is, there have been several transitions in this family of hormones.

Oxytocin acts on the smooth muscular system of the uterus, promoting contraction in labour,

Cys–Tyr–Ile–Gln–Asn–Cys–Pro–Arg–Gly–NH_2 (with a bracket linking the two Cys residues)

and it also stimulates milk ejection in the lactating mammary gland. It possesses the biological properties of vasopressin but to a lesser extent (antidiuretic activity, 0.5 per cent; pressor activity, 1 per cent).

In 1953 the elucidation of the structure of oxytocin was described independently by DU VIGNEAUD *et al.* and by TUPPY and MICHL. A year later DU VIGNEAUD published a total synthesis: it was the first synthesis of a biologically active peptide. The protected nonapeptide amide was obtained from three fragments. A 2 + (3 + 4) coupling was preferred:

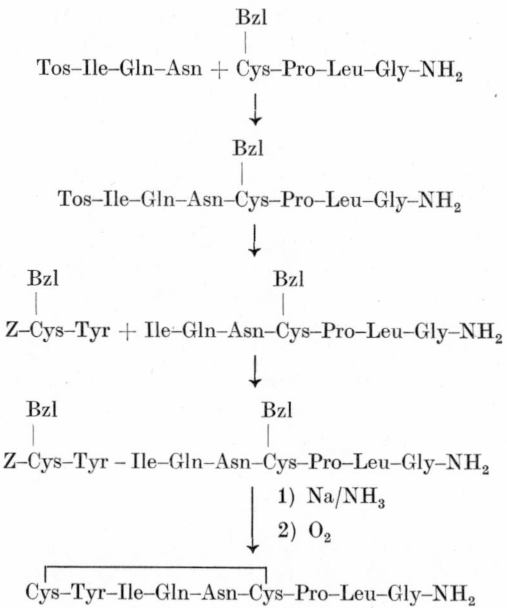

In 1955 further syntheses of oxytocin were published independently by RUDINGER *et al.* and BOISSONNAS. In the meantime a number of improved syntheses have been devised which are used for the production of oxytocin.

About 300 analogues of oxytocin have been synthesised (largely by the groups of DU VIGNEAUD, RUDINGER and BOISSONNAS) to investigate the relationship between structure and biological activity. The objectives of these syntheses of analogues are diverse but the major aim is to produce analogues with better pharmacological properties, such as higher biological activity, more prolonged effectiveness and better specificity of action for clinical use and for hormone-receptor studies. Many interesting results have been obtained but it is not possible to describe them here. An excellent review by RUDINGER[8] on peptide hormone analogues contains an extensive bibliography.

Although many synthetic analogues with higher biological potency than the natural hormone have been obtained, [4-threonine] oxytocin (synthesised by MANNING *et al.*[21], using the MERRIFIELD technique) was the first analogue in which the increase in activity was obtained simply by exchange of an amino acid. Apart from its clinical potential, this compound is also phylogenetically interesting.

As a result of extensive NMR investigations, WALTER and co-workers[22] have suggested that there is an 'active conformation' of oxytocin (Figure 24). This concept of an 'active conformation' enables a great number of the biological effects of synthetic analogues to be interpreted.

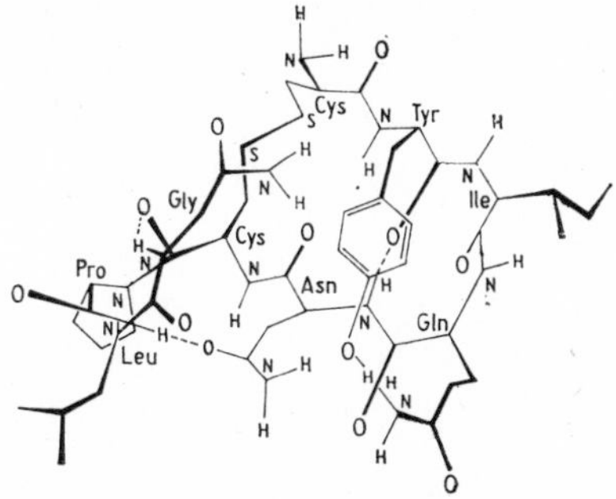

Figure 24. Proposed conformation of the biologically active conformation of oxytocin as proposed by WALTER *et al.*[22]

Vasopressin[23] is an antidiuretic and is hypertensive because it causes constriction of arterioles and capillaries. It also stimulates the smooth muscle system of the small intestine. It possesses some of the activity of oxytocin (uterine contraction, 3 per cent; milk ejection, 15 per cent). Under normal physiological conditions its effect on the kidney is the most important. The antidiuretic action involves the promotion of the resorption of water and the salt concentration of the urine is increased. Vasopressin is present as Arg^8-vasopressin in all the mammalian species studied, except in the pig and the hippopotamus, in which arginine is replaced by lysine in position 8.

By the synthesis of certain analogues it has been possible to change the main biological effects (hypertensivity and antidiuresis) in such a way that one is enhanced while another is suppressed. In [Phe^2, Ile^7, Orn^8]-vasopressin, for example, the hypertensivity/antidiuresis ratio is 220 : 1; in the [$Tyr(Me)^2$–

Lys^8]-vasopressin it is 1 : 115. As a result of the structural and topochemical modifications the first compound has been made unrecognisable to the 'anti-diuresis-specific' receptor, whereas the second compound is specially adapted to it, and two compounds show the opposite behaviour at the receptor responsible for the hypertensive effect.

2.3.2.7. *Releasing Hormones*[14, 24–26]

The releasing hormones are produced in different areas of the hypothalamus and are transported in the pituitary gland via the hypothalamic–hypophyseal portal vessel system. In the anterior lobe of the pituitary gland they stimulate the specific release of the hormones of the hypophysis via the adenylate cyclase system. In the synthesis of growth hormone, prolactin and melanotropin there is no negative hormonal feed-back mechanism from peripheral tissue. Instead, the release of these hormones is additionally controlled by release-inhibiting hormones.

Thyrotropin-releasing Hormone (TRH)

TRH is produced in the hypothalamus and stimulates the secretion of thyrotropin after transport into the anterior lobe of the pituitary gland. TRH was the first releasing hormone to be isolated, structurally elucidated and synthesised. The groups of Guillemin and Schally overcame extraordinary difficulties in the isolation and structure determination of the hormone, which was finally formulated as:

$$\text{Glu–His–Pro–NH}_2$$

The hypothalami of 400000 slaughtered animals provided only a fraction of a milligram of pure TRH for structure elucidation work. Only chemical synthesis provided the final evidence for the structure and material for pharmacological studies. TRH is used to diagnose hypothalamic and hypophyseal disturbances of function. More recently it has been found to have antidepressive action and this promises further clinical use, possibly by oral application.

Gonadotropin-releasing Hormone (GnRH)

The hypothalamus controls the release of luteinising hormone (LH) and follicle-stimulating hormone (FSH) with neurohumoral-releasing hormones. Schally *et al.* showed that the following decapeptide is responsible for both effects:

$$\text{Glu–His–Trp–Ser–Tyr–Gly–Leu–Arg–Pro–Gly–NH}_2$$

Although there are questions with regard to identity, the term 'gonadotropin-releasing hormone' will have to be used for the time being.

GnRH has found clinical application in the treatment of female sterility, and may also play a significant role in the development of new contraceptives, because the GnRH controls the release of the steroid sex hormones by the production of LH and FSH. A competitive GnRH antagonist might inhibit ovulation by combining with the GnRH receptor without evoking a biological response but excluding the natural GnRH. Such a compound, with the structure ⌐Glu–Trp–Ser–Tyr–Gly–Leu–Arg–Pro–NH–C_2H^5 has been synthesised.

Corticotropin-releasing Hormone (CRH)

Although extracts of hypothalamus show CRH activities, the structure of this releasing hormone, which stimulates the release of ACTH, is not yet known, but it is thought to be a peptide hormone.

Growth-hormone-releasing Hormone (GH-RH) and Growth-hormone-release-inhibiting Hormone (GH-RIH)

According to SCHALLY *et al.*, the releasing hormone which controls the secretion of growth hormone has the following primary structure:

Val–	His–	Leu–	Ser–	Ala–	Glu–	Glu–	Lys–	Glu–	Ala
1	2	3	4	5	6	7	8	9	10

This decapeptide and the Gln^9-analogue have been synthesised by HIRSCHMANN, DENKEWALTER *et al.* because the nature of the amino acid residue in position 9 was uncertain (Glu or Gln). Both peptides are similar to the N-terminal sequence of the β-chain of porcine haemoglobin:

Val–His–Leu–Ser–Ala–Glu–Glu–Lys – – – –

This fact suggests that haemoglobin could be the biogenetic precursor (prohormone) of the GH-RH. HIRSCHMANN *et al.* also synthesised the N-terminal decapeptide of the β-chain of human haemoglobin, Val–His–Leu–Thr–Pro–Glu–Glu–Lys–Ser–Ala, which may be therapeutically important in the medical treatment of hormonal disturbances of growth.

The growth-hormone-release-inhibiting hormone (somatotropin-release-inhibiting hormone, SR-IH, *somatostatin*) has the following sequence:

(disulfide bridge Cys 3 – Cys 14)

Ala–	Gly–	Cys–	Lys–	Asn–	Phe–	Phe–	Trp–	Lys–	Thr–	Phe–	Thr–	Ser–	Cys
1	2	3	4	5	6	7	8	9	10	11	12	13	14

Prolactin-releasing Hormone (PRH) and Prolactin-release-inhibiting Hormone (PRIH)

These hormones regulate the production of prolactin and its release in the anterior lobe of pituitary gland. It is possible that PRH is chemically identical with the thyrotropin-releasing hormone. The structure of the release-inhibiting hormone is not yet certain.

Melanotropin-releasing Hormone (MSH-RH) and Melanotropin-release-inhibiting Hormone (MRIH)

R. Walter *et al.* have suggested that MSH-RH corresponds to the partial sequence of oxytocin Cys–Tyr–Ile–Gln–Asn. The mechanism of MSH release and inhibition is not yet understood. The enzymic formation of Pro–Leu–Gly–NH_2 (MRIH) from oxytocin was the first example of a well-defined peptide hormone serving as a precursor or second-order prohormone for a smaller peptide with entirely different hormonal activities (R. Walter, 1971).

2.3.2.8. Insulin[27–31]

The polypeptide hormone insulin is produced in the β-cells of the islets of Langerhans of the pancreas. The physiological signal for the synthesis and secretion of this hormone is increased blood glucose level (hyperglycaemia). Insulin influences the whole of intermediate metabolism. It causes an increase of cell permeability for glucose and other monosaccharides as well as for fatty acids and amino acids. It accelerates glycolysis, the pentosephosphate cycle and the synthesis of glycogen in the liver. Furthermore, insulin promotes protein biosynthesis and fatty acid synthesis. Insulin deficiency causes diabetes mellitus, a metabolic malfunction disease characterised by hyperglycaemia, glycosuria and a tendency to ketosis. Diabetes mellitus may arise because of impaired biosynthesis of insulin or of the proinsulin–insulin conversion, or by enhanced inactivation (by degradation) or by increased release of hormone antagonists. The unsolved problem of diabetes is attracting increasing efforts by medical and pharmaceutical research workers, especially since it is on the increase all over the world, apparently being connected with increasing standard of living.

Insulin was discovered by F. G. BANTING and C. H. BEST (Nobel Prize 1923) in 1922, and successfully purified and crystallised by J. J. ABEL in 1926. The elucidation of the primary structure was published in 1958 by F. SANGER, who was awarded the Nobel Prize for the work. Finally, in 1969 DOROTHY HODGKIN and her collaborators established the three-dimensional structure by X-ray analysis (cf. Figure 28). Insulin consists of two polypeptide chains, the A-chain containing 21 residues and the B-chain containing 30 residues. The two chains are covalently linked by two disulphide bridges, and in addition the A-chain has an intrachain disulphide bridge (cf. Figure 26). It is interesting that the insulins of different species show about the same biological activity in the common biological test systems (mouse convulsion test, rat adipose tissue test) despite their structural differences. The sequence analyses of these insulins show that the differences occur mainly in the positions 8–10 of the A-chain, and it is probable that this fragment is not important for the biological activity. From the pancreas of rats and some fishes, two different insulins have been isolated. They contain, for instance, a methionine residue which is absent in other insulins. Guinea-pig insulin differs from rat insulin in 17 positions. Generally, the greatest differences in primary structure are shown between species which are furthest apart in their physiological development.

On the basis of its molecular weight of about 6000 and its number of amino acid residues, insulin is classed as a polypeptide; its aggregation to form dimers or hexamers depending on conditions justifies also its classification as a protein. In acidic solution in the absence of Zn^{2+} ions and in neutral solution, it forms dimers. In presence of Zn^{2+} ions in neutral solution, three dimers come together to form a hexamer, containing two bound Zn^{2+} ions. Insulin also shows a tendency to form complexes: the complex of insulin, zinc and protamine is therapeutically important.

At present insulin is produced industrially from porcine and bovine pancreas by extraction processes. About 2000 IU per kilogram are obtained from bovine pancreas, 10000 IU from calf pancreas and 75000 IU from fish pancreas. Although the chemical synthesis of insulin for medicinal purposes is not at the moment economically feasible, the present stage of development of chemical approaches will now be discussed. In the beginning of the 1960s synthetic peptide chemistry had reached the stage when the synthesis of insulin could be seriously attempted. About ten groups worked in this area, and groups in Aachen, Pittsburgh and the People's Republic of China reported successful syntheses after several years of work. The strategy of ZAHN and also of the other groups consisted of separate synthesis of the two peptide chains of insulin followed by their statistical oxidative combination to form a biologically active product. The possibility of forming insulin in this way was demonstrated by the work of DIXON and WARDLAW[33], who reduced the disulphide bridges of natural insulin, separated both chains and then obtained a product with 1–2 per cent insulin activity by oxidative recombination of the two chains. This was encouraging in view of the fact that this recombination should, in theory, produce 12 isomeric insulins, 4 intramolecular linked monomers of the A- and B-chains (Figure 25) and a number of polymeric products. Calculations showed that the yield of natural insulin with the correct disulphide links would be only 4.53 per cent of the total if probability were the only controlling factor.

Figure 26 shows the subdivision of the A- and B-chains into suitable ragments and the coupling methods for the fragment condensations which

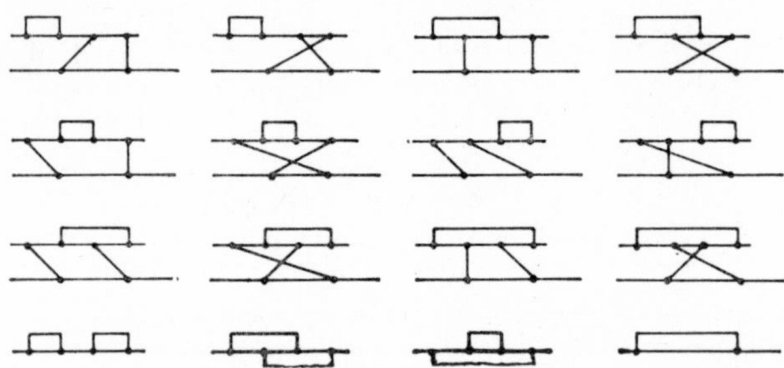

Figure 25. Schematic representation of insulin isomers and intramolecularly linked monomers of the A- and B-chains of insulin (according to ZAHN)

were used in the total synthesis by the Aachen group[34]. The synthetic schemes of KATSOYANNIS and of the Chinese group differ only in methodological details.

The benzyloxycarbonyl group was used as N(α)-protecting group. The cysteine and histidine side chains were protected as benzyl derivatives, the side chains of lysine and arginine by tosyl groups and the γ-carboxy group of glutamic acid by its *tert*-butyl ester.

The protecting groups of the A- and B-chains were removed by treatment of an equimolar mixture of the chains with sodium in liquid ammonia. The oxidative combination of the chains provided a protein mixture with 0.6–0.7 per cent of the activity of crystalline bovine insulin (27 IU/mg). The insulin content of the synthetic product was about the same as the insulin content of the product obtained from recombination of the natural chains. The total yield of the protected A-chain (related to the C-terminal amino acid) was 2.9 per cent and the yield of the B-chain 7.0 per cent. For the synthesis of the A-chain 89 steps and for the synthesis of the B-chain 132 steps were necessary. An additional three reaction steps were required for the combination of the chains.

In 1963 KATSOYANNIS *et al.*[35] successfully synthesised the A-chain of sheep insulin, which was combined with the natural B-chain by G. DIXON.

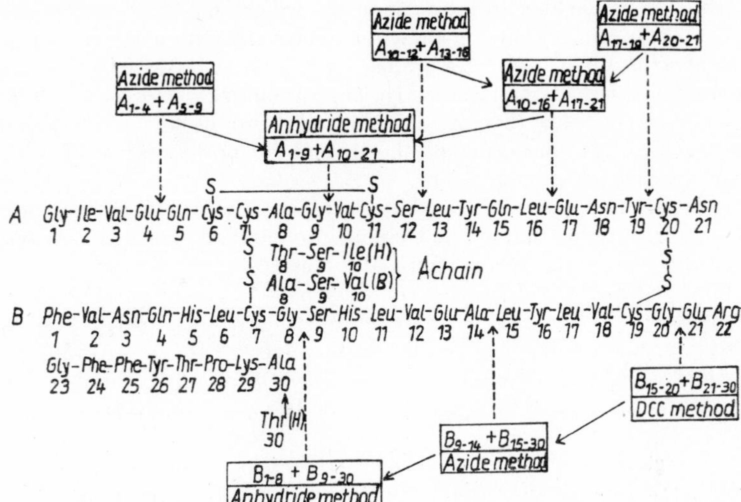

Figure 26. Primary structure of sheep insulin [human insulin (H) and bovine insulin (B) are different from it in the sequence A_{8-10}; human insulin also in B_{30}] and subdivision of the chains for the first total synthesis (Aachen, 1963)

This resulted in a product with an insulin content of 0.6 per cent. In March 1964 Katsoyannis described the total synthesis of sheep insulin and in 1966 the total synthesis of human insulin. Some investigators suggested that these syntheses confirm only the sequences of the A- and B-chains elucidated by Sanger but are not a complete proof of structure of insulin, since this would require construction of the disulphide bridges in a controlled manner. This criticism was weakened by the total synthesis of bovine insulin in China in 1965. This synthesis resulted in a product which contained 1.2–1.5 per cent insulin and from this crude product insulin was crystallised for the first time. The comparison of the synthetic product with natural insulin showed coincidence in all properties[36].

The A- and B-chains have also been synthesised on polymeric supports, but the results obtained did not come up to expectations. Marglin and Merrifield obtained the chains in yields of 67 and 37 per cent, respectively, but on combination a biological activity of only 0.16 per cent was obtained (the comparable combination of the natural chains showed 3.2 per cent activity). Weitzel *et al.* have synthesised a number of A- and B-chains with modified sequences mainly by the Merrifield method and have combined these chains with native chains to form hybrid insulins which have yielded interesting results concerning structure–activity relationships.

An alternative approach involves the synthesis of the separate chains followed by the directed linkage of the disulphide bridges by use of SH-protecting groups which can be removed independently of one another. This has been explored with model peptides by Hiskey, who has also been successful in the synthesis of a protected insulin A-chain which contains the intact intrachain disulphide bridge. The cysteine side chains in positions 7 and 20 are protected by benzhydryl and trityl groups so that the requirements for selective formation of the disulphide bridges with the insulin B-chain are met (cf. p. 162).

In 1974 P. Sieber *et al.*[37] described a total synthesis of human insulin involving direct formation of the three disulphide bonds at different stages of the fragment-condensation approach. This first genuine total synthesis of insulin was facilitated by the application of two new methods for the de-protection of S-Trt- and N(α)-Trt-protecting groups. In the first case, two S-Trt-protected cysteine residues are converted to the intramolecular disulphide bond 6–11 without affecting S-Acm-protected cysteine residue in position 7 in the A-chain fragment 6–13. The second new approach consists in a very mild, pH-controlled, acidolysis of N(α)-Trt-protecting groups, leaving intact N(α)-Bpoc and other acid-labile protecting groups.

A further possibility for the synthesis of insulin has resulted from the discovery of proinsulin by Steiner and co-workers[38]. It was found in 1967 that insulin is formed from a larger polypeptide precursor, which was termed proinsulin, in the β-cells of the islets of Langerhans. Proinsulin was isolated from human pancreas tumour cells[39]. Since then proinsulins have also been isolated from other species such as monkey, pig and codfish. It is also a constituent of commercial insulin. Figure 27 shows the primary structure of porcine proinsulin, which has 84 residues. The 33 amino acids of the C peptide form a connecting link between the N-terminal glycine of the A-chain and the C-terminal alanine of the B-chain. Both ends of the C-chain of all proinsulins contain basic amino acids which may be important for the enzymic release of insulin. However, such an enzyme system is not yet known. Proinsulin can also be activated by trypsin, which produces des-Ala-insulin from bovine and porcine proinsulins and des-Thr-insulin from human proinsulin. The connecting peptide (C peptide) consists of 33 residues in porcine proinsulin, 30 residues in bovine proinsulin and 35 residues in human proinsulin.

Insulin synthesis from a single-chain precursor is also a possible approach for a chemical synthesis. Steiner reported that reduced proinsulin can be reoxidised to form proinsulin in high yield (80 per cent); in other words, the polypeptide chain spontaneously takes up an optimal disulphide connection.

Recently some groups have begun to synthesise proinsulin even though the enzyme system which is responsible for releasing insulin is not yet known: H. Zahn *et al.* propose to synthesise six main fragments (A 10–21, A 2–9, B 24–30, B 1–23, C 1–16, C 17–A 1) and to connect these fragments to produce protected proinsulin.

The elucidation of the three-dimensional structure of insulin (Figure 28) has led to new thinking on methods of synthesis. The X-ray structure shows that the N-terminal residue of the A-chain and the C-terminal residue of the B-chain are close together. The distance between the amino group of glycine in position A^1 and that of lysine in position B^{29} is only about 10 Å.

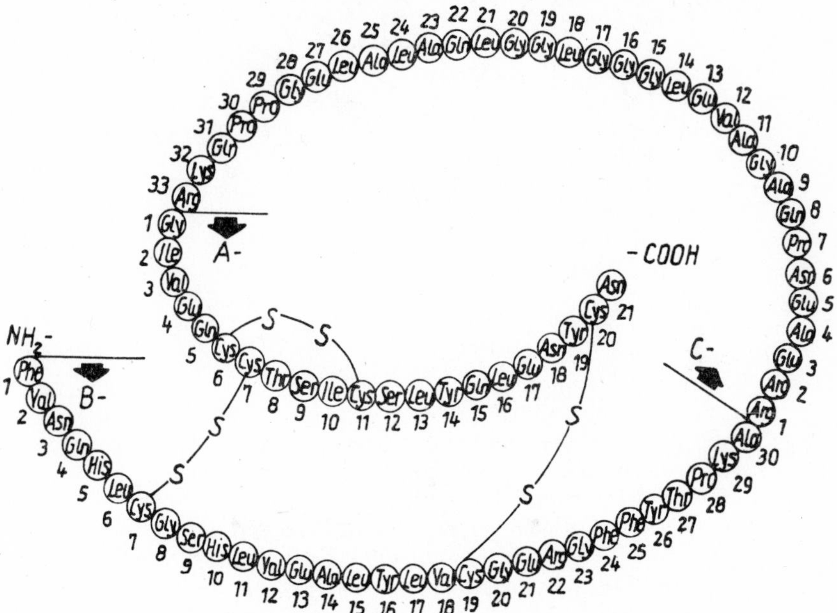

Figure 27. Primary structure of porcine proinsulin (according to CHANGE, ELLIS and BROMER[40])

Therefore, an intramolecular cross-linkage of the α-amino function of glycine with the ε-amino function of lysine should produce a proinsulin analogue. The connecting agent would perform the same function as the C-peptide and would hold the A- and B-chains in the position necessary for a correct combination of the disulphide bridges. Connection of the A- and B-chains using dicarboxylic acids of different chain lengths[42, 43] has proved feasible but it has so far not been possible to remove them selectively. Diaminodicarboxylic acids are more suitable as bridging agents because they can be removed by the EDMAN degradation. BRANDENBURG, SCHERMUTZKI and ZAHN have described the $N^{\alpha A1}$-, $N^{\varepsilon B29}$ cross-linked diaminosuberoyl insulin as a potentially suitable intermediate for a chemical synthesis of insulin[44]. Of course, the phenylalanine in position B^1 is also eliminated by the removal of the bridging group by the EDMAN degradation,

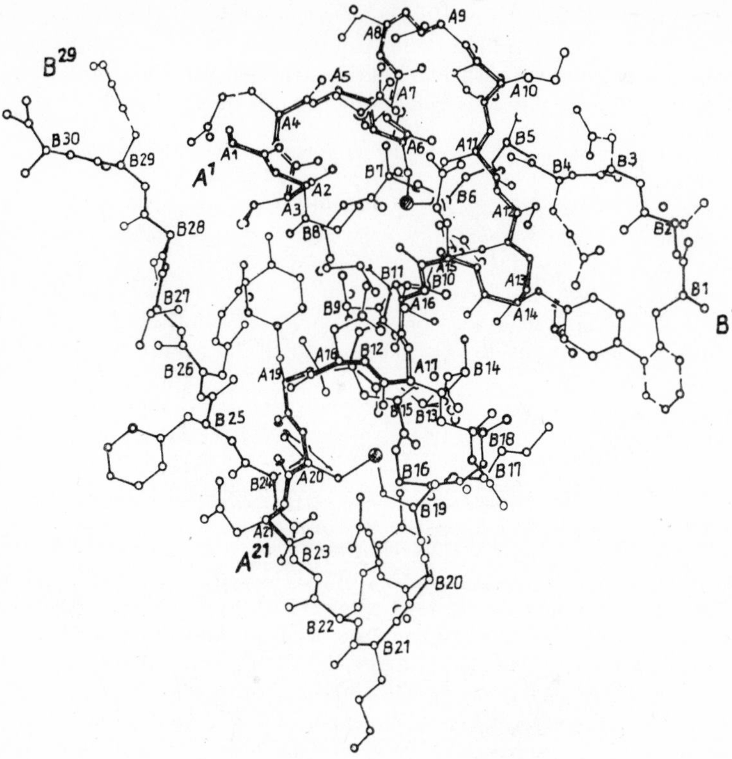

Figure 28. Model of the structure of monomeric insulin
(according to BLUNDELL *et al.*[41])

but the resulting des-Phe^{B1}-insulin has complete biological activity. Similar investigations were carried out simultaneously by GEIGER and OBERMEIER[45].

2.3.2.9. Glucagon

Glucagon, which is an insulin antagonist, is also produced in the pancreas, in the α-cells of the islets of LANGERHANS. The blood-sugar-increasing effect arises through stimulation of glycogen degradation in the liver and muscle. Glucagon also possesses lipolytic properties: like the other peptide hormones discussed, its actions involve the adenylate cyclase system. It is a polypeptide containing 29 residues with the following primary sequence (BROMER *et al.*, 1956):

His–Ser–Gln–Gly–Thr–Phe–Thr–Ser–Asp–Tyr–Ser–Lys–Tyr–Leu–Asp–
1 2 3 4 5 6 7 8 9 10 11 12 13 14 15

Ser–Arg–Arg–Ala–Gln–Asp–Phe–Val–Gln–Trp–Leu–Met–Asn–Thr
16 17 18 19 20 21 22 23 24 25 26 27 28 29

Of particular interest is the similarity of certain parts of this sequence to that of secretin.

Wünsch *et al.*[46] have described a total synthesis of glucagon. The synthesis of glucagon (which contains all the proteinogenic amino acids except cysteine, isoleucine and glutamic acid) is made particularly difficult by the great number of hydroxyamino acids, the N-terminal histidine and the absence of unracemisable amino acids. Moreover, the initially planned fragment condensation approach using the azide method had to be abandoned because of the low yield.

In the last phase of the synthesis the protected sequences 1–6 [Adoc–His(Adoc)–Ser(Bu^t)–Gln–Gly–Thr(Bu^t)–Phe] and 7–29 [Thr(Bu^t)–Ser(Bu^t)–Asp(OBu^t)–Tyr(Bu^t)–Ser(Bu^t)–Lys(Boc)–Tyr(Bu^t)–Leu–Asp(OBu^t)–Ser(Bu^t)–Arg(HBr)–Arg(HBr)–Ala–Gln–Asp(OBu^t)–Phe–Val–Gln–Trp–Leu–Met–Asn–Thr(Bu^t)–OBu^t] were coupled via the carbodiimide/*N*-hydroxysuccinimide method in a yield of 84 per cent. The sequence 7–29 was synthesised by the same method from the sequences 7–15, 16–21 and 22–29. Because of the great difference between the solubilities of the starting products and of the end product, the isolation and purification of the intermediate sequences was not difficult. It was possible to remove all the protecting groups in one step from the complete protected nonacosapeptide with trifluoroacetic acid. The crude glucagon was purified by gel filtration on Sephadex G50. A crystalline product was obtained which possessed complete activity and had properties identical with those of natural glucagon.

2.3.2.10. Calcitonin[47]

The production of calcitonin in the thyroid gland (thyrocalcitonin) is stimulated by increasing blood calcium level. It causes depression of blood calcium and phosphate levels: parathyroid hormone is an antagonist.

In 1968 three different groups described the determination of the structure and the synthesis of the calcitonin from porcine thyroid glands[48–51]. In the same year the sequence analysis and synthesis of human calcitonin (isolated from extracts of human thyroid carcinoma) were carried out[52].

The highest hypocalcaemic activity so far observed is shown by calcitonin isolated from the ultimobranchialic glands of the salmon. The synthetic product obtained by Guttmann *et al.*[53] was twenty times more active than porcine calcitonin and thirty-five times more active than human calcitonin.

The calcitonins show a striking uniformity in the amino acid sequence 1–7 and in the C-terminal proline amide. In the other positions considerable differences from the human calcitonin sequence are observed.

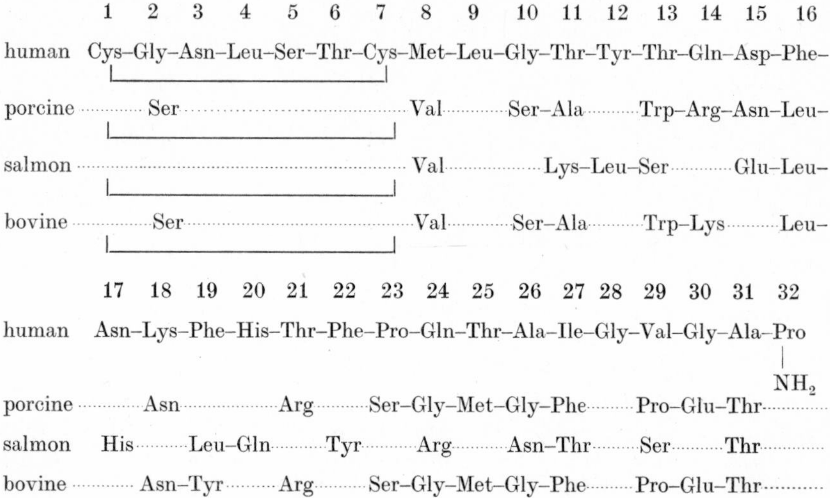

Human calcitonin, which only occurs in low concentrations, is accompanied by its anti-parallel dimer, calcitonin D:

```
                              1       7                  32
                            H–Cys.....Cys................Pro–NH2
                               |       |
                               |       |
H2N–Pro.....................Cys.......Cys–H
    32'                      7'       1'
```

The calcitonins have been synthesised by fragment condensation using the azide method or carbodiimide-combination methods. The use of the partial sequence 1–10 with a preformed disulphide bridge is an interesting feature of the synthetic work.

2.3.2.11. Parathyroid Hormone[54]

Parathyroid hormone also regulates calcium and phosphate metabolism. The increase of the blood calcium which it causes is the result of the activation of the osteoclast cells and the phosphate which is produced is excreted via the kidney. It is produced in the subthyroid gland: its antagonist is calcitonin.

The amino acid sequence of parathyroid hormone was worked out in 1970. It is a single-chain polypeptide with 84 residues. The primary structure of the porcine hormone differs from that of the bovine hormone in seven positions. Not all 84 residues are essential for biological activity. This was shown by the results of partial hydrolysis with dilute acids, which were carried out by POTTS *et al.* in 1968: these studies indicated that the N-terminal sequence of 30–40 residues contains the biologically active part of the hormone. POTTS *et al.* then synthesised the bovine N-terminal (1–34)-tetratriacontapeptide

Ala–	Val–	Ser–	Glu–	Ile–	Gln–	Phe–	Met–	His–	Asn–	Leu–	Gly–	Lys–	His–	Leu–	Ser–	Ser–
1	2	3	4	5	6	7	8	9	10	11	12	13	14	15	16	17
Met–	Glu–	Arg–	Val–	Glu–	Trp–	Leu–	Arg–	Lys–	Lys–	Leu–	Gln–	Asp–	Val–	His–	Asn–	Phe
18	19	20	21	22	23	24	25	26	27	28	29	30	31	32	33	34

on a polymeric support (a graft copolymer from styrene and trifluorochloroethylene)[55]. The synthetic product showed all the specific physiological and biochemical properties of the native hormone (about 80 per cent of the activity on a weight basis).

In 1973 R. H. ANDREATTA *et al.* described the synthesis of the sequence 1–34 of human parathyroid hormone:

Ser–	Val–	Ser–	Glu–	Ile–	Gln–	Leu–	Met–	His–	Asn–	Leu–	Gly–	Lys–	His–	Leu–	Asn–	Ser–
1	2	3	4	5	6	7	8	9	10	11	12	13	14	15	16	17
Met–	Glu–	Arg–	Val–	Gln–	Trp–	Leu–	Arg–	Lys–	Lys–	Lys–	Gln–	Leu–	Val–	His–	Asn–	Phe
18	19	20	21	22	23	24	25	26	27	28	29	30	31	32	33	34

It was prepared by condensing the fragments 25–28 with 29–34 and then joining on the sections 18–24, 13–17, 4–12 and 1–3.

2.3.2.12. Gastrin[56]

Gastrin is produced in the mucous membrane of the stomach: it stimulates the stomach and the pancreas to secrete hydrochloric acid and digestive enzymes, and also stimulates the gastrointestinal muscular system to increase intestinal motility.

Porcine (G) and human (H) gastrin are very similar linear heptadecapeptides (R. A. GREGORY, G. W. KENNER, R. C. SHEPPARD, 1964–66). Both exist in two forms I and II, the only difference between each pair being at position 5:

Glu–Gly–Pro–Trp–Met–$(Glu)_5$–Ala–Tyr–Gly–Trp–Met–Asp–Phe–NH_2 G I
1 2 3 4 5 6–10 11 12 13 14 15 16 17

SO_3H
|
..........Tyr.......... G II

..........Leu.......... H I

SO_3H
|
..........Leu..........Tyr.......... H II

Sheppard *et al.* and Morley *et al.* have synthesised a wide range of gastrins and numerous analogues.

Investigations of structure–activity relationships have shown that the C-terminal tetrapeptide Trp–Met–Asp–Phe–NH_2 possesses the complete biological activity of the natural hormone. More active than the tetrapeptide is the pentapeptide Boc–β-Ala–Trp–Met–Asp–Phe–NH_2, which is used in diagnosistic investigations of stomach secretion (Morley *et al.*, 1965). Also of interest in connection with these compounds is the accidental discovery that Asp–Phe–OMe has an intense sweet taste (160 times sweeter than sucrose in aqteous solution).

The gastrin-inhibiting hormone (enterogastrone) exerts an inhibiting effect on hydrochloric acid secretion. In 1970, Brown *et al.* isolated a polypeptide from the duodenum (pig) with the physiological properties of enterogastrone. It is a single-chain polypeptide with 43 residues which is remarkably similar to secretin and glucagon:

Tyr–Ala–Glu–Gly–Thr–Phe–Ile–Ser–Asp–Tyr–Ser–Ile–Ala–Met–Asp–
1 2 3 4 5 6 7 8 9 10 11 12 13 14 15

Lys–Ile–Arg–Gln–Gln–Asp–Phe–Val–Asn–Trp–Leu–Leu–Ala–Gln–Gln–
16 17 18 19 20 21 22 23 24 25 26 27 28 29 30

Lys–Gly–Lys–Lys–Ser–Asp–Trp–Lys–His–Asn–Ile–Thr–Gln
31 32 33 34 35 36 37 38 39 40 41 42 43

Although absolute proof of the identity of this compound with enterogastrone is still lacking, the name 'gastrin-inhibiting polypeptide (GIP)' has been suggested.

2.3.2.13. Secretin

Secretin is produced by the duodenal mucous membrane: it stimulates the pancreas to secrete an $NaHCO_3$-containing digestive liquid. In addition, the liver is stimulated to produce bile and the production of hydrochloric acid in the stomach is inhibited. The primary structure of secretin was elucidated by JORPES and MUTT in 1966:

His–Ser–Asp–Gly–Thr–Phe–Thr–Ser–Glu–Leu–Ser–Arg–Leu–Arg–Asp–
1 2 3 4 5 6 7 8 9 10 11 12 13 14 15

Ser–Ala–Arg–Leu–Gln–Arg–Leu–Leu–Gln–Gly–Leu–Val–NH_2
16 17 18 19 20 21 22 23 24 25 26 27

Because of the similarity of the sequence to that of glucagon a common biogenetic precursor is likely.

In the same year as the sequence determination BODANSZKY *et al.*[57] successfully synthesised this heptacosapeptide amide by stepwise elongation from the C-terminal end (cf. p. 177) and by fragment condensation. In the meantime E. WÜNSCH *et al.* synthesised it by fragment condensation with DCC/*N*-hydroxysuccinimide and DCC/1-hydroxybenzotriazole, respectively, as coupling agents. The activity of the crude product was 25 per cent, but after purification by gel filtration, electrophoresis and ion exchange chromatography this increased to 60 per cent of the natural activity.

2.3.2.14. Cholecystokinin–Pancreozymin (CCK–PZ)

CCK–PZ is produced in the intestinal mucous membrane: it stimulates contraction of the gall-bladder and secretion of enzymes by the pancreas. The polypeptide chain of the CCK–PZ consists of 33 residues but the C-terminal sequence 22–33 is primarily responsible for the biological activity:

SO_3H (on Tyr 27)

—Ile–Ser–Asp–Arg–Asp–Tyr–Met–Gly–Trp–Met–Asp–Phe–NH_2
22 23 24 25 26 27 28 29 30 31 32 33

The C-terminal dodecapeptide amide is produced on partial degradation by trypsin, the octapeptide amide by complete digestion. Structure–activity investigations have shown that the C-terminal heptapeptide amide is the shortest peptide possessing significant biological activity. The biological activity on a weight basis increases with elongation of the chain and attains at the decapeptide level ten times the activity of the native hormone: on the molar basis this corresponds to two and a half times the activity of the complete CCK–PZ. The synthetic work necessary for these results was mainly provided by SHEEHAN, PLUSCEC *et al.*

The rather complicated name of this hormone arises from the history of its discovery. In 1928 IVY and OLDBERG reported on the discovery of a gall-bladder contracting substance and named it cholecystokinin (CCK). Fifteen years later HARPER and RAPER discovered a substance in the duodenum which stimulated enzyme production by the pancreas. They called this pancreozymin. Finally, JORPES and MUTT in 1964 isolated a polypeptide from the same intestinal tissue which was shown to have the activity of both hormones hence: its rather unwieldy name.

The C-terminal pentapeptide of the CCK–PZ is identical not only with the corresponding C-terminal sequence of gastrin but also with the corresponding sequence of caerulein. Caerulein is a decapeptide with the primary structure:

```
                           SO3H
                            |
Glu–Glu–Asp–Tyr–Thr–Gly–Trp–Met–Asp–Phe–NH2
 1   2   3   4   5   6   7   8   9   10
```

It was isolated in 1967 by ANASTASI, ERSPARMER and ENDEAN from the skin of the amphibian *Hyla caerulea*. It has gastrin, CCK–PZ and blood-pressure-lowering activities.

2.3.2.15. Angiotensin[58, 59]

Angiotensin (formerly called angiotonin or hypertensin) is released from a plasma protein of the α_2-globulin fraction by a two-step enzymatic process. First the biologically inactive decapeptide angiotensin I is produced from angiotensinogen (renin substrate) by the endopeptidase renin, which occurs in the kidney. In a second step an activating enzyme (a chloride-dependent, EDTA-sensitive exopeptidase) releases the vasoactive octapeptide angiotensin II with the concomitant formation of the dipeptide His–Leu:

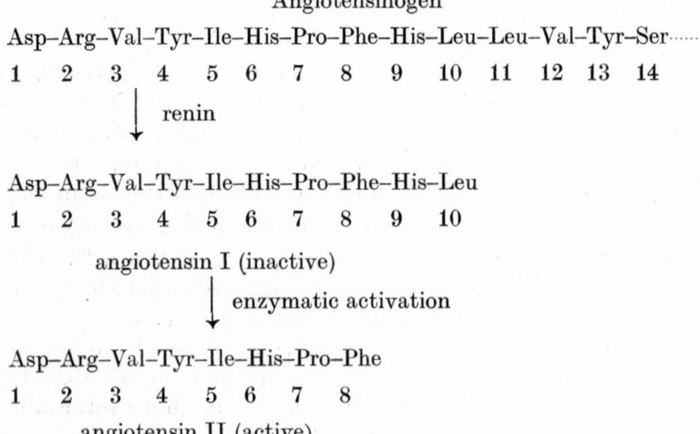

The [Ile^5]-angiotensin II of the horse and pig, shown in the scheme, is no different from the bovine [Val^5]-angiotensin II in its biological activity.

Structure–activity studies of numerous groups have revealed the essential nature of the Tyr, His, Pro, Phe and Val (or Ile) residues in position 5. In contrast, structural variations in the N-terminal sequence 1–3 are possible without drastic influence on the activity. The sequence 3–8 appears to be necessary for well-defined pressor activity. Although angiotensin is one of the most thoroughly investigated peptide hormones, interest in it shows no sign of diminishing. Several competitive antagonists have been found, such as [4-phenylalanine, 8-tyrosine]-angiotensin II, [1-sarcosine, 8-leucine]-angiotensin II, and [8-isoleucine]- angiotensin II. Angiotensin itself has been crystallised. Impressive progress has been made in understanding its physiological roles from the point of view of conformation and receptor interactions.

The influence of angiotensin on blood pressure and circulation is associated with its action on vascular muscle. Angiotensin causes a rapid increase of short duration in blood pressure. It also influences the smooth muscle of the uterus and of the intestinal tract and, in addition, increases the production and secretion of aldosterone.

2.3.2.16. Bradykinin[60, 61]

The name 'kinin' is applied to peptide tissue hormones which affect smooth muscle and have vasoactive properties with hypertensive or hypotensive action. The term 'plasma kinin' has been recommended for peptide hormones with bradykinin-like properties.

In 1949 Rocha e Silva *et al.* discovered a substance which stimulated smooth muscle. This substance was shown to be derived from a plasma protein by cleavage with proteolytic enzymes such as trypsin or those from snake venoms. The sequence of this substance, which proved to be a nonapeptide and was named bradykinin, is indicated below:

Arg–Pro–Pro–Gly–Phe–Ser–Pro–Phe–Arg
1 2 3 4 5 6 7 8 9

Bradykinin was first synthesised in 1960 by Boissonnas *et al.* using conventional procedures. In 1963 Merrifield described a solid-phase synthesis with an over-all yield of 68 per cent. In the course of structure–activity studies many bradykinin analogues have been synthesised in various laboratories.

It is interesting to note that enzymic digests of various proteins have produced fragments which possess bradykinin-potentiating activity and various peptides with bradykinin-potentiating activity have been synthesised.

A bradykinin-potentiating factor with the sequence Glu–Lys–Trp–Ala–Pro has been isolated from the venom of *Bothrops jararaca.*

Bradykinin and the related peptides kallidin, methionyl-lysyl-bradykinin, phyllokinin and polistes kinin all act upon intestinal and bronchial smooth muscles and cause contraction of the uterus. They increase capillary permeability and cause vasodilation and therefore have a hypotensive effect in mammals.

Kallidin, a decapeptide, is liberated from a plasma protein termed kallidinogen by trypsin cleavage. WERLE *et al.* established the following sequence by application of the EDMAN technique:

Lys–	Arg–	Pro–	Pro–	Gly–	Phe–	Ser–	Pro–	Phe–	Arg
1	2	3	4	5	6	7	8	9	10

Its physiological properties are similar to those of bradykinin ($^1/_3$ of the activity of bradykinin on guinea-pig ileum, $^2/_3$ on rat uterus; it is twice as active on the blood pressure of the rabbit as bradykinin). Many kallidin analogues have been synthesised, especially in the laboratories of SCHRÖDER, NICOLAIDES and BODANSZKY.

The structures of bradukinin-like peptides from ampkibians, such as *phyllokinin* (I) and *polistes kinin* (II), are indicated below:

I Arg–Pro–Pro–Gly–Phe–Ser–Pro–Phe–Arg–Ile–Tyr(SO_3H)

Arg–	Pro–	Pro–	Gly–	Phe–	Ser–	Pro–	Phe–	Arg–	Ile–	Tyr (SO_3H)
1	2	3	4	5	6	7	8	9	10	11

II <Glu–Thr–Asn–Lys–Lys–Leu–Arg–Gly–Arg–Pro–Pro–Gly–Phe–Ser–Pro–Phe–Arg (pyroglutamyl ring)

Glu–	Thr–	Asn–	Lys–	Lys–	Leu–	Arg–	Gly–	Arg–	Pro–	Pro–	Gly–	Phe–	Ser–	Pro–	Phe–	Arg
1	2	3	4	5	6	7	8	9	10	11	12	13	14	15	16	17

2.3.2.17. Substance P[62]

In 1931 VON EULER and GADDUM discovered in brain extracts a hypotensive principle which stimulated smooth muscule. The name substance P was suggested for the dried powder (P for Powder) obtained from the alcoholic extract. In 1971 CHANG, LEEMAN and NIALL[63] solved the sequence of substance P:

Arg–	Pro–	Lys–	Pro–	Gln–	Gln–	Phe–	Phe–	Gly–	Leu–	Met–NH_2
1	2	3	4	5	6	7	8	9	10	11

The sequence was confirmed by solid-phase synthesis[64], which after purification gave material indistinguishable from the natural substance P. Diverse biological and pharmacological activities have been reported for substance P. It was recently isolated in a pure form from bovine hypothalami and from equine intestine.

2.3.2.18. Eledoisin

In 1952 ERSPARMER isolated from extracts of salivary glands of cuttlefishes (*Eledone moschata* and *Aldrovandi*) a compound which he called eledoisin. Eleven years later its structure[65] was established as:

$$\text{Glu–Pro–Ser–Lys–Asp–Ala–Phe–Ile–Gly–Leu–Met–NH}_2$$

1 2 3 4 5 6 7 8 9 10 11

In the same year SANDRIN and BOISSONNAS reported the first synthesis[66]. The structure–activity relationships of eledoisin have been extensively studied. The carboxyl terminal hexapeptide ('hexeledoisin') has about 15–30 per cent of the activity of the natural peptide. The corresponding octa-, nona- and decapeptides have higher activities than eledoisin. Many other analogues have been synthesised which possess higher activity than the native hormone.

Eledoisin causes smooth muscle preparations to contract and causes marked peripheral vasodilation leading to a decrease in blood pressure.

2.3.2.19. Physalaemin

Physalaemin has an action similar to that of eledoisin, but it is 3–4 times more potent. In contrast to eledoisin, it is inactivated more rapidly, so that its effects are more short-lived. The isolation of physalaemin from the skin of *Physalaemus fuscumaculatus* and the determination of its structure were carried out by ERSPARMER *et al.*:

$$\text{Glu–Ala–Asp–Pro–Asn–Lys–Phe–Tyr–Gly–Leu–Met–NH}_2$$

1 2 3 4 5 6 7 8 9 10 11

The first synthesis was published by BERNARDI *et al.* in 1964[67].

2.3.2.20. Bombesin[68]

Bombesin was found in extracts of the skin of the European frogs *Bombina bombina* and *Bombina variegata* (ANASTASI, ERSPARMER and BUCCI, 1971):

$\ulcorner$Glu–Gln–Arg–Leu–Gly–Asn–Gln–Trp–Ala–Val–Gly–His–Leu–Met–NH_2
1 2 3 4 5 6 7 8 9 10 11 12 13 14

The most important properties of bombesin are its actions on vascular and extravascular smooth muscle and on the kidney and stomach.

The tetradecapeptide, *alytesin*, extracted from the skin of *Alytes obstetricans*, and the undecapeptide, *ranatensin*, extracted from the skin of an American frog (*Rana pipiens*) have very similar structures. Bombesin and alytesin differ only in two positions: the Gln^2 residue is replaced by glycine and the Asn^6 residue by threonine.

Ranatensin has the following sequence:

$\ulcorner$Glu–Val–Pro–Gln–Trp–Ala–Val–Gly–His–Phe–Met–NH_2
1 2 3 4 5 6 7 8 9 10 11

2.3.3. *Peptide Antibiotics*[69, 70]

The antibiotic polypeptides consist mainly of metabolic products of microorganisms. They are often resistant to degradation by the proteolytic enzymes of normal cells. The reasons for this are attributable not only to their cyclic structure but also to the fact that they frequently contain structural features not usually found in proteins such as D-amino acids[71], uncommon amino acids and non-peptide inter-residue links.

The peptide antibiotics can be classed according to chemical structure as homomeric homodetic cyclic peptides such as the gramicidins, tyrocidins and bacitracins or heteromeric heterodetic cyclic peptides such as the therapeutically important polymyxins or the cyclic depsipeptides, which will be dealt with in Section 2.3.4. To date about 250 different peptide antibiotics have been discovered in nature.

2.3.3.1. *Gramicidins*

Gramicidin S, which is active against Gram-positive bacteria, was isolated by Soviet investigators from cultures of *Bacillus brevis* (C. F. Gause, M. G. Brazhnikova) in 1942. R. L. M. Synge *et al.* determined its structure and showed it to be a 30-membered cyclic decapeptide. In 1965 Schwyzer and Sieber[72] confirmed the structure by total synthesis:

cyclo(Val–Orn–Leu–D–Phe–Pro–Val–Orn–Leu–D–Phe–Pro)

The linear protected decapeptide was synthesised from pentapeptide moieties and the cyclisation of the decapeptide was performed using the *p*-nitrophenyl ester method:

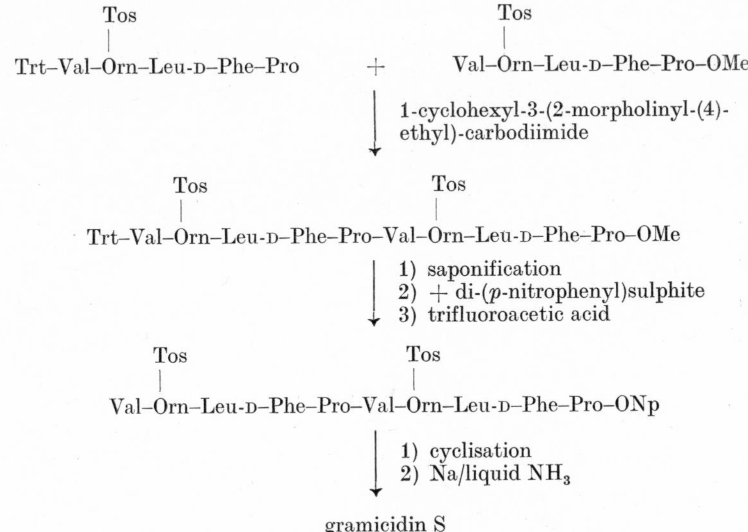

According to WAKI and IZUMIYA (1967) in their synthesis of gramicidin S by combined dimerisation and cyclisation of pentapeptide *p*-nitrophenyl esters, a cyclic pentapeptide is formed as by-product (32 per cent). The separation of this biologically inactive 'cyclosemigramicidin S' was achieved by column chromatography on Sephadex LH 20.

Structure–activity studies have shown that the proline residue of gramicidin S can be replaced by glycine or sarcosine without loss of activity.

SARGES and WITKOP[73] showed that gramicidin A, in contrast to gramicidin S, is a linear peptide. Its structure proved more difficult to elucidate because the N-terminal amino group is blocked by a formyl group and the C-terminal carboxylic group by an ethanolamine residue:

HCO–Val–Gly–Ala-D–Leu–Ala-D–Val–Val-D–Val–Trp-D–Leu–Trp-D–Leu–Trp-D–Leu–

1 2 3 4 5 6 7 8 9 10 11 12 13 14

Trp–NH–CH_2–CH_2OH

15

2.3.3.2. *Tyrocidins*

The tyrocidins are closely related to gramicidin S, and they, too, are produced by *Bacillus brevis*.

Tyrocidin A was isolated by A. R. Battersby and L. C. Craig in 1952 and its structure was determined and confirmed by synthesis by M. Ohno and N. Izumiya in 1966. In tyrocidin B the L-phenylalanine residue in position 6 is replaced by L-tryptophan (T. P. King and L. C. Craig, 1955). Tyrocidin C contains D-tryptophan in position 7 instead of the D-phenylalanine of tyrocidin B (Ruttenberg, King and Craig, 1965):

cyclo-(Val–Orn–Leu-D–Phe–Pro–Phe-D–Phe–Asn–Gln–Tyr) 1 2 3 4 5 6 7 8 9 10	tyrocidin A
cyclo-(.........................Trp.........................)	tyrocidin B
cyclo-(.........................Trp-D–Trp–Asn.................)	tyrocidin C
cyclo-(...Phe)	tyrocidin D
cyclo-(......................................Asp..........Phe)	tyrocidin E

The tyrocidins C and E were synthesised by Izumiya *et al.* in 1970[74].

The tyrocidins are active against Gram-positive bacteria. A mixture of tyrocidins (80 per cent) and gramicidins (20 per cent) is used therapeutically against infections of the mouth and pharynx as well as against infections of the skin and mucous membranes.

2.3.3.3. *Bacitracins*

In 1945 a group of new polypeptides was discovered in culture filtrates of *Bacillus licheniformis*. Seven of them were isolated by countercurrent distribution (Craig *et al.*) and named bacitracins A–G.

Examination of the main component, bacitracin A, showed it to be based on a cyclic hexapeptide structure in which ring closure is brought about via a peptide bond formed by the β-carboxylic group of an aspartic acid residue and the ε-amino group of lysine. The α-amino group of lysine is joined to a tetrapeptide sequence and the α-carboxylic group of the aspartic acid is connected to a residue of asparagine. The N-terminal isoleucine of the tetrapeptide moiety is coupled to the mercapto group of the neighbouring cysteine residue to form an unusual thiazolidine ring system:

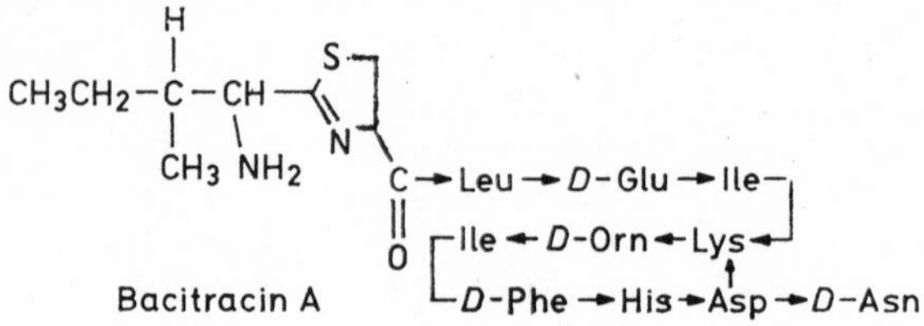

```
                H
                |              S─┐
CH3CH2─C─CH─<   │
                |    \         N─┘
               CH3  NH2          C → Leu → D-Glu → Ile ─┐
                                 ‖                      │
                                 O ┌ Ile ← D-Orn ← Lys ←┘
                                   │               ↑
                                   └ D-Phe → His → Asp → D-Asn
```

Bacitracin A

Bacitracin F is formed from Bacitracin A via oxidative deamination of the NH_2-terminal isoleucine and dehydrogenation of the thiazoline to a thiazole:

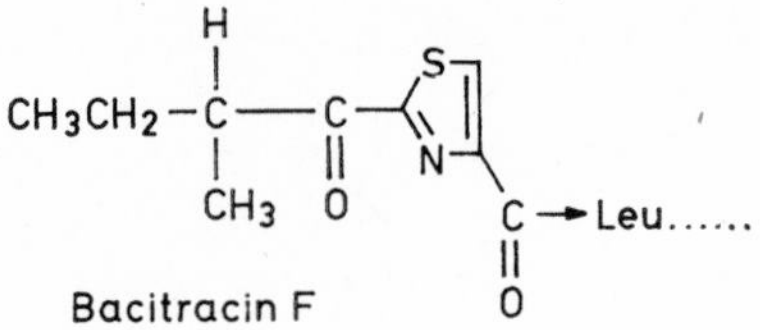

$$CH_3CH_2-\underset{CH_3}{\overset{H}{C}}-\underset{O}{\overset{\|}{C}}-\text{(thiazole)}-\underset{O}{\overset{\|}{C}}\rightarrow \text{Leu}\ldots\ldots$$

Bacitracin F

Bacitracins are active against Gram-positive bacteria. Commercial bacitracin contains about 70 per cent of the A-component and is used for the treatment of skin infections.

2.3.3.4. Polymyxins

Polymyxins are fatty-acid-containing cyclic peptides produced by *Bacillus polymyxa*. Unlike the peptide antibiotics mentioned above, the polymyxins have a high content of α,γ-diaminobutyric acid (Dbu) and are active against Gramnegative bacteria, particularly *Pseudomonas pyocyanea*.

Of the known polymyxins, polymyxin B and *D* have received most attention. They were separated into several components (B_1, B_2, D_1, D_2) by countercurrent distribution. Work on these was very difficult and the total synthesis of polymyxin B_1 was only finished in 1965 by Vogler *et al.*

Polymyxin B_1 is a decapeptide consisting of a tripeptide which is coupled to the γ-amino group of a diaminobutyric acid residue of a cyclic heptapeptide. The α-amino group of the terminal Dbu-residue is acylated by (+)-6-methyloctan-1-oic acid (MOA), [(+)-isopelargonic acid].

Polymyxin B_2 contains 6-methylheptanoic acid (IOA) instead of MOA. Polymyxins are widely used in the treatment of dysentery, gastroenteritis and meningitis, and other infections which cannot be treated successfully with other antibiotics. Their efficacy may be enhanced when they are used in combination with antibiotics effective against Gram-positive bacteria.

Additional antibiotics which belong to the polymyxin group are the circulins A and B and the colistins A, B and C: (see p. 220).

Circulin was discovered in 1948 by Murray and Tetrault: it is the active principle of the nonhaemolytic strain of *Bacillus circulans*. The main constituents were named circulin A and B. Their structures were determined in 1965 by Suzuki *et al.* and confirmed by the total synthesis

$$R_1-Dbu-Thr-R_2-\overline{Dbu-Dbu-R_3-R_4-Dbu-Dbu-Thr}$$

peptides of the polymixin group	R_1	R_2	R_3	R_4
polymyxin B_1	MOA	Dbu	*D*-Phe	Leu
polymyxin B_2	IOA	Dbu	*D*-Phe	Leu
polymyxin D_1	MOA	*D*-Ser	*D*-Leu	Thr
polymyxin D_2	IOA	*D*-Ser	*D*-Leu	Thr
polymyxin E_1	MOA	Dbu	D-Leu	Leu
polymyxin E_2	IOA	Dbu	*D*-Leu	Leu
circulin A	MOA	Dbu	*D*-Leu	Ile

of circulin A[75], which disproved the earlier proposal of KOFFLER that circulin A was a homodetic cyclic heteromeric decapeptide without a side chain.

In 1963 SUZUKI *et al.* isolated by countercurrent distribution three antibiotics from culture filtrates of *Bacillus colistinus,* which were named colistins A, B and C. It was later shown that colistins A and B are identical with polymyxins E_1 and E_2 (S. A. WILKINSON and L. A. LOWE, 1964; T. SUZUKI, 1965). The synthesis of colistin A (polymyxin E_1) has been described by R. STUDER (1965).

2.3.3.5. Additional Peptide Antibiotics

Fungisporin, formerly thought to be a cyclooctapeptide, is also antibiotically active. It was isolated from spores of several strains of *Penicillium* and *Aspergillus* (K. MIYAO, 1960):

$$\overline{\text{Phe-D-Val-Val-D-Phe}}$$

Other peptide antibiotics include:

Malformin, a metabolic product of the fungus *Aspergillus niger* (S. MARUMO and R. W. CURTIS, 1961; synthesis: A. SCHÖBERL, M. RIMPLER and E. CLAUSS, 1970): it can cause curvatures on bean plants and on corn roots. In a reinvestigation of the chemistry of malformin it could be proved that the structure which was believed to be that of malformin was wrong and that in fact the following structure represents the correct one (M. BODANSZKY *et al.*, 1973):

→D-Cys–D-Cys–Val-D–Leu–Ile—

Albomycin, from *Aclinomyces subtropicus* (SORM and co-workers[76]), is structurally similar to ferrochrome:

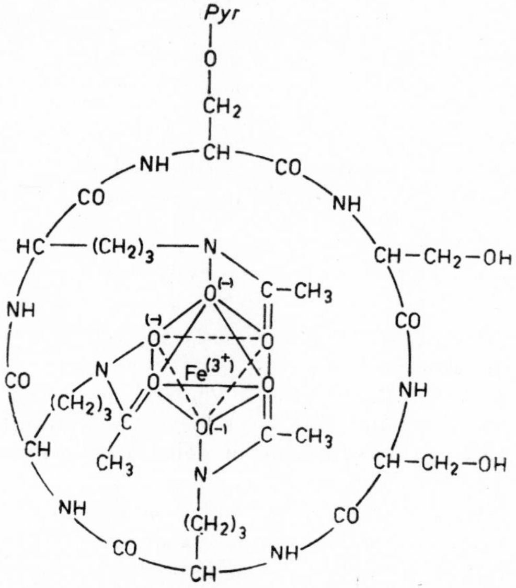

Albonoursin (I) from *Streptomyces noursei* (synthesised in 1967 by SHIN *et al.*[77]) and *echinulin,* which has the dioxopiperazine structure (II), from strains of the fungus *Aspergillus glaucus* (G. CASNATI *et al.*, 1963):

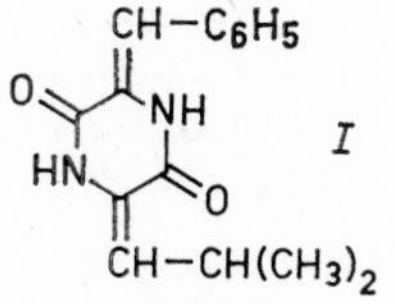

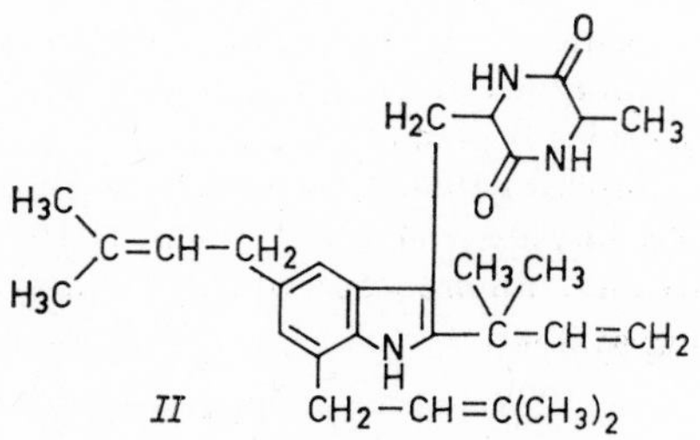

Penicillin (a product of *Penicillium notatum* discovered by FLEMING in 1928) can be related biogenetically to the peptide antibiotics. It is derived from valine and cysteine linked together via a thioether and a lactam ring in a bicyclic system. The individual penicillins differ from one another only in the residue R, which is a benzyl group in the best-known member of the series:

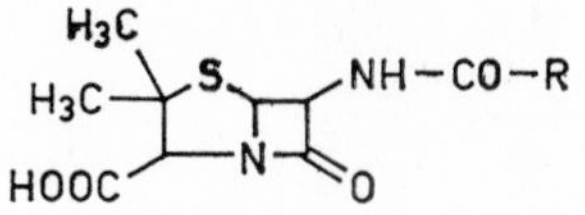

```
     H3C
H3C─┤  S   NH─CO─R
HOOC    N──
            O
```

2.3.4. *Depsipeptides*[78–81]

The name 'depsipeptide' is a combination of the terms depside (ester of a hydroxy acid) and peptide. It was proposed by SHEMYAKIN in 1953 to cover compounds which contained both ester and peptide bonds:

```
     R1         R2         R1         R2         R1
     |          |          |          |          |
—NH—CH—CO—O—CH—CO—NH—CH—CO—O—CH—CO—NH—CH—CO—
```

R_1 = residue of a amino acid
R_2 = residue of a hydroxy acid

Cyclic compounds of this sort have been discovered in large numbers as metabolic products of microorganisms. They show a surprisingly high degree of antibiotic activity. Their widespread application in medicine is prevented in most cases by the sensitivity of the ester bonds to hydrolysis.

The depsipeptides which contain exclusively amino acids (often also *N*-methyl- and D-amino acids) and hydroxy acids include the enniatins, amidomycin, valinomycin, the sporidesmolides, serratomolide and esperin.

In other depsipeptides the hydroxy groups of hydroxy amino acids such as serine and threonine are involved in the ester bonds. They can also contain complex structural units of types never found in proteins. To this group belong the actinomycins, etamycin and echinomycin. SCHRÖDER and LÜBKE have classified the depsipeptides according to their structure into *O*-peptides, peptolides and peptide lactones. A survey of the most important depsipeptides is given in Table 22.

The simplest naturally occurring depsipeptide is a phytopathogenic toxin from *Pseudomonas tabaci* (WOOLLEY, 1955). It consists of lactic acid and α,α′-diamino-adipinic acid which are linked together to form a dioxomorpholine ring system:

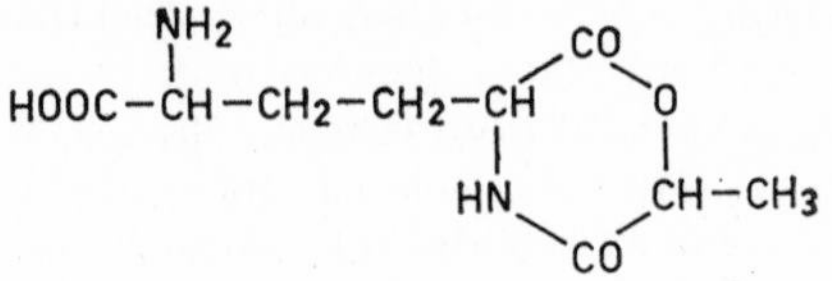

```
       NH2                    CO
       |                    /    \
HOOC—CH—CH2—CH2—CH       O
                    |        |
                    HN       CH—CH3
                      \    /
                        CO
```

The *enniatins* (which act upon membranes) have been proved to be cyclohexadepsipeptides by comparison of the natural products with synthetic materials. Their main activity is against various mycobacteria whose growth is totally inhibited by low depsipeptide concentrations.

SHEMYAKIN *et al.* have shown that the antibacterial activity of the enniatins and their analogues depends on the size of the ring system and on the configuration of the constituents. Exchange of amino acid residues is possible to a limited extent but the exchange of hydroxy acid residues leads to a great loss of activity. The most important discovery is that the antibacterial activity of the depsipeptides parallels their ability to promote the transport of potassium ions across biological membranes. The ion-transporting ability of the cyclodepsipeptide antibiotics depends on metal–ion–ligand interactions between the potassium ion and the amide and ester links.

The structure of *amidomycin* is not clear at present. It is composed of D-hydroxyisovaleric acid and D-valine, and is an antibiotic with specific activity against the growth of certain plant pathogenic organisms (corn rust, etc.) and some types of yeast.

Valinomycin, so called because of its high valine content, is a cyclic dodecapeptide which is potent against *M. tuberculosis*. The primary structure determined by BROCKMANN has been confirmed by synthesis (SHEMYAKIN *et al.* 1963). Conformational studies of this cyclic depsipeptide have provided strong evidence that valinomycin adopts different conformations in polar and non-polar solvents and yet a third conformation is involved when a K^+ ion is complexed. It also seems probable that the conformations in solution and in the crystal are different. Valinomycin and other depsipeptides are able to transport alkali metal ions across biological and artificial membranes.

In contrast to the depsipeptides discussed above, in the *sporidesmolides* the distribution of amino acid and hydroxy acid residues is less symmetrical. Instead of the common D-antipodes, these biologically inactive cyclohexadepsipeptides contain 'normal' peptide bonds and L-hydroxyisovaleric acid.

Serratomolide, a cyclic tetradepsipeptide, and *esperin* are depsipeptides containing β-hydroxy acids. The total hydrolysis of serratomolide showed the presence of L-serine and D-β-hydroxy-decanoic acid. Esperin was shown to be a heptapeptide lactone by Japanese investigators, using mass spectrometry. The fatty acid component is a mixture of homologues (cf. Table 22, $R = C_{12}H_{25}$ (40 per cent), $R = C_{11}H_{23}$ (35 per cent), $R = C_{10}H_{21}$ (20 per cent)) and the C-terminal leucine is partially replaced by valine.

The *actinomysins*—sometimes called chromopeptides because they contain both a chromophore and a peptide moiety—are highly toxic antibiotics produced by a variety of *Streptomyces* species. More than 20 different such compounds have now been isolated and crystallised. They differ only in the amino acid sequences of the two lactone ring systems which are

Table 22. Some important depsipeptides

Compound	Formula	Occurrence	Isolation, Synthesis (S)
Enniatin A	cyclo-[*D*-Hyv–MeIle]$_3$	*Fusarium orthocerus*	PLATTNER, 1947/48; (S) 1963
Enniatin B	cyclo-[*D*-Hyv–MeVal]$_3$	*Fusarium oxysporum* and *avenaceum*	PLATTNER, 1947/48; (S) 1963
Amidomycin	cyclo-[*D*-Hyv–*D*-Val]$_n$	*Streptomyces* sp.	VINNING and TABER, 1957
Valinomycin	cyclo-[*D*-Val–Lac–Val–*D*-Hyv–]$_3$	*Streptomyces fulvississinus*	BROCKMANN, 1955; (S) SHEMYAKIN, 1963
Sporidesmolide I	cyclo-(Hyv–*D*-Val–*D*-Leu–Hyv–Val–MeLeu)	*Spirodesmium bakeri*	RUSSEL 1960; (S) SHEMYAKIN, 1963
Serratamolide	cyclo-[*D*-Hyd–Ser–]$_2$	*Serratia marcescens*	WASSERMANN, 1960; (S) SHEMYAKIN, 1963
Esperin	OC–Glu–Leu–Leu–Val–Asp–Leu–Leu, with OC linked to H_2C–CH(R)–O– linked to Asp	*Bacillus mesentericus*	THOMAS and ITO, 1969
Actinomycin C_3	–CO–Thr–*D*-aIle–Pro–Sar–MeVal– / –CO–Thr–*D*-aIle–Pro–Sar–MeVal– (each Thr linked to MeVal of its chain)	*Streptomyces*	(S) BROCKMANN, LACKNER, 1960
Etamycin	cyclo-(Hypic–Thr–Leu–aHyp–Sar–DiMeLe–Ala–PheSar)	*Streptomyces*	SHEEHAN; ARNOLD, 1958
Echinomycin	quinoxaline–OC–Ser–Ala–N(CH_3)---C(–S–)–CO–MeVal–CO; Ser–O–CO–MeVal–OC–C(–H_2C–S)–N(CH_3)–Ala---Ser---CO–quinoxaline; MeVal–CO–O–Ser; ring: C–S–H_2C–C–S–CH_2–C	*Streptomyces echinatus*	KELLER-SCHIERLEIN, 1959

connected to actinocin by amide bonds. *Actinomycin* C_1 shows the highest antibacterial activity, and has been tested clinically as a cytostatic. It is active against HODGKIN's disease, lymphatic sarcoma and metastatic WILMS tumour.

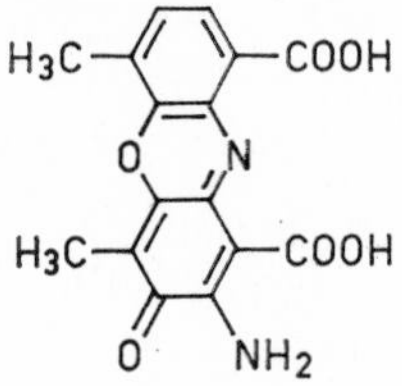

Actinocin

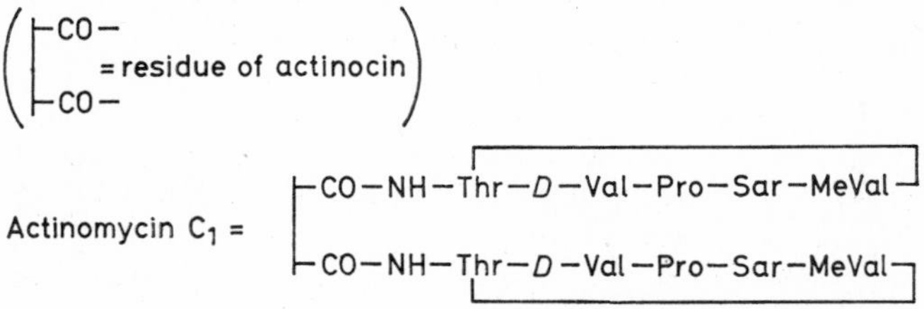

(⊢CO– / ⊢CO– = residue of actinocin)

Actinomycin C_1 = ⊢CO–NH–Thr–*D*–Val–Pro–Sar–MeVal┘ / ⊢CO–NH–Thr–*D*–Val–Pro–Sar–MeVal┐

In vivo the actinomycins act on and inhibit the DNA-dependent RNA-polymerase system. The actinomycin concentration necessary for the inhibition depends on the base composition of the nucleic acids, increasing with decreasing guanidine content.

Actinomycin C_3 was synthesised in 1960 by H. BROCKMANN and H. LACKNER (formula in Table 22). In a recent synthesis of actinomycin C_1 (J. MEIENHOFER, 1970) the crucial ring formation was achieved by cyclisation between the proline and sarcosine residues.

YOSHIOKA *et al.*[82] have isolated two compounds from cultures of *Streptomyces griseo verticillatus* which are active against tuberculosis: they called them *tuberactinomycin A and B* (*TUM A and B*). TUM B is identical with viomycine. Two additional tuberactinomycins are produced if the microorganisms are treated with nitrosoguanidine (TUM N and O). The compounds differ only in the arrangement and number of the hydroxyl groups.

Footnote to Table 22.

Abbreviations:

Hyd = β-hydroxydecanoic acid; Hyv = hydroxyisovaleric acid; Lac = lactic acid; Hypic = 3-hydroxypicolinic acid; Sar = sarcosine; PheSar = L-α-phenylsarcosine; MeLeu, MeIle, MeVal = N-methylamino acids; DiMeLeu = L-β-N-dimethylleucine.

Tuberactinomycins (TUM)

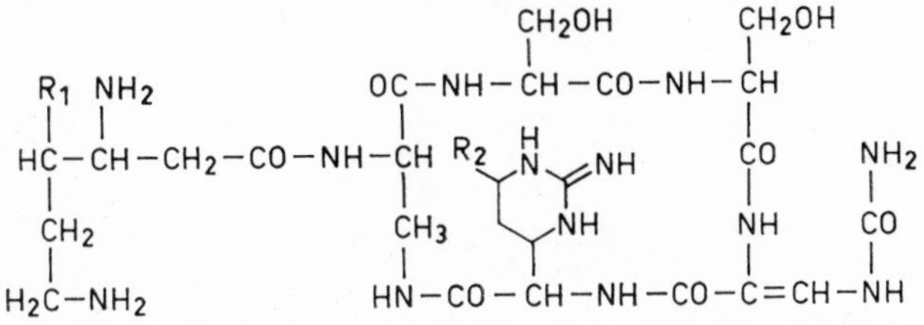

```
                                  CH2OH       CH2OH
                                   |           |
 R1 NH2                 OC-NH-CH-CO-NH-CH
 |  |                    |                      |
HC-CH-CH2-CO-NH-CH R2  H                CO      NH2
|                       |   N   NH                 |        |
CH2                    CH3       NH       NH      CO
|                       |         |        |        |
H2C-NH2                HN-CO-CH-NH-CO-C=CH-NH
```

TUM	R_1	R_2	TUM	R_1	R_2
A	OH	OH	N	OH	H
B	H	OH	O	H	H

Etamycin (viridogrisein) is also a peptide antibiotic. It contains 3-hydroxy-picolinic acid as hydroxy component. It is active against Gram-positive bacteria and *M. tuberculosis.*

In *staphylomycin* the ring closure takes place through the hydroxy group of threonine to form a peptide lactone. It is used in treating local infections by Gram-positive bacteria, particularly staphylococci.

Echinomycin, which is also active against Gram-positive bacteria, contains a 1,4-dithiane ring, a dilactone ring system and two 2-quinoxaline carboxylic acid residues.

The basic approaches to the synthesis of depsipeptides (Figure 29) are not very different from those of peptide synthesis. It is possible to synthesise a regular depsipeptide chain step by step alternately forming ester and peptide bonds (I) or alternatively by fragment condensation (II). By appropriate choice of the partial sequences used the fragment condensation steps can be limited to the formation of amide or ester bonds.

The use of compounds which are converted into amino acid or hydroxy acid residues during the course of the synthesis is a third possible approach to depsipeptide synthesis. The reaction of N-protected amino acid residues with diazoacetic acid derivatives is in this class. It was developed by Curtius and later used by Gibian and Lübke (1960) for the synthesis of depsipeptides of glycolic acid:

$$\text{Y-NH-AA-COOH} + |\text{N}{\equiv}\overset{\oplus}{\text{N}}-\overset{\ominus}{\bar{\text{C}}}\text{H-COOR}' \rightarrow \text{Y-NH-AA-COO-}\underset{}{\overset{\overset{\oplus}{\text{N}}\equiv\text{N}|}{\overset{|}{\text{C}}}}\text{H-COOR}'$$

$$\xrightarrow{-\text{N}_2} \text{Y-NH-AA-CO-O-CH}_2\text{COOR}' \quad \text{etc.}$$

I Stepwise synthesis of depsipeptides

$$Y-NH-AA-COOH+HO-HA-COOR$$
$$\downarrow$$
$$Y-NH-AA-\boxed{CO-O}-HA-COOR$$
$$\downarrow -R$$
$$Y-NH-AA-CO-O-HA-COOH+H_2N-AA-COOR$$
$$\downarrow$$
$$Y-NH-AA-CO-O-HA-\boxed{CO-NH}-AA-COOR$$
$$\text{etc.} \downarrow -R$$

II Synthesis of regular depsipeptides by fragment condensation
IIa (only formation of peptide bonds)

$$Y-NH-AA-CO-O-HA-COOH+H_2N-AA-CO-O-HA-COOR$$
$$\downarrow$$
$$Y-NH-AA-CO-O-HA-\boxed{CO-NH}-AA-CO-O-HA-COOR$$
$$\text{etc.} \downarrow -R$$

IIb (only formation of ester bonds)

$$X-O-HA-CO-NH-AA-COOH+HO-HA-CO-NH-AA-COOR$$
$$\downarrow$$
$$X-O-HA-CO-NH-AA-\boxed{CO-O}-HA-CO-NH-AA-COOR$$
$$\text{etc.} \downarrow -R$$

Figure 29. The most important synthetic approaches for depsipeptides: Y = amino-protecting group; X = O-protecting group; R = carboxyl-protecting group; $H_2N-AA-COOH$ = amino acid; $HO-HA-COOH$ hydroxy acid

Similarly, the PASSERINI reaction (in which carboxylic acids, carbonyl compounds and isonitriles are converted in a one-step reaction into α-acyloxy-carboxylic acid amides) has been used by UGI and FETZER for the synthesis of depsipeptides of glycolic and hydroxyisovaleric acids:

$$\text{Y-NH-}\underset{\displaystyle\text{CH}}{\overset{\displaystyle R_1}{|}}\text{-COOH} + R_2\text{-CHO} + \text{CN-}\underset{\displaystyle\text{CH}}{\overset{\displaystyle R_3}{|}}\text{-COOR}'$$

$$\rightarrow \text{Y-NH-}\underset{\displaystyle\text{CH}}{\overset{\displaystyle R_1}{|}}\text{-CO-O-}\underset{\displaystyle\text{CH}}{\overset{\displaystyle R_2}{|}}\text{-CO-NH-}\underset{\displaystyle\text{CH}}{\overset{\displaystyle R_3}{|}}\text{-COOR}'$$

These methods are only useful for the synthesis of racemic depsipeptides because in both reactions asymmetric centres either have to be formed or have to take part in the reaction. In the conventional synthesis of depsipeptides the coupling methods used are the same as those used in ordinary peptide synthesis, but the relative importance of the individual methods is different. The acid chloride method and the mixed anhydride are more important in depsipeptide work than in the synthesis of straightforward peptides, but the carbodiimide method and activated esters are less useful.

Because of the sensitivity of the ester bonds to base the number of protecting groups that can be used is restricted.

The combination of a protecting group that is resistant to hydrogenation with a protecting group that is susceptible to hydrogenation is particularly useful in depsipeptide synthesis. A common combination is that of a *tert*-butyloxycarbonyl group with a benzyl or *p*-nitrobenzyl group. The acidolytic removal of the Boc-group with HCl/ether or HCl/ethyl acetate does not affect the depsipeptide ester group. The nitroso group which can be cleaved with acid is suitable for *N*-methylamino acids and the combinations Boc-/–NH–NH–Z or Z-/–NH–NH–Boc enable the azide method to be used.

A detailed review of the problems of depsipeptide syntheses has been published by LOSSE and BACHMANN[80].

2.3.5. *Strepogenin Peptides*[83]

Strepogenins stimulate the growth of microorganisms, especially of lactic acid bacteria. They are found, for example, in liver extracts, tomato juice and in partial enzymic hydrolysates of, for example, insulin, casein, ribonuclease, etc. The standard unit of strepogenin activity is the increase of the growth of *Lactobacillus casei* produced by 1 mg of a standard liver extract (Table 23).

Further investigations[83] have shown that the strepogenin activity of synthetic products is connected with the presence of cysteine. The highest activities are found when cysteine is either connected N-terminally with leucine or is situated between two leucine residues. The oxidation of cysteine to give the corresponding cysteic acid derivative results in the loss of activity.

Table 23

Origin of the strepogenin peptides	Amino acid sequence	Strepogenin activity (U/mg)
Insulin hydrolysates	Ser–His–Leu–Val–Glu	86
	Ser–His–Leu–Val–Glu–Ala–Leu	98
	[Leu–Val–Cys–Gly–Glu–Arg]$_2$	200
Synthetic products	Leu–Cys–Leu–Val–Glu	400
	Leu–Cys–Leu–Ala–Glu	400
	Leu–Cys–Leu–Ala	400
	Cys–Leu–Ala–Glu	200
	Leu–Cys–Leu	130
	Leu–Cys(SO_3H)Leu–Val–Glu	0
	Leu–Val–Glu	0

2.3.6. Peptide Toxins

The structures of the toxins of the death cap toadstool (*Amanita phalloides*) have been elucidated by WIELAND and co-workers[84, 85]. These toxins are divided into the *phallatoxins* (phalloidin, phalloin and phallacidin) and the more toxic but slower acting *amatoxins* (α-, β-, γ-amanitins).

Structure of the phallatoxins

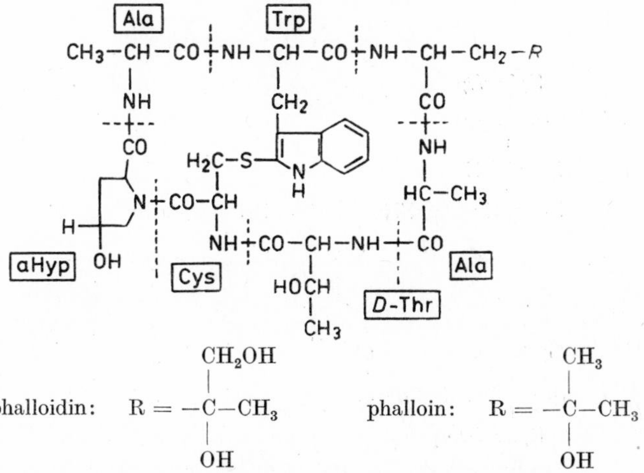

phalloidin: R = $-C(CH_2OH)(OH)-CH_3$ phalloin: R = $-C(CH_3)(OH)-CH_3$

Phallacidin: instead of the Ala-D–Thr unit in phalloidin phallacidin contains Val-D-erythro-β-hydroxyaspartic acid

Toxicity of the most important toxic components of *Amanita phalloides*, LD_{50} mg/kg white mouse:

γ-amanitin	0.2	phalloin	1.35
α-amanitin	0.35	phalloidin	1.85
β-amanitin	0.97	phallacidin	2.50

Investigations of the relation between structure and toxicity indicate that the cyclic heptapeptide structure and the thioether bridge of these compounds are essential.

The amatoxins, which consist of L-amino acids only, are also cyclic heptapeptides. In these the thioether group is replaced by a sulphoxide group. The presence of the γ-hydroxy group in the isoleucine side chain is essential for toxicity. It is absent in amanullin, which is structurally very similar to γ-amanitin but which is nevertheless non-toxic.

More than 90 per cent of the fatal cases of poisoning by fungi are due to *A. phalloides* and related species. The first symptoms (vomiting and diarrhoea) usually appear some 10–24 hours after eating the fungus. The fatal stage involves entry of the toxins into the liver and irreversible liver damage. The discovery that the green bulbous agaric contains the cyclic decapeptide *cyclo*-(–Pro–Phe–Phe–Val–Pro–Pro–Ala–Phe–Phe–Pro–) as well as the highly toxic compounds was of great interest because this decapeptide, *antamanide*, is an antidote for the toxic effect of phalloidin and α-amanitin (Th. WIELAND, 1968). The structure of antamanide was determined by mass spectrometry and has been confirmed by total synthesis.

Structure of the amatoxins

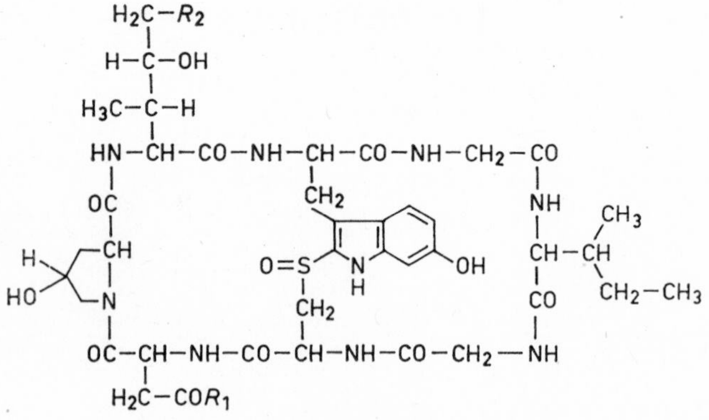

```
         H2C-R2
           |
         H-C-OH
           |
        H3C-C-H
           |
        HN-CH-CO-NH-CH-CO-NH-CH2-CO
        |           |              |
        OC          CH2            NH     CH3
        |                          |     /
  H    CH                          CH-CH
   \   |         O=S     N      OH |     \
  HO   N             |   H         CO     CH2-CH3
       |             CH2           |
       OC-CH-NH-CO-CH-NH-CO-CH2---NH
          |
         H2C-COR1
```

α-amanitin: $R_1 = NH_2$, $R_2 = OH$; β-amanitin: $R_1 = OH$, $R_2 = OH$
γ-amanitin: $R_1 = NH_2$, $R_2 = H$

Complete protection against the action of the toxins is only obtained if the protective dose of antamanide is taken at the same time as the toxin. The protective action of antamanide and some of its analogues depends on their behaviour at membranes. They form complexes with K^+ and Na^+ ions analogous to those formed by the enniatins and the valinomycins, which have been extensively investigated at the Shemyakin Institute of Bioorganic Chemistry in Moscow (Ovchinnikov *et al.*, 1971).

In 1966 F. Habermann *et al.* published the results of their structural work on the *melittins*, which are, together with the polypeptide *apamin*, the *MCD-peptide* and histamine as well as the enzymes hyaluronidase phospholipase A, the main components of bee sting venom. This work showed that uncommon amino acids or ring structures are not invariably features of peptide toxins. An interesting characteristic of the melittins is the predominance of amino acids with hydrophobic character towards the N-terminus and marked hydrophilic character near the C-terminus of the polypeptide chain. Melittins I and II consist of 26 and 27 amino acid residues, respectively.

Fifty per cent of the bee sting venom (*Apis mellifera*, Hymenoptera) consists of melittin I, which is responsible for the haemolytic action of the toxin (LD_{50} 4 mg/kg mouse).

Gly–Ile–Gly–Ala–Val–Leu–Lys–Val–Leu–Thr–Thr–Gly–Leu–Pro–Ala–Leu–
1 2 3 4 5 6 7 8 9 10 11 12 13 14 15 16

Ile–Ser–Trp–Ile–Lys–Arg–Lys–Arg–Gln–Gln–NH_2 mellitin I
17 18 19 20 21 22 23 24 25 26

It is thought that the haemolytic properties are due to the ability of the toxin to destroy the structure of lipid membranes. The melittins are also active against Gram-positive bacteria and give (in mice) protection against radiation after subcutaneous injection.

Melittin II contains the same amino acid sequence 1–20 as melittin I, then

–Ser–Arg–Lys–Lys–Arg–Gln–Gln–NH_2
21 22 23 24 25 26 27

Apamin is a peptide that acts on the central nervous system. It is composed of 18 amino acids, four of which are cysteine residues forming two disulphide bridges:

Cys–Asn–Cys–Lys–Ala–Pro–Glu–Thr–Ala–Leu–Cys–Ala–Arg–Arg–Cys–Gln–Gln–His–NH_2
1 2 3 4 5 6 7 8 9 10 11 12 13 14 15 16 17 18

MCD-peptide ('peptide 401') is composed of 22 amino acids and similar to apamin:

```
    ┌──────────────────────────────────────────────────┐
Ile–Lys–Cys–Asn–Cys–Lys–Arg–His–Val–Ile–Lys–Pro–His–Ile–Cys–Arg–Lys–Ile–Cys
1   2   3   4   |5  6   7   8   9   10  11  12  13  14  15  16  17  18  |19
                └───────────────────────────────────────────────────────┘
Gly–Lys–Asn–NH2
20  21  22
```

The name MCD-peptide was suggested from the ability of this substance *m*ast *c*ells to *d*egranulate.

In the separation of the components of bee sting venom gel filtration with Sephadex G 50 combined with other chromatographic purification techniques proved invaluable. Syntheses of melittin and some of its partial sequences have been described by E. SCHRÖDER and K. LÜBKE.

2.3.7. Peptide Insecticides

In recent years some naturally occurring insecticidal cyclic peptides have been discovered. *Asprochacin*, which is highly toxic to silkworms, is a metabolic product produced by *Aspergillus ochraceus*. It consists of *N*-methylalanine, *N*-methylvaline, ornithine and an octatrienoic acid side chain residue[86], making up the first naturally occurring cyclic tripeptide to be discovered.

Asprochacin

```
       HN—CH2—CH2—CH2      O
       |           |       ||
     O=C           HC—NH—C—R
       |           |
   H3C—CH          C=O
       |           |
   H3C—N——C——CH——N—CH3
          |  ||
          O  CH
            /  \
         H3C    CH3
```

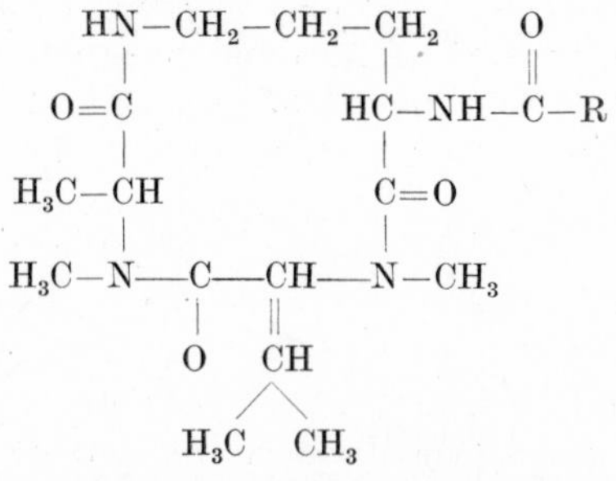

$R = -CH{=}CH-CH{=}CH-CH{=}CH-CH_3$

In 1970 SUZUKI *et al.*[87] isolated the insecticidal cyclodepsipeptides *destruxin* C and D from cultures of *M. anisopliae*:

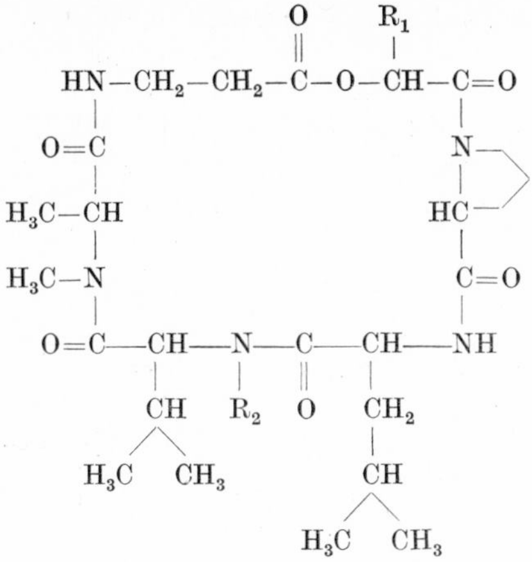

```
                       O       R1
                       ||      |
HN—CH2—CH2—C—O—CH—C=O
 |                                  |
O=C                                N—
 |                                  |    >
H3C—CH                              HC—
 |                                  |
H3C—N                               C=O
 |                                  |
O=C——CH——N——C——CH——NH
       |     |    ||   |
       CH    R2   O    CH2
      / \              |
   H3C   CH3           CH
                      / \
                   H3C   CH3
```

	R_1	R_2
Destruxin C	$HO-CH_2-CH(CH_3)-CH_2-$	CH_3-
Destruxin D	$CH_3-CH(COOH)-CH_2-$	CH_3-
Desmethyl destruxin	$CH_3-CH(CH_3)-CH_2-$	$H-$

While on the subject of insecticidal peptides, the investigations of Poduška, Slama *et al.* into the juvenile hormone activity of simple peptides must be mentioned. L-Isoleucyl-L-alanyl-*p*-aminobenzoic acid ethyl ester, for example, is structurally related to the analogues of juvenile hormones which have been synthesised from aliphatic monoterpenes and *p*-substituted aromatic compounds. The biological activity of the peptide was increased if isoleucine was replaced by a Boc-group. The most active compound found so far is the α-chloroisobutyric derivative, 1 mg of which is enough to destroy 2000 kg of insects of the family of Pyrrhocoridae at the stage when the final larval instar changes to the chrysalis.

2.3.8. Peptoides

Apart from the chromopeptides, which have been well studied, relatively few peptoides are known and these are found in rather diverse sources. The most important peptoides (for a definition see p. 76) are the nucleo-, phospho-, lipo-, glyco- and chromo-peptides.

The nucleotide and peptide components of the nucleopeptides are interesting in connection with protein biosynthesis, as they are generally joined together in the same way as the aminoacylnucleic acids (cf. p. 62). Nucleopeptides have been isolated from animal organs (pig and bovine liver) and from yeast. The purine base, adenine, and the pyrimidine base, uracil, have been observed as constituents. The *nucleopeptides* that have been identified in pancreatic extracts are especially stable to bases. It is thought that this arises as a result of a phosphonamide bond between the phosphoric acid residue and the N-terminal amino acid. In 1960 J. W. DAVIES and G. HARRIS successfully elucidated the structure of a nucleopeptide from brewery yeast.

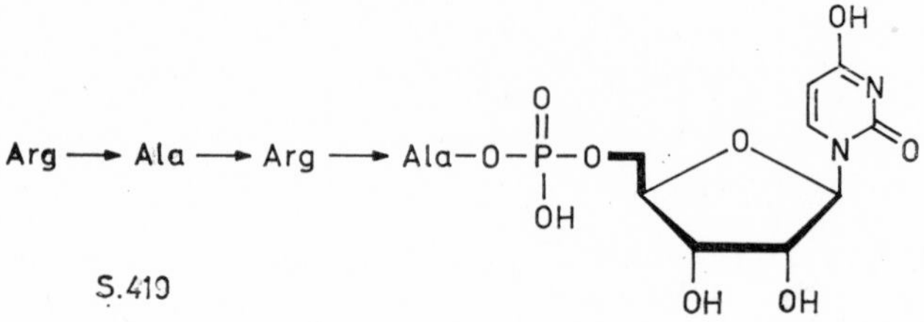

Arg ⟶ Ala ⟶ Arg ⟶ Ala–O–P(=O)(OH)–O–[ribose(OH)(OH)]–[pyrimidine(OH)(=O)]

S.410

Peptides containing organically bound phosphorous are called *phosphopeptides*. They are of considerable interest in relation to the phosphoproteins of eggs and milk: some simple members of the group have been synthesised.

Lipopeptides are composed of amino acids and fatty acids. They are fat-soluble, and sometimes contain natural amino acids, although more often the D-configuration is found. The few known representatives include *fortuitine* from *M. fortuitum* (see p. 282) and *peptidolipin NA* isolated in 1963 from *Nocardia asteroides*. The latter compound contains a depsipeptide bound to D-3-hydroxyeicosanoic acid to form a macrocyclic lactone. The fatty acid is connected with the heptapeptide Thr–Val–D-Ala–Pro–D-aIle–Ala–Thr via the hydroxy and carboxylic groups[88].

Glycopeptides are produced by the enzymic or partial acid hydrolysis of glycoproteins, such as globulins or ovalbumins. It is common in these compounds for the peptide component to be esterified to one of the carbohydrate hydroxy residues (A=CH_2–O–CO–CH(R_1)–NH–CO–CH(R_2)–NH..., B=OH) or alternatively bound via an amide bond to the amino group of an amino sugar (A = CH_2OH, B = –NH–CO–CH(R_1)–NH–CO–CH(R_2)–NH...) or even via an amide bond to the carboxylic group of a sugar acid (A = CO–NH–CH(R_1)–CO–NH–CH(R_2)–CO..., B = OH):

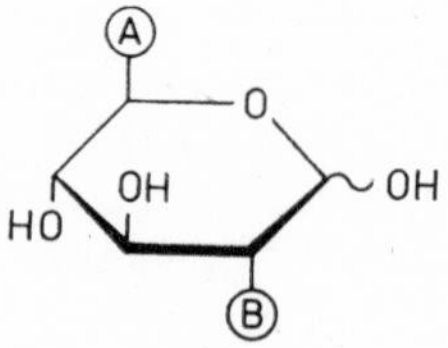

[Structure: pyranose ring with substituents A, B, HO, OH, OH]

Chromopeptides contain chromophores covalently linked to peptides: the most important natural representatives are the actinomycins (p. 223).

Ferrichrome belongs to a group of compounds which was found to be present in fungal cultures as biologically active ferric complexes. It has the following structure[89]:

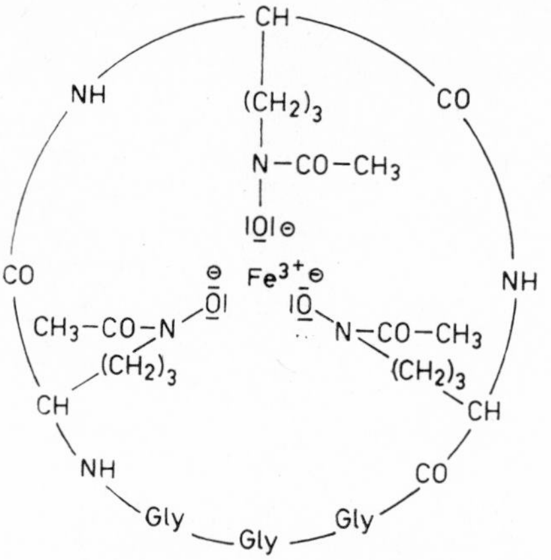

Ferrichrome has been synthesised by cyclisation of a hexapeptide containing three nitronorvaline residues and three glycine residues. After the reduction of the nitro groups to hydroxylamine functions and acetylation the trihydroxamic acid was isolated as its Fe^{3+} complex. It possessed all the properties of the natural product.

2.3.9. *Peptide Alkaloids*[90]

In addition to the ergot alkaloids, the structure of which is set out in the scheme,

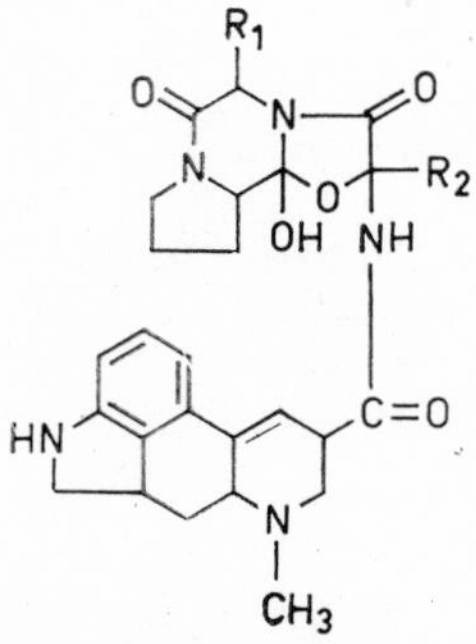

Basic structure of the ergot alkaloids

	R_1	R_2
ergotamine	$-CH_2-C_6H_5$	$-CH_3$
ergosine	$-CH_2-CH(CH_3)_2$	$-CH_3$
ergocristine	$-CH_2-C_6H_5$	$-CH(CH_3)_2$
ergocryptine	$-CH_2-CH(CH_3)_2$	$-CH(CH_3)_2$
ergocornine	$-CH(CH_3)_2$	$-CH(CH_3)_2$

some peptide alkaloids have also been found to be constituents of *Rhamnacea* (R. Tschesche *et al.*, 1967): the structure of some of these compounds has been elucidated with the aid of mass spectrometry. A 14-membered ring system with a *p*-alkoxy-styrylamine residue in the ring is characteristic of the class.

Frangulanine, the main alkaloid of the bark of *Rhamnus frangula* L., consists of leucine, *N,N*-dimethylisoleucine (Me_2Ile), β-hydroxy-leucine (3-Hyle) and *p*-hydroxystyrylamine. Additional peptide alkaloids which contain an ether bridge in the ring system are *intergerrenein,* which has β-hydroxyphenylalanine instead of the 3-Hyle residue in the frangulanine formula; *integerresine,* which has β-hydroxyphenylalanine and phenylalanine

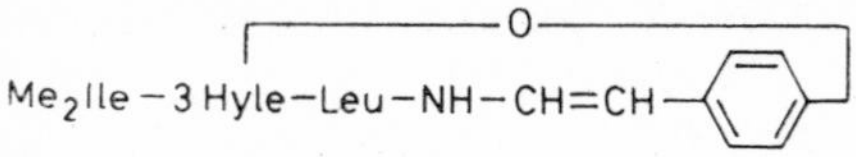

Me_2Ile – 3 Hyle–Leu–NH–CH=CH–(p-phenylene, O-bridged to 3 Hyle)

Frangulanine

Me_2Phe–Pro–3 Hyle–Phe–NH–CH=CH–(p-phenylene, O-bridged to 3 Hyle)

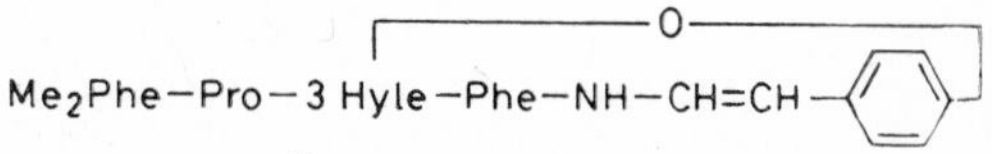

Scutianine

instead of the 3-Hyle and Leu residues; and *scutianine,* which is produced in the bark of the South American three *Rhamnacea scutia buxifolia* Reiss.

Zizyphine is an alkaloid with a linear peptide component. Its structure was elucidated by chemical degradation, NMR spectroscopy and mass spectrometry (E. Zbiral *et al.*, 1965).

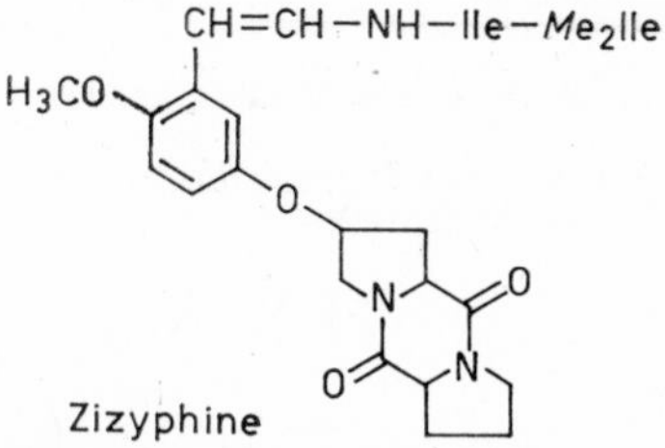

CH=CH–NH–Ile–Me_2Ile
H_3CO
O
N
O
O
N

Zizyphine

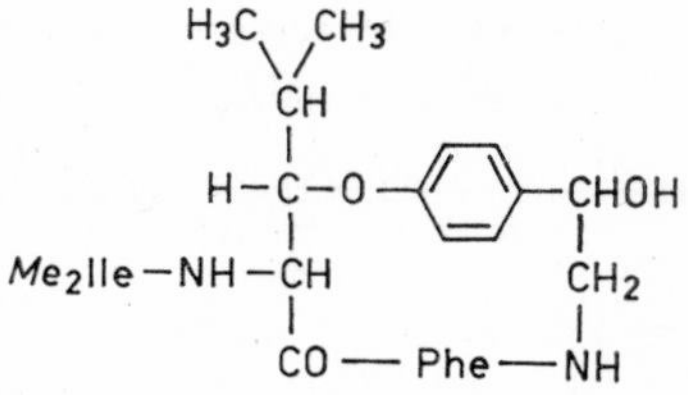

H_3C CH_3
CH
H–C–O–(p-phenylene)–CHOH
Me_2Ile–NH–CH CH_2
CO — Phe — NH

Pandamine

Pandamine, the peptide alkaloid of *Panda oleosa,* contains isoleucine, phenylalanine, β-hydroxyleucine and a hydroxyethylamine derivative (M. X. Pais *et al.*, 1964).

References

1 Bricas, E., and Fromageot, C. (1953). *Advan. Protein Chem.*, **8**, 4

2 Schwyzer, R. (1958). *Chimia*, **12**, 53

3 Waley, S. G. (1966). *Advan. Protein Chem.*, **21**, 2

4 SCHRÖDER, E. and LÜBKE, K. (1966). *The Peptides*, Vol. 2, Academic Press, New York

5 JESCHKEIT, H., LOSSE, G. and KNOPF, D. (1962). *Pharmazie*, **18**, 658

6 *Advances in Experimental Medicine and Biology* (1968). BLACK, N., MARTINI, L. and PAOLETTI, R., eds.: Vol. 2, *Pharmacology of Hormonal Polypeptides and Proteins*, Plenum Press, New York

7 BERSON, S.A. and YALOW, R.S. (1973). *Peptide Hormones*, North-Holland, Amsterdam

8 RUDINGER, J. (1971). *Drug Designs*, Vol. 2, ARIENS, E. J., ed., Academic Press, New York

9 WALTER, R., GRIFFITHS, E. C. and HOOPER, K. C. (1973). *Brain Research*, **60**, 449

10 WALTER, R. (1973). *Proc. 12th Europ. Peptide Symp.*, 1972, p. 363, North-Holland, Amsterdam

11 SCHWARTZ, I. L. and WALTER, R.: ref. 7, p. 179

12 SCHWYZER, R. (1973). *Proc. 12th Europ. Peptide Symp.*, p. 424, North-Holland, Amsterdam

13 GARREN, L. D., GILL, G. N., MASUI, H. and WALTON, G. M. (1971). *Rec. Progr. Hormone Res.*, **27**, 433

14 GOTH, E. and FÖVENYI, J. (1971). *Polypeptide Hormones*, Akademie Kiado, Budapest

15 SCHWYZER, R. and SIEBER, P. (1965). *Helv. Chim. Acta*, **49**, 134

16 SCHWYZER, R. (1966). *Naturwissenschaften*, **53**, 189

17 LI, C. H. and YAMASHIRO, D. (1970). *J. Amer. Chem. Soc.*, **92**, 7608

18 PECILE, A. and MÜLLER, E. E., eds. (1972) *Growth and Growth Hormone*, Excerpta Medica, Amsterdam

19 LI, C. H. *et al.* (1969). *Nature*, **224**, 696

20 DU VIGNEAUD, V. (1970). *Perspectives in Biological Chemistry*, OLSON, R. E., ed., p. 133, Dekker, New York

21 MANNING, M., COY, E. and SAWYER, W. H. (1970). *Biochemistry*, **9**, 3925

22 WALTER, R., *Structure-Activity Relationships of Protein and Polypeptide Hormones*, Vol. 2, MARGOULIES, M. and GREENWOOD, F. C., eds., p. 181, Excerpta Medica

23 THORN, N. A. (1970). *Advan. Metabolic Disorders*, **4**, 40

24 SCHALLY, A. V. and KASTIN, A. J. (1970). *Advan. Steroid Biochem. Pharmacol.*, **2**, 41

25 GUILLEMIN, R. (1971). *Advan. Metabolic Disorders*, **5**, 1

26 MARTINI, L., MOTTA, M. and FRASCHINI, F. (1971). *The Hypothalamus*, Academic Press, London

27 KLOSTERMEYER, H. and HUMBEL, R. E. (1966). *Angew. Chem.*, **78**, 871

28 ZAHN, H. (1970). *Fortschr. Med.*, **88**, 163

29 LÜBKE, K. and KLOSTERMEYER, H. (1970). *Advan. Enzymol.*, **33**, 445

30 STEINER, D. F. *et al.* (1969). *Rec. Progr. Hormone Res.*, **25**, 207

31 GRODSKY, G. M. (1970). *Vitamins and Hormones*, **28**, 37

32 ADAMS, M. J., BLUNDELL, T. L., DODSON, E. J., DODSON, G. G., VIJAYAN, M., BAKER, E. N., HARDING, M. M., HODGKIN, D. C., RIMMER, B. and SHEAT, S. (1969). *Nature*, **224**, 491

33 Dixon, G. H. and Wardlaw, A. C. (1960). *Nature*, **188**, 721

34 Meienhofer, J., Schnabel, E., Bremer, H., Brinkhoff, O., Zabel, R., Okuda, T., Sorka, W., Klostermeyer, H., Brandenburg, D. and Zahn, H. (1963). *Z. Naturforsch.*, **18**b, 1120

35 Katsoyannis, P. G., Tometsko, A. and Fukuda, K. (1963). *J. Amer. Chem. Soc.*, **85**, 2863

36 Kung, Y.-t., Du, Y.-c., Huang, W.-t., Chen, C.-c., Ke, L.-t., Hu, S.-c., Jiang, R.-q., Chu, S.-q., Niu, C.-i., Hsu, J.-z., Chang, W.-c., Cheng, L.-l., Li, H.-s., Wang, Y., Loh, T.-p., Chi, A.-h., Li, C.-h., Shi, P.-t., Yie, Y.-h., Tang, K.-l. and Hsing, C.-y. (1965). *Scientia Sinica*, **14**, 1710

37 Sieber, P., Kamber, B., Hartmann, A., Jöhl, A., Riniker, B. and Rittel, W. (1974). *Helv. Chim. Acta*, **57**, 2617

38 Steiner, D. F., Cunningham, D., Spiegelman, L. and Aten, B. (1967). *Science*, **157**, 697

39 Mehlis, B. and Köller, G. (1970). *Pharmazie*, **25**, 669

40 Change, R. E., Ellis, R. M. and Bromer, W. W. (1968). *Science*, **161**, 165

41 Blundell, T. L., Dodson, G. G., Dodson, E., Hodgkin, D. C. and Vijayan, M. (1971). *Res. Progr. Hormone Res.*, **27**, 1

42 Lindsay, D. G. (1972). *FEBS Lett.*, **21**, 105

43 Brandenburg, D. (1972). *Hoppe-Seyler's Z. Physiol. Chem.*, **353**, 869

44 Brandenburg, D., Schermutzki, W. and Zahn, H. (1973). *Hoppe-Seyler's Z. Physiol. Chem.*, **354**, 1521

45 Geiger, R. and Obermeier, R. (1973). *Biochem. Biophys. Res. Comm.*, **55**, 60

46 Wünsch, E. and Wendelberger, G. (1968). *Chem. Ber.*, **101**, 3659; Wünsch, E., Jaeger, E. and Scharf, R. (1968). *Chem. Ber.*, **101**, 3664

47 *Proc. 2nd Internat. Symp. Calcitonin*, London, 1969: *Calcitonin 1969*, Tayler, S. and Forster, G., eds., Heinemann Medical, London (1970)

48 Neher, R., Riniker, B., Zuber, H., Rittel, W. and Kahnt, F. W. (1968). *Helv. Chim. Acta*, **51**, 917

49 Rittel, W., Brugger, M., Kamber, B., Riniker, B. and Sieber, P. (1968). *Helv. Chim. Acta*, **51**, 924

50 Guttmann, St., Pless, J., Sandrin, E., Jaquenoud, P. A., Bossert, H. and Willems, H. (1968). *Helv. Chim. Acta*, **51**, 1155

51 Bell, P. H., Barg, W. F. and Collucci, W. F. (1968). *J. Amer. Chem. Soc.*, **90**, 2704

52 Riniker, B. *et al.* (1968). *Helv. Chim. Acta*, **51**, 1738, 1900, 2057

53 Guttmann, St., Pless, J., Huguenin, R. E., Sandrin, E., Bossert, H. and Zehnder, K. (1969). *Helv. Chim. Acta*, **52**, 1789

54 Behrens, O. K. and Grinnan, E. L. (1969). *Ann. Rev. Biochem.*, **38**, 83

55 Potts, J. T., Jr., Tregear, G. W., van Rietschoten, J., Niall, H. D., and Keutmann, H. T. (1973). Jakubke, H.-D., eds., *Proc. 12th Europ. Peptide Symp.*, 1972, p. 191, North-Holland, Amsterdam

56 Grossman, M. I. (1970). *Nature*, **228**, 1147

57 Bodanszky, M. *et al.* (1966). *Chem Ind.*, 1757

58 Fisher, J. W. (1971). *Kidney Hormones*, Academic Press, London

59 GROSS, F. (1971). In *Pharmacology of Naturally Occurring Polypeptides and Lipid-soluble Acids*, Vol. 1, p. 73, WALKER, J. M., ed., Pergamon Press, Oxford
60 ROCHA E SILVA, M. (1970). *Kinin Hormones*, C. C. Thomas, Springfield, Ill.
61 *Bradykinin, Kallidin and Kallikrein*, ERDÖS, E. G., ed., Springer-Verlag, Berlin (1970)
62 LEMBECK, F. and ZETTER, G. (1971). In *Pharmacology of Naturally Occurring Polypeptides and Lipid-soluble Acids*, Vol. 1, p. 29, WALKER, J. M., ed., Pergamon Press, Oxford
63 CHANG, M. M., LEEMAN, S. E. and NIALL, H. D. (1971). *Nature New Biology*, **232**, 86
64 TREGEAR, G. W., NIALL, H. D., POTTS, D. J., Jr., LEEMAN, S. E., and CHANG, M. M. (1971). *Nature New Biology*, **232**, 87
65 ERSPARMER, V. and ANASTASI, A. (1962). *Experientia*, **18**, 58
66 SANDRIN, E. and BOISSONNAS (1962). *Experientia*, **18**, 59
67 BERNARDI, L., BOSISIO, G., GOFFREDO, O. and DE CASTIGLIONE, R. (1964). *Experientia*, **20**, 492
68 DE CASTIGLIONE, R., ANGELUCCI, F., ERSPARMER, V., FALCONIERI, G., and NEGRI, L. (1973). *Proc. 12th Europ. Peptide Symp.*, 1972, p. 463, North-Holland, Amsterdam
69 SHEEHAN, J. C. (1963). *Pure Appl. Chem.*, **6**, 297
70 BODANSZKY, M. and PERLMAN, D. (1964). *Nature*, **204**, 804
71 BODANSZKY, M. (1968). *Nature*, **218**, 291
72 SCHWYZER, R. and SIEBER, P. (1957). *Helv. Chim. Acta*, **40**, 624
73 SARGES, R. and WITKOP, N. (1964). *J. Amer. Chem. Soc.*, **86**, 1861
74 KUROMIZU, K. and IZUMIYA, N. (1970). *Tetrahedron Letters*, **17**, 1471
75 STUDER, R. O., LERGIER, W. and VOGLER, K. (1966). *Helv. Chim. Acta*, **49**, 974
76 TURKOVA, J., MIKES, O. and SORM, F. (1964). *Czech. Chem. Comm.*, **29**, 280
77 SHIN, C., CHIGIRA, Y., MASAKI, M. and OHTA, M. (1967). *Tetrahedron Letters*, 4601
78 SHEMYAKIN, M. M. and KHOKHLOV, A. S. (1953). *The Chemistry of Antibiotics*, 2nd edn, State Publishing House on Chemistry, Moscow
79 SHEMYAKIN, M. M. (1960). *Angew. Chem.*, **72**, 342
80 LOSSE, G. and BACHMANN (1964). *Z. Chem.*, **4**, 241
81 SHEMYAKIN, M. and OVCHINNIKOV, Y. (1967). *Recent Developments in Chemistry of Natural Carbon-Compounds*, Vol. II, Hungarian Academy of Sciences, Budapest
82 YOSHIOKA, H. *et al.* (1971). *Tetrahedron Letters*, 2043
83 BANDET, P., BORECKA, I. and CHERBULIEZ, E. (1968). *Helv. Chim. Acta*, **51**, 1
84 WIELAND, TH. (1968). *Science*, **159**, 946
85 SCHMID, R. (1968). *Naturwiss. Rundsch.*, **21**, 514
86 MYOKEI, R., SAKURAI, A. and CHING-FUN CHANG (1969). *Tetrahedron Letters*, 695
87 SUZUKI, A., TAGUCHI, H. and TAMURA, S. (1970). *Agr. Biol. Chem.*, 813
88 GUINAND, M. and MICHEL, G. (1966). *Biochim. Biophys. Acta*, **125**, 76
89 KELLER-SCHIERLEIN, W. and MAURER, B. (1969). *Helv. Chim. Acta*, **52**, 603
90 SNIECKUS, V. A. (1972). *The Alkaloids*, SAXTON, J. E., ed. (Specialist Periodical Reports), Vol. 2, p. 271, Chemical Society, London

2.4. *Polyamino Acids*[1–4]

Polyamino acids are polymers containing amino acid moieties connected by peptide bonds. The terms 'polyamino acid' and 'polypeptide' are often wrongly supposed to be synonymous: the name 'polyamino acid': should be reserved for materials which can be obtained by polymerisation of simple amino acid derivatives.

Naturally occurring polyamino acids include the poly-γ-glutamic acids, which have been discovered in the outer membrane of anthrax bacteria and in *Subtilis* bacteria. Their structures and the synthesis of stereoisomeric materials with either the L- or the D-configuration or mixed L- and D-configuration have been intensively studied by V. Bruckner *et al.* and S. G. Waley.

Polyamino acids can be formed of identical monomeric units (I) or alternatively can be copolymers of different amino acids.

$$-(\underset{\displaystyle}{NH}-\underset{\displaystyle R_1}{\underset{|}{CH}}-CO-NH-\underset{\displaystyle R_1}{\underset{|}{CH}}-CO-NH-\underset{\displaystyle R_1}{\underset{|}{CH}}-CO-NH-\underset{\displaystyle R_1}{\underset{|}{CH}}-CO-)_n- \qquad (I)$$

Homopolyamino acids (I) can be synthesised easily up to a molecular weight of 10^3–10^6. One of the best methods is the polymerisation of 3-alkyloxazolidin-2,5-diones (Leuchs anhydrides). The anhydride ring is opened by nucleophiles so that polymerisation can be initiated by the presence of traces of moisture and by bases. Investigations by Goodman and Hutchison[5] show that the reaction follows the mechanism suggested by Bamford. By polymerisation of a mixture of *N*-carboxyanhydrides of various amino acids polymers with a statistical distribution of the amino acid residues can be produced (random copolymers: II). By reacting the *N*-carboxyanhydride of a single amino acid using a poly-α-amino acid initiator, block copolymers are produced (III):

$$-(NH-\underset{\displaystyle R_1}{\underset{|}{CH}}-CO-NH-\underset{\displaystyle R_2}{\underset{|}{CH}}-CO-NH-\underset{\displaystyle R_1}{\underset{|}{CH}}-CO-NH-\underset{\displaystyle R_2}{\underset{|}{CH}}-CO-)_n- \qquad (II)$$

$$-(NH-\underset{\displaystyle R_1}{\underset{|}{CH}}-CO-)_n-(NH-\underset{\displaystyle R_2}{\underset{|}{CH}}-CO)_m- \qquad (III)$$

However, polymers with a known constant repeating amino acid sequence, called 'sequence polymers' are more important protein models[6].

The first sequence polymer was produced by E. Fischer at the beginning of this century:

$$n\text{Ala–Gly–Gly–OMe} \rightarrow (\text{Ala–Gly–Gly})_n$$

Over the last 15 years improved methods for the synthesis of sequence polymers have been developed. Free peptides can be polymerised, for example, by the mixed anhydride method, with alkyl pyrophosphites or using dicyclohexylcarbodiimide. The reaction of Gly–Ser–Ala–NH–NH_2 with *N*-brominosuccinimide also produces sequence polymers.

However, the most suitable method involves the use of activated esters. The groups of De Tar, Kovacs and Stewart have preferred *p*-nitrophenyl esters. In this way more than 35 polymeric di-, tri-, tetra-, penta- and hexapeptides have been synthesised. Cyclisation is observed as an unwanted side reaction, especially in the polymerisation of activated dipeptide esters:

$$\text{HBr, Pro–Gly–ONp} \xrightarrow{tert\text{-base}} \begin{cases} (\text{Pro–Gly})_n & (10\%) \\ \text{dioxopiperazine} & (90\%) \end{cases}$$

In order to minimise this side reaction, the polymerisation is carried out at high concentration with DMF or DMSO as solvent. At the end of the polymerisation the polymeric product is precipitated with ethanol, dialysed against water and isolated by lyophilisation.

In contrast to homopolymeric amino acids, sequence polymers can be synthesised only up to a molecular weight of about 12000 to 15000. Preliminary rules for the nomenclature of poly-α-amino acids have been published by the IUPAC–IUB Commission[7].

Poly-α-amino acids are valuable synthetic protein models[8] for conformational studies in the solid state and in solution. The X-ray diffraction pattern of poly-(glycyl-alanine), for example, is similar to the silk fibroin of *Bombyx mori*. Some poly-α-amino acids and sequence polymers are suitable substrates for proteolytic and other enzymes and have been used to study enzymic attack on high-molecular-weight peptide chains. There have also been extensive applications in the immunological field.

While on the subject of sequence polymers, the so-called proteinoids have to be mentioned. These are produced by heating amino acids to 140–200 °C in presence of an excess of glutamic acid or aspartic acid. It appears that the amino acid residues in these substances are not distributed randomly but occur in definite arrangements dictated by structure and charge. Proteinoids have considerable interest in connection with the prebiotic evolution of proteins[9].

A remarkable process which leads to sequence polymers occurs in the so-called 'plastein reaction'. In this reaction oligopeptides are condensed by treatment with proteolytic enzymes such as pepsin and chymotrypsin to produce high-molecular-weight polypeptides. At one time only a few oligopeptides, formed by partial enzymatic hydrolysis, were thought to be plastein-active. The phenomenon is, however, now known to be more general than this, the work of, *inter alia*, Determan and Wieland having established the principle features of the structure–activity correlation.

The first synthetic pentapeptide to be discovered which could be condensed in the presence of pepsin was Tyr–Ile–Gly–Glu–Phe. Plastein activity was retained when the sequence of the amino acids in the peptide was changed, but the activity was lost if Phe was replaced by Ala. In the light of this evidence it was concluded that a lipophilic C-terminal amino acid is necessary. Shortening of the peptide chain to three amino acid residues also results in loss of activity, and so does incorporation of a peptide bond formed via the γ-carboxyl group of glutamic acid, possibly because the α-helical structure of the product is interrupted by such a link.

References

1 Bamford, C. A., Elliot, A. and Hanby, W. E. (1956). *Synthetic Polypeptides*, Academic Press, New York

2 Katchalski, E. and Sela, M. (1966). *Advan. Protein Chem.*, **13**, 243

3 Katchalski, E., Sela, M., Silman, H. I. and Berger, A. (1964). In *The Proteins*, Neurath, H., ed., 2nd edn, Vol. II, p. 406, Academic Press, New York

4 Blout, E. R. (1962). *Polyamino Acids, Polypeptides, and Proteins*, Stahman, M. A., ed., University of Wisconsin Press, Madison, Wisc.

5 Goodman, M. and Hutchinson, J. (1965). *J. Amer. Chem. Soc.*, **87**, 3524

6 Johnson, B. J. (1974). *J. Pharm. Sci.*, **63**, 313

7 IUPAC–IUB Commission on Biochemical Nomenclature (CBN) (1967). *Europ. J. Biochem.*, **3**, 129

8 Fasman, D. (1967). *Poly-α-Amino Acids, Protein Models for Conformation Studies:* Vol. 1, Biological Macromolecules Series, Dekker, New York

9 Fox, S. W., Krampitz, G. and Waehnefeldt, T. V. (1967). *Bild Wissenschaft*, 1014

3. Proteins[1–8]

3.1. Significance and Historical Aspects

Proteins are high-molecular-weight substances composed, either entirely or to a very large extent, of amino acids. The majority (80–90 per cent) of the organic compounds of living organisms are proteins. For instance, the cells of *Escherichia coli* contain over 3000 different protein molecules. The metabolism, structure and function of all cells is controlled by proteins. Virtually all chemical reactions that occur in living systems are catalysed by specific enzymes. Enzymes carry out reactions under the extremely mild conditions demanded by natural systems many times faster than they would otherwise proceed. Enzymes, too, do not affect the position of equilibrium but merely the speed of attaining the equilibrium. Other proteins such as the scleroproteins play structural roles. The immune response, the method by which living organisms protect themselves, depends on protein antibodies: the immunoglobulins. Proteins are also important constituents of biological membranes. Carrier proteins act as electron carriers during respiration and photosynthesis; others transport the metabolic products and gases resulting from respiration. The actin-myosin proteins in muscle cells are involved in muscular action. Proteins are components of the blood coagulation system. They take part in differentiation processes and have regulatory functions. Many proteins are hormones; others have toxic properties. This brief outline should serve to indicate the importance of proteins in diverse biochemical functions.

The construction of cell proteins is coded in the genetic material of the cell (DNA); the code determines the sequence of amino acids in the polypeptide chains. A species may contain between a thousand and a million different proteins, depending on its complexity; the individuality of each member of a species is ensured because this protein biosynthesis is controlled by specific, inherited genes.

The separation of a polypeptide chain from the ribosome after synthesis is followed by its arrangement into the correct three-dimensional shape to fulfil its specific biochemical function. The elucidation of these three-dimensional structures is necessary if we are to understand life processes on a molecular scale.

Proteins, carbohydrates and fats are the three major components of human food (see Section 1.1.2.). At the present time, the main supply of nutritional proteins for the developed world is obtained from animals. However, the growing population's demand for protein will outstrip the possibility of meeting the need with animal protein. Animal husbandry is in any case a very inefficient way of using primary food sources. Research into the chemical synthesis of protein is in progress. One area where some success has been achieved is in the microbiological production of protein—for example, from wastes containing carbohydrate, sulphite and melasses.

It has been found that certain yeasts (*Candida, Saccharomyces*) and also some bacteria can utilise petroleum hydrocarbons. Although it seems that some success has been achieved in solving the technical problems of biological deparaffination, the safety of the use of such types of proteins has not yet been proved fully.

Another source of protein which has not been fully exploited is the green parts of plants, especially the leaves. Proteins in leaves constitute about one-third of the dry weight and can be extracted from the plant material in a yield of 50–60 per cent. The utilisation of plant proteins by animals results in a degree of utilisation of only about 18 per cent. Leaf proteins have a high essential amino acid index. Their lysine content amounts to 5–6 per cent, which is comparable to that of soya protein, which is the basis of artificial foods at present. Japan has paid special attention to the use of soya protein, fish protein and algal protein. Other countries have directed their efforts to the cultivation of improved varieties of wheat and other cereal crops, but all these attempts will have to be reinforced in order to solve the world's food problems.

De Fourcroy (1784) was the first to state that proteins constituted a special class of substances. The name 'protein' comes from the Greek 'proteuo' ('I occupy the first place') and was originally given by Berzelius.

After the first experiments of Kühne in which he attempted to clarify the structure of proteins using enzymatic methods, Kossel succeeded, at the end of the nineteenth century, in isolating a number of proteins. At the same time, Hofmeister and E. Fischer recognised that proteins were chains of amino acid units, and this was proved by synthetic work carried out by the latter. Hofmeister, working with egg albumin, was the first to extract a protein in crystalline form. Abel (1925) succeeded in crystallising insulin and 10 years later Sumner described the crystallisation of urease. Stanley (1935) obtained tobacco mosaic virus in crystalline form. Th. Svedberg (1925–1930) succeeded in determining the molecular weight of various proteins with the aid of the ultracentrifuge. The development of new analytical techniques—for example, electrophoresis (Tiselius, 1937) and improved chromatographic methods—led to further advances in this field.

The determination of the amino acid sequence of insulin by Sanger (1951–1956) was a major advance. The methods which were developed in this work formed the basis for the systematic elucidation of the primary structure of many other proteins. More recently, the sequencer developed by Edman (1966), the application of mass spectroscopy for sequence analysis and the use of computers to evaluate the mass spectroscopic data have all played their part, and over 15000 publications on sequence-analysis have now appeared and the primary structures of more than 350 proteins have been determined.

After 1945, systematic research began on the secondary structure of proteins, their three-dimensional spatial shape. Following on the fundamental work of PAULING and COREY on the conformation of peptide chains, KENDREW and PERUTZ elucidated the structure of myoglobin and haemoglobin using X-ray diffraction. Structural data obtained *in vitro* were then able to provide clues to the biochemical functions of proteins *in vivo*.

PHILLIPS was the first to describe the structure of an enzyme, lysozyme: this served to clarify, at a molecular level, the mechanism of binding and of splitting of the substrate in the lysozyme reaction.

3.2. Classification and Natural Occurrence

As we have seen, some species have up to a million different protein molecules, so no complete systematic classification of these proteins on a structural basis is possible. Therefore, proteins are classified in a variety of ways —for example, according to the source of the proteins (plant proteins, animal proteins, virus proteins, bacterial proteins), or according to their occurrence in different organs (plasma proteins, muscle proteins, milk proteins, egg proteins, etc.) or in cell organelles (ribosomal proteins, nuclear proteins, microsomal proteins, membrane proteins, etc.). The general biological function can be used as another criterion (for example, enzyme proteins, immunoproteins, structure proteins, carrier proteins, storage proteins, etc.).

The earliest classification of proteins was based essentially on differences in solubility and molecular shape:

Globular proteins	*Fibrous proteins*
soluble in water and dilute salt solutions; have a globular molecular shape	insoluble in water and salt solutions; fibrous structures at the macroscopic level; resistant to acids, alkalis and proteases

An early classification based on chemical composition was into two types: simple or pure proteins and conjugated proteins (proteids). Total hydrolysis of a simple protein yields only amino acids, while total hydrolysis of conjugated proteins yields amino acids and other components.

Some of the original classifications as simple proteins have been proved wrong by modern analytical techniques, but we will nevertheless use this classification because protein nomenclature has developed alongside it.

References

1 *Advances in Protein Chemistry*, Academic Press, New York, annually from 1944
2 NEURATH, H. (1963). *The Proteins*, Academic Press, New York
3 HAUROWITZ, F. (1963). *The Chemistry and Function of Proteins*, Academic Press, New York
4 LEGGETT-BAILEY, J. (1969). *Techniques in Protein Chemistry*, Elsevier, Amsterdam
5 FASOLD, H. (1972). *Die Struktur der Proteine*, Verlag Chemie, Weinheim/Bergstraße
6 DICKERSON, R. E. and GEIS, I. (1969). *The Structure and Action of Proteins*, Harper and Row, New York
7 LEACH, S. J. (Ed.) (1970). *Physical Principles and Techniques of Protein Chemistry*, Part B, Academic Press, New York
8 LÜBKE, K., SCHRÖDER. E. and KLOSS, G. (1975). *Chemie und Biochemie der Aminosäuren, Peptide und Proteine*, Georg THIEME Verlag, Stuttgart

3.2.1. Simple or Pure Proteins

3.2.1.1. Albumins

Albumins are readily crystallisable, low-molecular-weight proteins which are easily soluble in water and diluted salt solutions in the pH range 4 to 8.5. They can only be precipitated by a high ammonium sulphate concentration (70–100 per cent saturation). Their acidic character is due to their high content (20–25 per cent) of glutamic and aspartic acids.

Important representatives of this group are serum alumin, α-lactalbumin, ovalbumin of egg-white and the plant albumins.

Serum albumin (M.W. 67000) is the main component of plasma proteins occurring to the extent of about 60 per cent. Its most important physiological function is the regulation of the osmotic pressure of blood. It exhibits a high binding capacity for K^+, Na^+ and Ca^{2+} ions as well as for hormones, fatty acids and drugs. Human serum albumin consists of a polypeptide chain with 579 amino acids and contains 17 disulphide bonds.

α-Lactalbumin occurs in the milk of mammals. It is characterised by its high heat-stability. Unlike serum albumin, α-lactalbumin contains an oligosaccharide which is attached to the polypeptide chain through the β-carboxyl group of an aspartic acid residue. Human α-lactalbumin (M.W. 14176) has 123 residues and contains 4 disulphide bridges. The primary structure not only coincides with the α-lactalbumins of other mammalian species but shows also a great similarity to the structure of lysozyme. Structural analysis shows an α-helix content of 26 to 30 per cent, while 14 per cent of the amino acid residues occur in β-structures. It has been shown that the α-lactalbumin polypeptide chain is folded in a similar way to that of lysozyme. It is possible that the α-lactalbumin arose by modification of lysozyme during evolution.

Ovalbumin (M.W. 44000) contains, in addition to 3.2 per cent carbohydrates, a phosphoric acid residue which is attached through a serine side chain.

Plant albumins occur in the seeds in small amounts, e.g. ricin (in castor seeds), leucosin (in the seeds of various cereals) and legumelin (in legominosae). Ricin and abrin (from the seeds of *Abrus precitorius*) are toxic proteins and are, therefore, of special interest.

Ricin (M.W. 65000) comprises 493 amino acid residues. It consists of two chains: the A-chain (M.W. 30000) and the B-chain (M.W. 35000) which are linked through disulphide bridges. Although the A-chain is the toxic component, the B-chain is significant as it acts as binding centre for the surface of the cell. Ricin and the structurally similar abrin act as inhibitors of protein biosynthesis and show antitumour activities (inhibition of the reproduction of EHRLICH ascites tumour cells).

3.2.1.2. Globulins

Globulins have higher molecular weights than the albumins and occur in practically all animal and plant cells. To this heterogeneous and extensive class of proteins belong numerous enzymes, plasma proteins, antibodies, milk proteins, many glycoproteins and plant storage proteins. Since many globulins contain carbohydrates firmly attached to them, there is no agreement over the exact classification of these compounds.

Globulins are insoluble or sparingly soluble in salt-free water, and a solubility minimum is observed at the isoelectric point. They are soluble in salt solutions and can be fractionally precipitated with ammonium sulphate.

β-Lactoglobin (M.W. 36726) is the most important protein component of cow's milk (3 g/l). It comprises two identical polypeptide chains of 162 amino acid residues: the β-lactoglobin of the pig, on the other hand, has one chain only. The α-helix content of β-lactoglobin is low (10–17 per cent). In addition, 24–42 per cent of the residues in β-lactoglobin occurs in β-structures. The reaction of β-lactoglobin with ϰ-casein results in the formation of a heat-stable casein micelle which prevents the precipitation of milk proteins (see p. 250).

The *plasma globulins* are classified, according to their electrophoretic behaviour, as α_1-, α_2-, β- and γ-globulins (see Figure 38). The γ-globulins act as antibodies.

Antibodies are immunoglobulins produced by the lymphoid cell system in response to invasion by a foreign species, the antigen. The immunoglobulins react specifically with the antigen. All immunological reactions involve this same primary step, the combination of antigen with antibody. Precipitation, agglutination or other more complicated reactions may occur in subsequent steps. The general organisation of the immunoglobulins is shown in Figure 30. The complete immunoglobulin is composed of a pair of light (L) chains, molecular weight 23000, and a pair of heavy (H) chains. There are five types of heavy chain (γ-, α-, μ-, ∂- and ε-types) and two types of light chain ($\varkappa$- and λ-types) normally present in immunoglobulins, which are therefore classified into five main groups (IgG, IgA, IgM, IgD and IgE), each of these groups being subdivided into two, depending on the light chain.

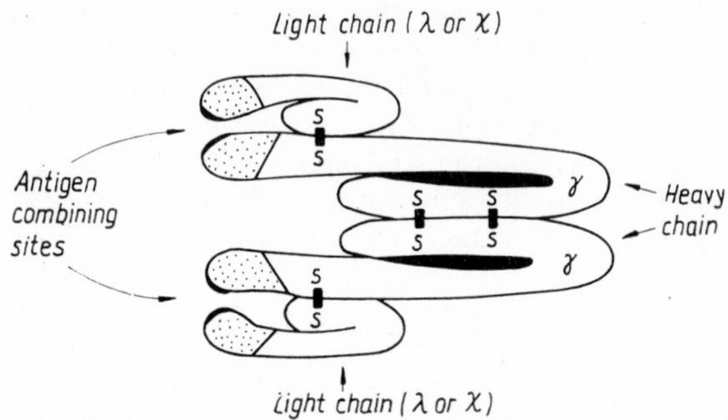

Figure 30. Structure of the immunoglobulin IgG

The α_2-*macroglobulin* (α_2-fraction) is of special interest as it is a natural inhibitor of plasmin; it also inhibits other enzymes, such as elastase, trypsin and chymotrypsin. It contains 8.2 per cent carbohydrates, has a molecular weight of 820000 and comprises 10 identical sub-units.

The *haptoglobin* is a tetramer consisting of two non-identical pairs of chains (α_2 and β_2) which are linked through disulphide bridges. The primary structures of the three genetic variants (Hp 1—1, Hp 2—2 and Hp 2—1) of the human haptoglobin are known. Haptoglobin forms a complex with oxyhaemoglobin. This results in a change in the conformation of the porphyrin ring, which can then be opened by encyme action.

Edestin, arachin, zein (maize) and tuberin (potatoes) are examples of vegetable globulins.

Edestin is a protein of hemp seeds with a molecular weight of 300000. It consists of six sub-units, each constructed of two non-identical polypeptide chains, attached to each other through disulphide bonds. *Arachin*, the peanut protein, has a hexameric structure. Its sub-units have two polypeptide chains bound covalently to each other. *Glycinin*, the biologically important storage protein of soya beans, is structurally similar to arachin. *Excelsin*, the protein of Brazil nut, is structurally related to edestin.

3.2.1.3. Histones

Histones are low-molecular-weight, basic proteins (high content of lysine and arginine) which are soluble in water and acids. These proteins are tissue-non-specific and occur in all eucaryotes, where they form reversible complexes with DNA (nucleohistones). They play an important role as non-specific gene-repressors.

The number of amino acid residues ranges between 101 and 212 and the molecular weight between 11200 and 21000. In nearly all histones various methylated and acetylated amino acid residues and also phosphate groups occur. Histone II b_1, for instance, is composed of 129 amino acids, has a molecular weight of 14000 and contains 14 lysine and 12 arginine residues; the N-terminal serine occurs in the acetylated form. Histone II b_2 contains 125 amino acids, including 20 lysine and 8 arginine residues: the N-terminal proline is not substituted. Out of the 116 amino acid residues of histone I, 45 are lysine and 16 are proline. Histone IV contains 102 amino acids, while histone III comprises 135 amino acids. In histone III two cysteine residues and various methylated and acetylated lysines have been detected.

3.2.1.4. Prolamines

The prolamines are characterised by a high content of glutamic acid (30 to 45 per cent) and proline (about 15 per cent). The nomenclature of this class of proteins is derived, as the name indicates, from their high proline content. They are insoluble in water and salt solutions, but soluble in 50–90 per cent ethanol. Gliadin (wheat and rye) and hordenin (barley) are important cereal protein representatives of this class: they contain low amounts of essential amino acids. Rice and oats do not contain any prolamines.

3.2.1.5. Protamines

These proteins are more basic than the histones as a result of their high content of arginine (80–85 per cent). They are soluble in water and acids. They are found in the cell nucleus combined with DNA. They function, like the histones, as gene-repressors. The known protamines have relatively low molecular weights ranging from 4000 to 4500. Protamines have been mainly isolated from fish-sperm. *Clupeine* (isolated from herring) consists of 30 amino acid residues. Their primary structures often show repetition of the same sequence. *Salmin A—I*, for example (isolated from salmon), has the following sequence: Pro–Arg_4–Ser_3–Arg–Pro–Val–Arg_5–Pro–Arg–Val–Ser–Arg_6–Gly–Gly–Arg_4.

3.2.1.6. Glutelines

The name of these proteins stems from their high content of glutamic acid (up to 45 per cent). They are insoluble in water, salt solutions and dilute ethanol, although they are soluble at both high and low pH values. *Glutenine* (wheat), *orzynine* (rice), *hordenin* (barley) are important representatives of this group. Glutenine and gliadin constitute the sticky protein (gluten) of wheat- and rye-flour. The ability of these flours to be baked into bread is due to their gluten content.

3.2.1.7. Fibrous Proteins

These proteins, which are also known as scleroproteins, are proteins which are practically insoluble in water and salt solutions. They are highly resistant to alkali, acids and most proteases. Collagen, elastin, keratin and fibroin are important representatives of these structural proteins (see Section 3.6.4.).

3.2.2. Conjugated Proteins

Conjugated proteins are classified according to their non-amino acid components, which are essential for their function and are attached to them by covalent, ionic or coordinate links.

3.2.2.1. Phosphoproteins

Phosphoproteins contain phosphoric acid groups, which are usually attached by ester bonds to the hydroxyl side chain functions of serine or, more rarely, of threonine.

The *caseins* are important phosphoproteins which occur at a high concentration in cows milk (25 g/l), constituting about 79 per cent of the milk proteins. The most important representatives are the α-, β- and $\varkappa$-caseins.

$\varkappa$-Casein is a single chain molecule with 169 amino acids and a molecular weight of 19093. On coagulation of milk only the Phe_{105}–Met_{106} bond is cleaved by the limited proteolytic activity of rennet enzymes to form *para*-$\varkappa$-casein and a macro-peptide. The sequence 1–105 is hydrophobic, while the C-terminal part comprising the other 64 amino acids is hydrophilic in character.

The six structurally similar α_S-caseins (α_{S0}–α_{S5}) are also single chain proteins, with the exception of α_{S5}-casein, which is composed of α_{S3}- and α_{S4}-monomers. Of the 199 amino acids present in α_{S1}-casein B, 16 are serine residues and 50 per cent of these are phosphorylated. The sequence of β-casein A^2 (M.W. 24400), which consists of 209 amino acids, has been elucidated. γ-Casein is formed from β-casein by the limited action of milk protease: it contains the 29–209 sequence of the β-casein. The low α-helix content of caseins is due to their high proline content. The micelles are formed from α_{S1}-, β- and $\varkappa$-caseins in the molar ratio 3: 2: 1. $\varkappa$-Casein is not precipitated by Ca^{2+} ions. Its C-terminal sequence forms the surface of the micelles. The $\varkappa$-casein functions as the stabiliser of milk on account of its heat stability (see p. 246) and its resistance towards the action of Ca^{2+} ions.

Phosphovitin and *vitellin* from egg-yolk are also phosphoproteins. Fifty per cent of the amino acid residues of phosphovitin are phosphoserine. Although vitellin is the principal protein of egg-yolk, it contains only one-tenth of the phosphate content of phosphovitin. Ovalbumin also belongs to the phosphoproteins (see p. 247).

3.2.2.2. Chromoproteins

These proteins have a chromophoric prosthetic group. The haemoproteins, which contain haem groups, are important representatives of this class.

Haemoglobin (see p. 307) consists of four haem groups and four polypeptide chains (two α-chains with 141 amino acids and two β-chains with 146 amino acids). Molecular oxygen transported by haemoglobin from the lungs is bound to the iron of the haem group and to the myoglobin in the muscles. *Myoglobin* contains one haem group and one polypeptide chain with 153 amino acids. Oxygen is stored by myoglobin until it is consumed in oxidative metabolic processes. Haemoglobin is also responsible for the reverse transport of the oxidation product, CO_2, to the lungs. The electron carrier protein *cytochrome c* contains a haem group and one polypeptide chain with 104 amino acids, but in contrast to myoglobin, the haem group in cytochrome c is covalently attached to the apoprotein by thioether bridges via two cysteine side chains.

The enzymes catalase and peroxidase are also haemoproteins.

3.2.2.3. Lipoproteins

Lipoproteins are physiologically important protein–lipid complexes. They occur primarily in blood plasma, egg-yolk, cell membranes and cell organelles. The lipoproteins of plasma are responsible for the transport of neutral fats and fat-like substances. Arteriosclerosis, found generally in old age, is attributed to the deposition of cholesterol in blood vessels. Therefore, the carrier proteins for cholesterol and cholesterol esters, as well as the metabolism of lipoproteins, are subjects of special clinical interest.

The largest lipoproteins, the chylomicrons, are 5000 Å in diameter. Their occurrence in blood causes lipoidaemia. The separation and isolation of lipoproteins can be accomplished by density-gradient centrifugation. Their lipid content, which varies from 40 to 95 per cent, is the cause of their relatively low density. With the exception of the very high-density lipoproteins they float and do not sediment on centrifugation, moving to the upper surface of the centrifuge tubes. Accordingly, the sedimentation coefficient is replaced by the floation coefficient, which is a characteristic standard for lipoproteins.

Studies on the primary structure of fat-free lipoproteins (that is, the apoliproteins) indicate that the protein components usually consist of two or more low-molecular-weight polypeptide chains.

3.2.2.4. Glycoproteins

Glycoproteins are carbohydrate–protein complexes in which short oligosaccharide chains are glycosidically attached through the side chain functions of certain amino acids. Several modes of attachment are known—for example, *O*-glycosidic linkage (to Ser, Thr), *N*-glycosidic linkage (to N-terminal amino groups, ε-amino groups of Lys, amide nitrogen of Asn) and ester-glycosidic attachment (Glu, Asp). The carbohydrate components are usually composed of hexoses (mainly D-galactose, D-mannose, D-glucose), with *N*-acetyl-hexosamines, as well as L-fucose or sialic acid (*N*-acetylneuraminic acid) as terminal components of the chains. The carbohydrate chains of glycoproteins are usually quite short, consisting of perhaps 8 or 10 individual saccharide residues.

Glycoproteins containing only one carbohydrate chain—for example, ovalbumin (see p. 247)—as well as those with 800 chains (sheep submaxillary glycoprotein) are known.

The carbohydrate chains are introduced by the action of glycosyl-transferases. The lack of genetic control on this stage sometimes results in microheterogeneity.

Glycoproteins are widely distributed in plants and animals. They are found as components of membranes, enzymes, antibodies, blood group substances, mucilages, complement factors, hormones and plasma proteins. They occur in all plasma fractions with the exception of the albumin fraction: their molecular weights range between 4000 and 1 million.

The acidic *α_1-glycoprotein* (*orosomucoid*) is one of the best-studied glycoproteins. Orosomucoid (M.W. 41000) consists of one polypeptide chain with 181 amino acids. It has a C-terminal sequence very similar to that of the α-chain of haptoglobin as well as to that of the H-chain of immunoglobin G. Orosomucoid has the highest carbohydrate content found among the plasma proteins. The five carbohydrate chains are attached to the asparagine residues at positions 15, 37, 53, 74 and 84.

The glycoproteins perform a number of different functions. On the cell surface they take part in the active transport across membranes, and act as carrier proteins for certain metal ions (Fe^{3+}, Cu^{2+}). The high viscosity of glycoprotein solutions gives them lubricating properties. They are the mucilagenous substances found in gland secretions, articular fluids and subartaneous connective tissue.

3.2.2.5. Nucleoproteins

Nucleoproteins are nucleic acid–protein complexes of a heteropolar nature which contain protamines, histones and non-histone proteins of the chromosomes as their protein components. DNA is the main nucleic acid in these proteins. Nucleoproteins are of prime importance in DNA-replication and gene control.

Tobacco mosaic virus provides the best-studied example of virus-envelope protein functioning as a nucleoprotein. Tobacco mosaic virus was the first virus to be isolated in crystalline form (W. STANLEY, 1937). It occurs in the form of cylindrical rods 300 nm in length and 17 nm in diameter.

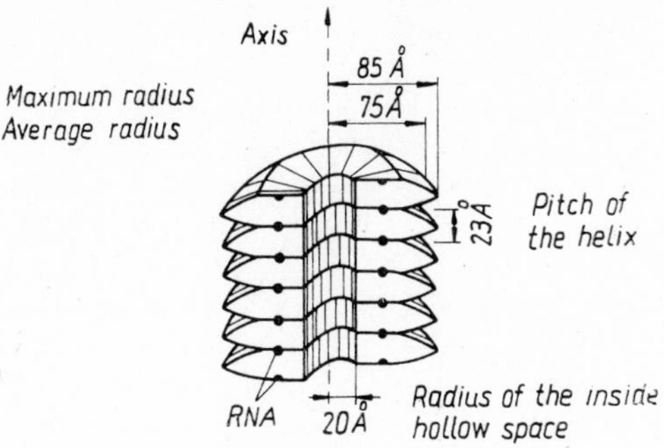

Figure 31. Tobacco mosaic virus according to R. FRANKLIN

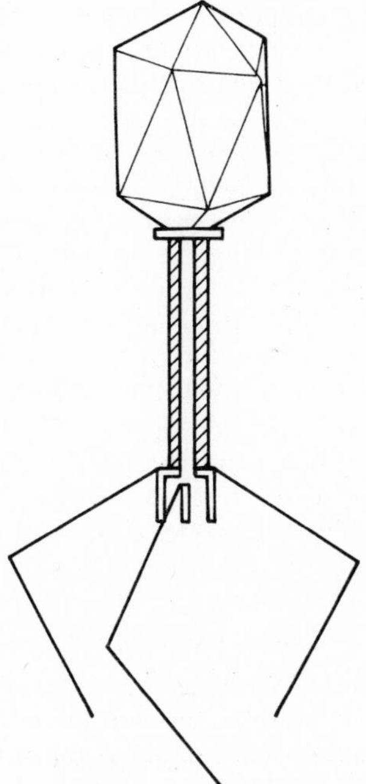

Figure 32. Bacteriophage T_4

The envelope protein of tobacco mosaic virus consists of a hollow cylinder. 2130 protein sub-units are arranged in a large spiral, wound around the hole in the cylindrical rod. Each sub-unit consists of a polypeptide chain with 158 amino acids. The ribonucleic acid is present as a long chain which is embedded between the turns of the spiral (Figure 31). Other viruses and phages, however, show more complicated structures than the tobacco mosaic virus.

In *bacteriophage* T_4 the envelope protein has the shape of an elongated icosahedron. The head of the bacteriophage is attached through its collar and neck to a tail of complex structure (Figure 32). The DNA of the head has a molecular weight of about 130×10^6 and occurs in a highly organised state. When bacterial cells are infected by the bacteriophage, the tail is attached to a specific receptor in the cell wall. The tip of the tail then penetrates the cell wall by a muscle-like contraction and the DNA enters the cell.

3.2.2.6. Metalloproteins

Certain metal ions are often bound to proteins in the form of complexes and are found as functional components in a great number of enzymes. Many oxidases (for example, phenoloxidases, cytochrome oxidase, ascorbic acid oxidase) are copper-containing protein. Mn^{2+} ion acts as a specific activator of the glutamic synthetase (from *Escherichia coli*) and also serves to stabilise its dodecameric structure. The glycolysis and proteolysis enzyme of tissues are manganese-containing proteins. In recent years, molybdenum enzymes have received a great deal of attention. Carbonic anhydrase is an example of a zinc-containing enzyme: it contains one equivalent of zinc.

Ferritin and *haemosidirin* are typical iron storage proteins of mammals. Together they contain 25 per cent of the iron of the whole organism. The iron-free apoferritin has a plate-shaped structure and consists of 24 sub-units. Iron occurs in ferritin in the form of hydroxide-oxide micelles, in which up to 4300 Fe^{3+} atoms can be included. In this form excess iron can be stored intracellularly (for example, in liver and bone-marrow) and mobilised when needed. In certain diseases iron is lost (for example, in the form of hemosidirin), mainly from the liver.

3.3. Physicochemical Characteristics[1–3]

3.3.1. Amphoteric Character

The amphoteric nature of proteins is chiefly the result of the number and distribution of the available basic and acidic side chain functions since the terminal amino and carboxyl groups in a long peptide chain, in contrast to amino acids, contribute relatively little dipolar character.

Proteins have a positive charge in acidic media and a negative charge in alkaline media and their degree of hydration and their solubility are greatest at pH extrema. The difference between the positive and negative absolute charges is the factor which determines the hydration of the molecule, not the charge itself. In contrast, the electrophoretic mobility is determined by the charge. At the isoelectric point (IEP) the protein has equal positive and negative charges which nullify each other. The minimum values of solubility and hydration are reached at the IEP. Determination of the IEP can be carried out (a) by determination of the solubility minimum in different buffer solutions, (b) electrophoretically at different pH values or (c) by electrofocusing. Proteins containing a high proportion of basic amino acids have their IEP in the alkaline range (for example, protamines, IEP = 11.8). Pepsin (IEP = 1) is a typical example of a protein with an IEP on the extremely acidic side. The IEP generally depends on the ionic strength and on the type of buffer used.

The physiologically important buffer effect of proteins depends on the equilibrium between

$$\text{Protein cation} \rightleftharpoons \text{Neutral protein} \rightleftharpoons \text{Protein anion}$$

Haemoglobin plays an important role in the pH-stabilisation of blood. A normal blood pH value is in the range 7.35 to 7.40: a decrease of 0.3 to 0.5 pH units puts life in danger.

3.3.2. Solubility

The solubility of proteins depends on several factors. These include pH, the nature of the solvent (dielectric constant), the electrolyte concentration (ionic strength) and the type of ions in the medium. The character of the protein as well as its structural features also play an important role in solubility. The different solubilities of globular and fibrous proteins originally formed a basis for their classification.

The addition of slightly polar or non-polar solvents like ethanol or acetone to an aqueous protein solution lowers the dielectric constant of the solvent system. As a result, a drop in hydration and solubility occurs, and if enough solvent is added, the protein is precipitated. The electrolyte concentration is of great importance in the solubility of proteins. A certain level of salt concentration is often needed to keep proteins with markedly asymmetric charge distribution (for example, serum albumins) in solution.

This *salting-in effect* is attributed to interference with the association or aggregation of protein particles. The ions of the salt accumulate on the protein molecule and strongly enhance its degree of hydration. This prevents reassociation of the protein molecules and enhances the solubility.

The *salting-out effect,* which leads to protein precipitation, is attributed to repression of protein hydration as a result of the hydration requirements of a large excess of electrolyte. Ammonium sulphate is generally used for salting-out because it is very soluble in water and its ions are more strongly hydrated than those of, for example, sodium chloride. Since different proteins are precipitated by different electrolyte concentrations, the salting-out method is an important, mild method for the preliminary separation of protein mixtures.

3.3.3. Denaturation[4]

Denaturation of proteins is defined as those changes brought about in native proteins by physical and/or chemical means which lead to a more or less complete loss of the biological activity or other individual characteristics of the protein in question while leaving the original primary structure unchanged. During denaturation hydrogen bonds are largely destroyed. In denaturation in presence of reducing agents, disulphide bonds are also cleaved. Denaturation is often reversible and, after removal of the denaturating effects, a return to the original native conformation is possible. This process is known as renaturation. Renaturation is also possible after reductive denaturation. However, there are many well-known cases in which irreversible denaturation takes place.

The transition from the native low-energy state to the denatured form is accompanied by increased disordering of the chain and, therefore, by an

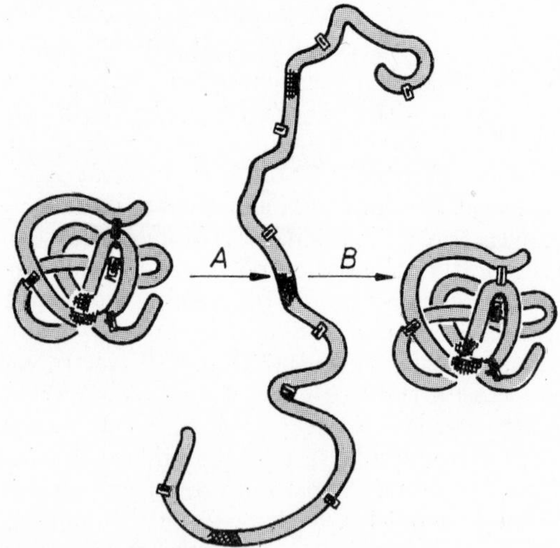

Figure 33. Renaturation as envisaged by C. B. Anfinsen *et al.* (1963): A, Unfolding of a native protein with four disulphide bonds (|||) in presence of 8 M urea and β-mercaptoethanol; B, refolding of the reduced protein (| = SH group) after removal of urea and β-mercaptoethanol and following spontaneous reoxidation

increase in entropy. However, the ordering of the neighbouring water molecules is increased as a result of hydration of the liberated hydrophobic groups of the amino acid side chains, so that the entropy increase is greatly over-compensated. During denaturation other changes also occur—for example, in solubility. Moreover, the isoelectric point may be shifted as other ionised groups become exposed. Unfolding of the peptide chains may lead to an increase in viscosity and also to changes in the UV absorption. The aggregation and precipitation which are often observed on denaturation of proteins is due to changes in hydration and solubility behaviour, and also to disulphide exchange. However, precipitation of a protein is no sure sign that it has been denatured, since salts and organic solvents can precipitate proteins from aqueous solution without affecting the secondary or tertiary structure. The most important physical methods of denaturation arestrong stirring, shaking, heating, UV, X-ray and radioactive irradiation and ultrasonic treatment. Chemical denaturation can be affected by compounds that are capable of breaking hydrogen bonds—for example, 6–8 M urea or 4 M guanidine solutions—by acid or base treatment (pH value under 3 or over 9) and also by treatment with 1 per cent sodium dodecylsulphate and other detergents. Individual proteins vary considerably in their sensitivity towards denaturation. Circular dichroism, rotatory dispersion and absorption spectrophotometry have all been used to follow the course of protein denaturation.

3.3.4. Molecular Weights[5–9]

The molecular weights of proteins vary from 10^4 to 10^5 in onechain molecules, from 5×10^4 to several millions in the majority of poly chain (oligomeric) proteins. A variety of physicochemical methods may be used to determine the molecular weight and molecular shape of a protein. These methods include determination of viscosity, estimation of diffusion and sedimentation rates in the ultracentrifuge, determination of electrophoretic and gel chromatographic migration rates, the measurement of light-scattering and the determination of osmotic pressure.

The light-scattering methods depend on the fact that an increase in particle size is accompanied by an increase in the Tyndall effect: the light-scattering is measured by comparison of the intensity of the incident light with that which is scattered at right angles—under ideal conditions the difference between the scattering of the pure solvent and the protein solution is directly proportional to the number and size of the protein molecules.

The osmotic pressure can be measured with a vapour pressure osmometer (M.W. $<$ *ca.* 20 000) or a membrane osmometer (M.W. $>$ *ca.* 20 000).

In the membrane osmometer, the osmotic pressure is shown by the difference of hydrostatic pressure between test and comparison capillaries. The time required to attain equilibration can be shortened by the use of dynamic osmometry, in which the rate of flow of water into or out of the osmometer is measured at different external pressures and can be automatically compensated. However, this method, although of historical importance, is only rarely used nowadays.

There are two methods of molecular weight determination using the *ultracentrifuge*: the sedimentation-velocity technique in which the rate of sedimentation is determined, and the sedimentation equilibrium technique, in which the sedimentation equilibrium is investigated.

The ultracentrifuge consists of a rotor which is driven by a high-speed electric motor and which holds cells containing the protein solution under investigation. The rotor compartment is refridgerated and may be evacuated, and an optical system is provided for measuring the protein concentration at each point in the cell throughout the duration of a run. The first ultracentrifuge was constructed by the Swedish chemist Th. SVEDBERG and his associates in 1925.

For the sedimentation-velocity technique, gravitational fields which ensure complete sedimentation are necessary. The sedimentation velocity is directly proportional to the molecular weight. In a sedimentation-velocity experiment the protein solution is subjected to a centrifugal field of about up to 500 000 times the force of gravity, which may necessitate a speed up to 70 000 rev/min for the ultracentrifuge rotor. The changes in concentration which occur during centrifugation are followed with the aid of optical methods, such as the Schlieren method or the RAYLEIGH interference method. The Schlieren method enables the refractive index gradient to be measured at each point at the cell. In consequence of the diffusion of the protein molecules, as the protein moves down the cell, the peak height diminishes and the peak width increases. A typical sedimentation pattern is shown schematically in Figure 34.

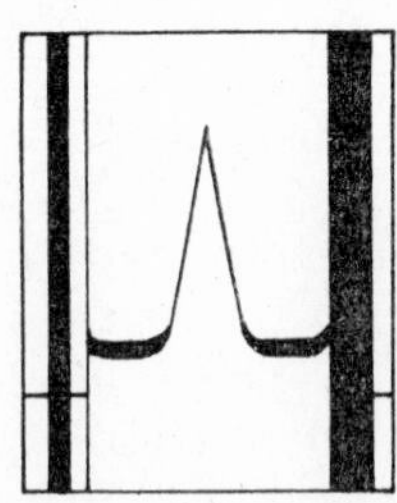

Figure 34. Sketch of an ultracentrifuge Schlieren photograph of a sedimenting peak

The RAYLEIGH interference method records concentration distributions within the cell in terms of interference fringes.

The basic equation employed to obtain the molecular weight is

$$M = \frac{R \cdot T \cdot S}{D(1 - \varrho L V_{\text{prot.}})}$$

where S = sedimentation constant or SVEDBERG unit (dimension $10^{-13}\,s^{-1}$) corrected with regard to the viscosity and density of water at 20 °C; R = gas constant; T = absolute temperature; $V_{\text{prot.}}$ = partial specific volume of protein; ϱL = density of the solvent; and D = diffusion constant.

The value of $V_{\text{prot.}}$ can be calculated from the density of the solution. In order to determine the molecular weight the diffusion constant D must be known. D can be calculated theoretically for globular molecules. However, since protein molecules are generally not globular, D is best determined by carrying out a second centrifugation at lower velocity and measuring the extension of the peak attributed to diffusion.

A knowledge of the diffusion constant D is not necessary when determining the molecular weight by the *sedimentation equilibrium technique.* Compared with the sedimentation velocity technique in which the protein solution is centrifuged in a field of about 400 000 to 500 000 g, a centrifugal field of only 10 000 to 15 000 g is sufficient for equilibrium centrifugation.

Under these conditions the rate of sedimentation of the protein molecules away from the centre of rotation is of the same order as the rate of diffusion of molecules from high-concentration regions near the cell bottom to the lower-concentration regions nearer the centre of rotation. Over a period of hours or days, this leads to an equilibrium state where the flow of particles is zero. The molecular weight can then be calculated from the concentration gradient which is set up. The disadvantage of the method is the long time taken to attain equilibrium. This can be avoided in the ARCHIBALD *method,* which is an example of a transient-state sedimentation method. In this low-speed method, the concentration gradient formed at the meniscus of the cell can be used for the estimation of molecular weight. The *zero meniscus concentration method* described by YPHANTIS (1964) enables the sedimentation equilibrium to be reached at high velocities. By this means, the centrifugation time can be shortened to about 2–4 hours. The *density gradient centrifugation method* of MARTIN and AMES (1961) is carried out in sugar gradients of increasing density under high-speed conditions. The distance of migration of the protein in these gradients is inversely proportional to its molecular weight. The molecular weight of an unknown protein can be determined with a reasonable degree of accuracy by comparison with standard proteins of known molecular weight.

An alternative means of measuring protein molecular weights is the *gel-filtration technique.* The most suitable materials are cross-linked polysaccharides, such as the different Sephadex types (G-50, G-100, G-150, G-200), and polyacrylamides (Bio-Gel P-100, P-150, P-300). The chromatography may be performed either in columns or on thin layers. According to the pore size of the gel used, the rate of diffusion of the molecules in the gel when treated with a specific solvent, depends on their particle size (molecular weight). Salts and low-molecular-weight compounds usually penetrate the gel easily, while proteins do not enter the pores easily, and are therefore found in earlier fractions of the eluate. Similarly, large proteins are eluted before smaller ones. Molecular weights can be cheaply and easily determined by this method using standard curves obtained from elution diagrams of proteins of known molecular weight. The accuracy obtained is about 5–10 per cent. *Thin layer gel chromatography* is similar in principle and has the advantage of requiring only small amounts of material (Figure 35).

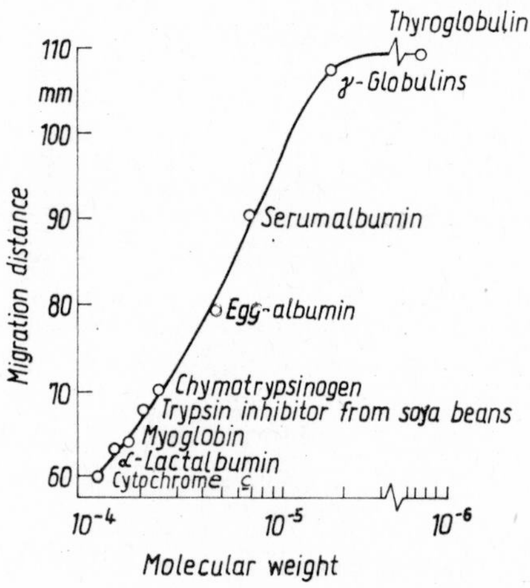

Figure 35. Molecular-weight determination of proteins by thin layer gel chromatography

Polyacrylamide gel electrophoresis is a variant of the method known as zone electrophoresis. The monomers of the gel are dissolved in buffer and are allowed to polymerise simultaneously with bisacrylamide in glass-tubes or in slabs to obtain a gel 7–10 cm in height. The sharp separations that may be obtained by this method are attributed to a 'molecular-sieve' effect. When this electrophoretic process is carried out in the presence of sodium dodecylsulphate, oligomeric proteins can be separated into their sub-units and their molecular weights can be estimated. The molecular weight is determined by comparison of the migration rate with that of standards and the accuracy is 5–10 per cent. The advantages of the method are its rapidity (2–4 hours) and sensitivity requiring only 10–50 μg of protein.

The polyacrylamide gel electrophoresis method can be modified to improve the sharpness of the bands by connecting short layers of polymeric material with greater mesh size. In this column technique, called *disc electrophoresis*, proteins are separated into sharp discs.

Many proteins tend either to aggregate or to dissociate into their subunits, and therefore the exact definition of molecular weight is not always easy. Interpretation of the results obtained using physicochemical means is often difficult. Normally the weights of the different fractions are added together and divided by the number of particles in the solution. Results obtained from ultracentrifuge measurements are average values.

3.3.5. Molecular Shape

There are various methods for the determination of the molecular shape of proteins. Since most proteins are not spherical, globular proteins, the axial ratio f/f_0 may be used to characterise their shape:

$$f/f_0 = a \Big/ \left(\frac{3V_{\text{prot.}} \cdot M}{4\pi N}\right)^{1/3}$$

where f = molar frictional coefficient;
f_0 = frictional coefficient of a globular molecule;
a = STOKES radius;
N = AVOGADRO's number;
$V_{\text{prot.}}$ = partial specific volume of protein;
M = molecular weight.

If the Stokes radius is known, the axial ratio can be calculated from the sedimentation coefficient obtained from ultracentrifuge measurements. The STOKES radius can be determined from K_D, the exclusion constant of gel chromatography, using the equation:

$$K_D = (1 - a/r)^2[1 - 2.104(a/r) + 2.09(a/r)^3 - 0.95(a/r)^5]$$

(r = effective diameter of pores)

The axial ratio of ellipsoidal protein molecules is in the range of 2 to 30 in case of most globular proteins, while being more than 30 in case of fibrillar proteins. The molecular shape can also be determined by measurement of viscosity. Direct estimation of the molecular shape by electron microscopy is also possible. The negative staining technique using osmium tetroxide or other heavy metals as contrast agents is frequently used. However, electron microscopy normally needs desiccated preparations, and this precludes the examination of proteins in their native hydrated state.

References

1 Leach, S. J. (1970). *Physical Principles and Techniques of Protein Chemistry*, Academic Press, New York

2 Tanford, C. (1961). *Physical Chemistry of Macromolecules*, Wiley, New York

3 Edsall, J. T. and Wyman, J. (1958). *Biophysical Chemistry*, Vol. 1, Academic Press, New York

4 Tanford, C. (1968). *Advan. Protein Chem.*, **23**, 121; (1969). **24**, 1

5 Creeth, J. M. and Pain, R. H. (1967). *Progr. Biophys. Mol. Biol.*, **17**, 217

6 Schachman, H. K. (1963). *Biochemistry*, **2**, 887

7 Schachman, H. K. (1959). *Ultracentrifucation in Biochemistry*, Academic Press, New York

8 Svedberg, Th. and Pedersen, K. O. (1940). *The Ultracentrifuge*, Oxford University Press, London

9 Ackers, G. K. (1970). *Advan. Protein Chem.*, **24**, 343

3.4. Isolation and Purification of Proteins[1–10]

The isolation of insoluble fibrous proteins presents fewer problems than the isolation of soluble proteins from animal or plant tissues, bacterial cultures or other cell suspensions, which is usually complicated by the presence of many other proteins, carbohydrates, nucleic acids and lipids. The separation is further complicated by the risk of denaturation by temperature and other factors and the sensitivity to the action of proteolytic enzymes. In a few favourable cases—for example, the isolation of haemoglobin from erythrocytes, of casein from milk or of albumin from egg-white—the protein content is large enough for simple precipitation to be used for the isolation. In the majority of cases, however, tedious purification processes involving several steps are usually necessary.

The first step is to break down the structure of the biological material since protein molecules, in general, cannot penetrate cell membranes. Cell walls can be destroyed mechanically in a homogeniser or by a variety of other methods including ultrasonic disruption, freezing and thawing, shaking with glass beads, grinding frozen material in a mortar, osmotic shock, osmotic lysis using distilled water and treatment with detergents. Cell organelles which do not contain the desired protein may be separated by differential centrifugation. The first step towards concentrating the protein is usually extraction with water or diluted salt solutions, followed by fractional precipitation using ammonium sulphate. Protamine sulphate can be used for the separation of nucleic acids.

Fractional precipitation of proteins can also be carried out with organic solvents such as acetone, methanol, ethanol, dioxan, etc., at low temperatures (0 to $-10\,^\circ C$). Cohn and co-workers[11] were able to separate plasma proteins using this method, which has also been used for the separation of other proteins. Figure 36 illustrates the fractionation of proteins by varying amount of solvent added and the pH of the medium.

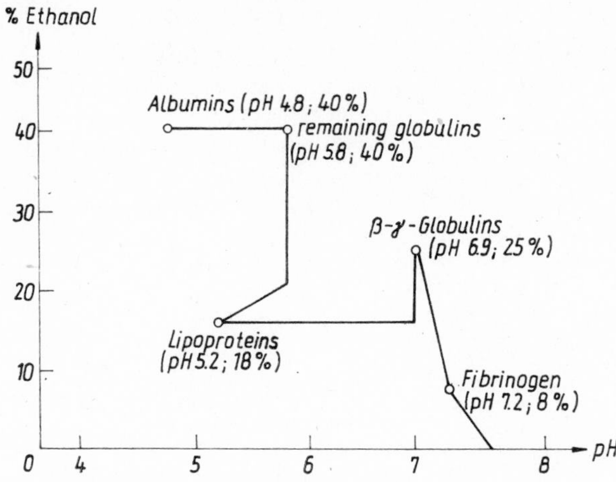

Figure 36. Fractionation of plasma proteins (E. Cohn)

After the desired protein has been concentrated in a certain fraction by these methods, it is finally purified by one of the chromatographic techniques. The most important of these are gel chromatography (molecular sieve chromatography), ion exchange chromatography and adsorption chromatography. The separation of protein molecules according to their size by the gel chromatography technique has already been described in Section 3.3.4. Since there is negligible adsorption onto the column materials, the sensitive structure of proteins is usually unaffected in gel chromatography. The most commonly used materials for adsorption chromatography of proteins include alumina gel, starch, calcium phosphate gel, diatomaceous earth (Celite) and hydroxyapatite. For similar reasons, cellulose or dextran derivatives are used as supporting materials in *ion exchange chromatography*. These substances may have diethylaminoethyl (DEAE), aminoethyl (AE), triethylammoniumethyl (TEAE) side chains attached and therefore form the basis of anion exchange resins. Cation exchange resins are formed by linking carboxymethyl (CM), sulphomethyl (SM), sulphoethyl (SE) and phosphate functions to a cellulose backbone. When these columns are used for ion exchange chromatography, they are developed by elution with buffers of increasing ionic strength or varying pH, as necessary.

In many cases a second chromatographic separation, perhaps using an alternative method, will be needed to obtain pure material. Preparative electrophoresis is another valuable technique for obtaining pure proteins. Two electrophoretic techniques, *moving boundary electrophoresis* and *zone electrophoresis*, are commonly used. In the moving boundary method, the protein solution is first placed in the apparatus and then carefully covered with a layer of protein-free buffer solution. Between the two solutions a sharp boundary is formed. The electrodes are then immersed in the buffer solution and the progress of the run is followed by a suitable optical system. Complete separation of the proteins is not usually achieved and this method is of limited use for the isolation of individual components of protein mixtures. Zone electrophoresis involves the use of a support (starch, paper, cellulose, polyacrylamide gels) which is impregnated with a buffer solution of suitable pH. Starch can be used in form of blocks or gels. The protein to be examined is applied to a very narrow zone in the centre, between the buffer vessels in which the electrodes are immersed. Depending on their charge, the components of a protein mixture migrate towards the anode or cathode. Since proteins, generally, have an over-all negative charge, their separation by different migration rates in an electrophoretic field is usually achieved in buffers with a pH range of 7 to 9. The choice of the support used in zone electrophoresis is largely determined by the scale required. For

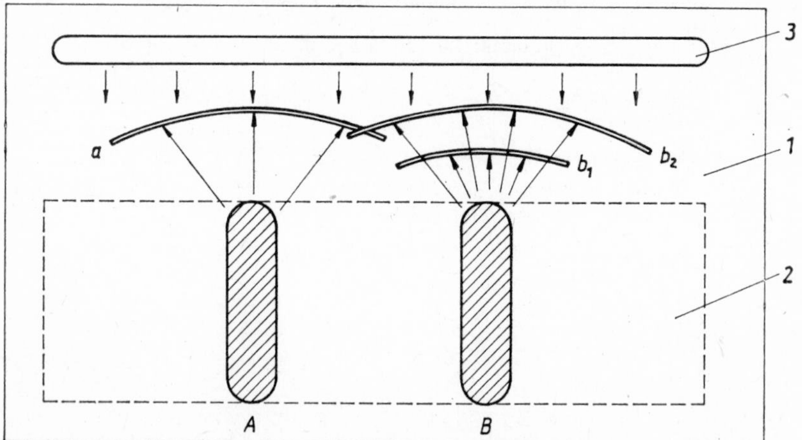

Figure 37. Principle of immunoelectrophoresis (from S. M. Rapoport, *Medizinische Biochemie*, Verlag Volk und Gesundheit, Berlin, 1968). 1, Agar layer; 2, electrophoretically separated serum (antigen mixture); 3, antiserum (mixture of antibodies); A, serologically uniform electrophoretic fraction; a, precipitation band corresponding to A; B, electrophoretic fraction consisting serologically of two components; b_1 and b_2, precipitation bands corresponding to B

example, the starch block technique is particularly useful for preparative purposes, because large quantities of material can be handled. At the end of a run, the starch block is divided into sections and the individual protein components are eluted. *Polyacrylamide gel electrophoresis* enables a fractionation of proteins according to their size, by a molecular sieving effect (see p. 260).

Electrofocusing is an electrophoretic separation technique on a column based on differences in isoelectric points using a pH gradient.

A very high degree of resolution has been attained by *immunoelectrophoresis*. Figure 37 illustrates the principle of immunoelectrophoresis, which

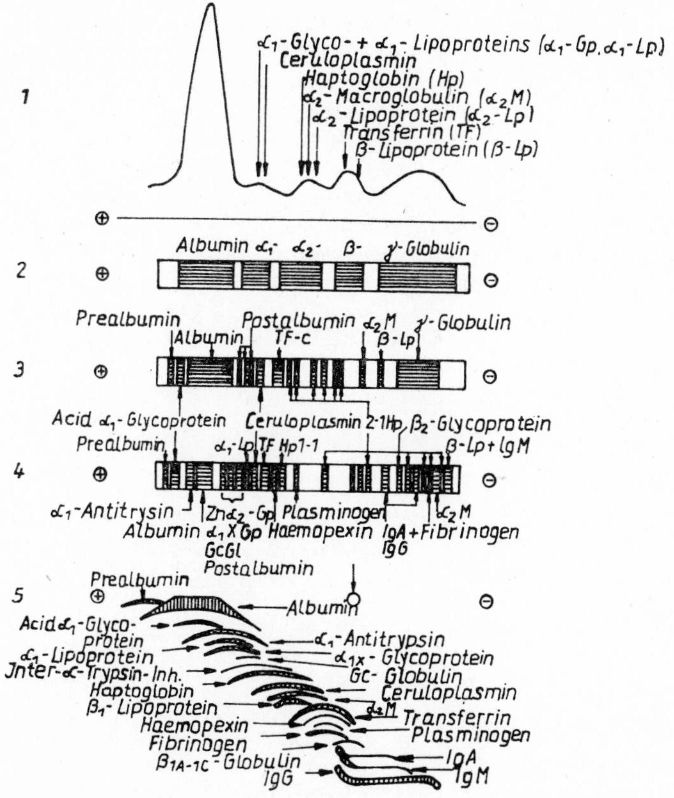

Figure 38. Electrophoretic fractionation of human plasma proteins at pH 8 by various methods: 1, free electrophoresis according to TISELIUS; 2, electrophoresis on paper; 3, electrophoresis on starch gel; 4, polyacrylamide electrophoresis in a discontinuous gel gradient; 5, immunoelectrophoresis on starch gel according to R. KLEINE. The arrow marks the start

is of particular importance in medical diagnosis. In this technique the protein mixture is first subjected to agar gel electrophoresis. The antiserum to the protein mixture is then allowed to diffuse in the agar strip opposite to the antigens. The resulting precipitin zones correspond to a definite antigen–antibody reaction depending on the heterogeneity of the protein mixture. In the case of a homogeneous protein only one precipitin band is formed.

Under certain conditions, preparative ultracentrifugation can be used to purify proteins.

Ultrafiltration is a method of concentrating protein solutions under mild conditions. It also enables proteins to be separated from low-molecular-weight impurities. Fractionation of proteins according to their molecular weights is performed using membrane filters of different pore sizes.

Several methods have been developed for the separation and purification of enzymes. These include fractional precipitation using polyethyleneglycol, ion filtration chromatography and affinity chromatography. In the *affinity chromatography* technique[12] a low-molecular ligand (competitive inhibitor, modified substrate, specific antigen, etc.) is linked, via a spacer, to an insoluble inert support such as Sepharose. Using this technique, it is possible to isolate a specific enzyme or an antibody from a protein mixture in one step. A specific protein–ligand complex forms and retains the desired protein, while other components of the mixture are unaffected and pass through the column. The desired protein must then be recovered from the complex, and this may be successfully carried out by varying the pH or the ionic strength of the medium. Several enzymes, isoenzymes, antibodies and antigens (immunoadsorption chromatography) have been isolated using this technique, and the method may also be applied to the isolation of receptor proteins.

Ion filtration chromatography combines the advantages of ion exchange chromatography and gel chromatography. It enables the separation of proteins on DEAE-Sephadex A-25 columns in a short time without the use of salt gradients.

The countercurrent distribution method of Craig (1944) has also been applied to the purification of proteins[13,14]. At every stage of distribution, the equilibrium is governed by the Nernst distribution law $K = c_1/c_2$, where K is the distribution coefficient, and c_1 and c_2 are the concentrations of components to be distributed in the upper and lower phases. When two components are to be separated, the higher the separation factor the greater the separation effect. The separation factor can be calculated from the ratio K_1/K_2. The Craig distribution method has proved useful for the purification of both biologically active native peptides and synthetic intermediates. Solvent systems commonly used include butanol/acetic acid, trichloroacetic acid/*p*-toluenesulphonic acid, chloroform/benzene/methanol/phenol/water, and other combinations. The choice of solvents to be used in protein separation presents problems because of the ever-present risk of denaturation. Craig *et al.* succeeded in isolating ribonuclease and also lysozyme without loss of biological activity using the system ethanol/water/ammonium sulphate after 3746 and 3420 distribution steps, respectively. The same authors[15] succeeded in separating into four fractions a serum albumin, previously thought to be a uniform protein with molecular weight 68000 from its ultracentrifuge behaviour. After 401 distribution steps in the system butanol/trichloroacetic acid/acetic acid/ethanol, the pure serum albumin was obtained with $K = 0.875$ (see Figure 39).

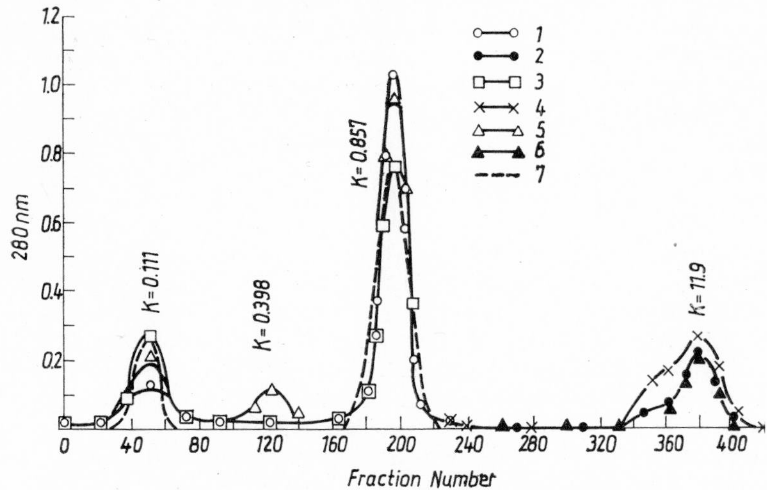

Figure 39. Countercurrent distribution of a serum albumin (HAUSMANN and CRAIG, 1958)

Polymeric solvents able to form two-phase systems with water such as polyethyleneglycol, polypropyleneglycol, etc., have also been suggested for the countercurrent distribution of high-molecular proteins. The disadvantages are the length of time required for the two phases to separate and the great difficulty of recovering proteins from the polymeric solvents.

There are various criteria for establishing the purity of a protein obtained from any of the different purification techniques. The homogeneity of a protein can be established by the occurrence of only one protein band on electrophoresis, or by its characteristic sedimentation behaviour in the ultracentrifuge. Another criterion of purity of a protein is the solubility curve below saturation. A pure protein shows a linear relationship between the total amount of protein and that actually dissolved. At the saturation point a sharp break appears and, above the saturation point, a line of slope zero (Figure 40).

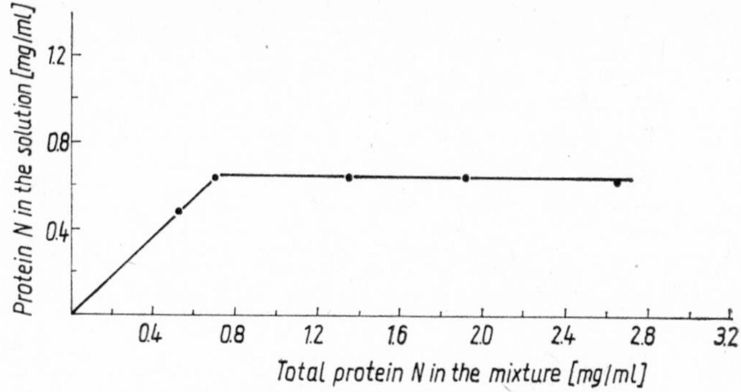

Figure 40. Solubility curve of a pure protein

The crystallinity of a protein is undoubtedly further evidence of its purity, but this cannot be considered an absolute criterion of purity, since there are many examples of heterogeneous crystalline preparations. Also, the crystallisation of proteins is not always successful, and the concentration, solvent and temperature conditions are critical and vary from case to case. The crystallisation of proteins is usually carried out in the presence of ammonium sulphate.

The biological activity of a protein (for example, kinetic parameters of enzymes, the hormonal activity of different proteohormones) is further evidence of its purity.

It is generally advisable to establish the purity and homogeneity of a protein by several independent criteria.

References

1 Alexander, P. and Block, R. J. (1961). *Analytical Methods of Protein Chemistry*, Pergamon Press, Oxford

2 Leggett-Bailey, J. (1969). *Techniques in Protein Chemistry*, Elsevier, Amsterdam

3 Work, T. S. and Work, E. (Eds.) (1970–1972). *Laboratory Techniques in Biochemistry and Molecular Biology*, Vols. 1–3, Elsevier, North-Holland, New York

4 Hrapia, H. (1965). *Einführung in die Chromatographie*, Akademie-Verlag, Berlin

5 Determann, H. (1967). *Gelchromatographie*, Springer-Verlag, Berlin

6 Fischer, L. (1972). *An Introduction to Gel Chromatography*, Elsevier, North-Holland, New York

7 Gordon, A. H. (1971). *Electrophoresis of Proteins in Polyacrylamide and Starch Gels*, Elsevier, North-Holland

8 Helferich, F. (1959). *Ionenaustauscher*, Verlag Chemie, Weinheim/Bergstraße

9 Clausen, J. (1971). *Immunochemical Techniques for the Identification and Estimation of Macromolecules*, Elsevier, North-Holland, New York

10 GRABAR, P. and BURTIN, P. (Eds.) (1964). *Immunoelektrophoretische Analyse*, Elsevier, North-Holland, Amsterdam
11 COHN, E. (1946). *J. Amer. Chem. Soc.*, **68**, 459
12 CUATRECASAS, P. (1972). *Advan. Enzymol.*, **36**, 29
13 TAVEL, P. V. and SIGHER, R. (1956). *Advan. Protein Chem.*, **11**, 237
14 SCHWENKE, K. D. (1965). *Z. Chem.*, **5**, 322
15 HAUSMANN, W. and CRAIG, L. C. (1958). *J. Amer. Chem. Soc.*, **80**, 2703

3.5. Detection and Quantitative Estimation of Proteins

There are several classic protein colour reactions. These colour reactions mostly depend on the presence of particular amino acid residues but are not always specific. The most important are summarised in Table 24.

Table 24. Colour reactions for the qualitative detection of proteins

Reaction	Procedure	Reacting component(s)	Colour
Biuret	Strongly alkaline solution treated with dilute $CuSO_4$ solution	Peptide bonds	Violet
MILLON	Solution heated with $Hg(NO_3)_2$ and conc. H_2SO_4	Tyr	Brownish-red precipitate
PAULY	Alkaline solution treated with diazo benzene sulphonic acid	Tyr and His	Red (yellowish on the addition of acid)
HOPKINS–COLE	Mixing with glyoxylic acid and underlayering with conc. H_2SO_4	Trp	A violet ring
SAKAGUCHI	Treatment with alkaline α-naphthol solution and hypobromite	Arg	Red
Xantho-protein	Treatment with conc. HNO_3	Tyr Phe	Yellow (orange on the addition of alkali)

Other qualitative detection methods depend on precipitation reactions resulting from the denaturation of proteins. These comprise treatment with trichloroacetic acid, picric acid, perchloric acid, phosphomolybdic acid, heavy metal salts (Cu, Pb, Zn, Fe, etc.), as well as heating at the isoelectric point. In the presence of exess reagent, the precipitation is usually quantitative and can be used to remove proteins from biological fluids. The biuret reaction (see Table 25) is suitable for the quantitative estimation of proteins. This reaction depends on the formation of a violet copper complex which absorbs between 540 and 560 nm and can be estimated colorimetrically.

The LOWRY method, which is a combination of the biuret method and the FOLIN *method* for tyrosine and tryptophan residues, is more sensitive, and is the technique most commonly used for the colorimetric micro determination of proteins. The blue-coloured complex formed (absorption maximum at 750 nm) is sufficiently stable for quantitative measurements. Serum albumin is usually used as the standard protein. The lower limit for detection is 5–10 μg protein/ml. Tris-buffer, guanidine, and SH-containing compounds interfere with the estimation.

The KJELDAHL *method* is the classic method for the quantitative estimation of proteins. The sample to be analysed is boiled with concentrated sulphuric acid with the addition of a catalyst. This treatment causes total destruction of the natural material, and organic nitrogen is converted to ammonium sulphate. On the addition of excess alkali, the ammonia liberated can be distilled into N/100 H_2SO_4 and estimated by back-titration.

Direct estimination of proteins can be carried out by measuring their UV absorbance at 280 nm (which is due to Tyr and Trp residues). If the extinction coefficient $E^{1\%,1cm}_{280nm}$ is known, the protein content of an unknown solution can be calculated.

Proteins can also be estimated, with varying degrees of accuracy and specificity, using other physical methods such as refractometry, specific density measurements and nephelometry.

3.6. Protein Structure

For the discussion of protein structure LINDERSTROM-LANG recommended the terms 'primary', 'secondary' and 'tertiary' structure. The 'primary structure' of a protein is the number and sequence of amino acids attached to each other through peptide bonds. The term 'secondary structure' is the conformation of the chains resulting from the formation of hydrogen bonds between the oxygen of the carbonyl group and the nitrogen of the amide group of the polypeptide backbone. The term 'tertiary structure' denotes the spatial structure of a polypeptide chain resulting from the intramolecular interaction of the side chains. The term 'quaternary structure' was introduced by J. D. BERNAL in 1958 to describe the association of protein monomers into definite oligomeric complexes.

3.6.1. Primary Structure

For the elucidation of the primary structure of a protein, the amino acid composition and sequence have to be determined. A succesful analysis can be achieved only if uniform proteins or polypeptide chains are investigated. Proteins consisting of more than one peptide chain are first treated with a concentrated solution of urea in order to break up the tertiary structure. Disulphide bridges are usually cleaved by reducing processes, and the

Val–Leu–Ser–Glu–Gly–Glu–Trp–Gln–Leu–Val–Leu–His–Val–Trp–Ala–Lys–
1 2 3 4 5 6 7 8 9 10 11 12 13 14 15 16

Val–Glu–Ala–Asp–Val–Ala–Gly–His–Gly–Gln–Asp–Ile–Leu–Ile–Arg–Leu–
17 18 19 20 21 22 23 24 25 26 27 28 29 30 31 32

Phe–Lys–Ser–His–Pro–Glu–Thr–Leu–Glu–Lys–Phe–Asp–Arg–Phe–Lys–His–
33 34 35 36 37 38 39 40 41 42 43 44 45 46 47 48

Leu–Lys–Thr–Glu–Ala–Glu–Met–Lys–Ala–Ser–Glu–Asp–Leu–Lys–Lys–His–
49 50 51 52 53 54 55 56 57 58 59 60 61 62 63 64

Gly–Val–Thr–Val–Leu–Thr–Ala–Leu–Gly–Ala–Ile–Leu–Lys–Lys–Lys–Gly–
65 66 67 68 69 70 71 72 73 74 75 76 77 78 79 80

His–His–Glu–Ala–Glu–Leu–Lys–Pro–Leu–Ala–Gln–Ser–His–Ala–Thr–Lys–
81 82 83 84 85 86 87 88 89 90 91 92 93 94 95 96

His–Lys–Ile–Pro–Ile–Lys–Tyr–Leu–Glu–Phe–Ile–Ser–Glu–Ala–Ile–Ile–
97 98 99 100 101 102 103 104 105 106 107 108 109 110 111 112

His–Val–Leu–His–Ser–Arg–His–Pro–Gly–Asn–Phe–Gly–Ala–Asp–Ala–Gln–
113 114 115 116 117 118 119 120 121 122 123 124 125 126 127 128

Gly–Ala–Met–Asn–Lys–Ala–Leu–Glu–Leu–Phe–Arg–Lys–Asp–Ile–Ala–Ala–
129 130 131 132 133 134 135 136 137 138 139 140 141 142 143 144

Lys–Tyr–Lys–Glu–Leu–Gly–Tyr–Gln–Gly
145 146 147 148 149 150 151 152 153

Figure 41. The primary structure of sperm whale myoglobin (A. B. EDMUNSON, 1965)

resulting free SH groups are blocked reversibly to prevent reoxidation and exchange of disulphide groups. The next step is to split the polypeptide chains into smaller fragments, and the investigation of each fragment then allows the primary structure of the whole chain to be determined. SANGER and co-workers carried out the pioneer work on systematic investigation of protein primary structure from 1945 to 1954. Up to the present time more than 12 000 publications on primary structures of proteins have appeared: all known structures are listed in the *Atlas of Protein Sequences and Structures*, edited by M. O. DAYHOFF.

3.6.1.1. Specific Cleavage of Polypeptide Chains

Proteolytic enzymes—notably trypsin, chymotrypsin, pepsin, papain, subtilisin, elastase and thermolysin—are used for the specific cleavage of peptide chains[1]. In practice, a combination of several enzymes with different substrate specificity is used, resulting in fragments of various size and composition. The amino acid sequence of the fragments usually overlaps, and this allows their arrangement in the original peptide chain to be deduced.

Trypsin is the most specific of the enzymes mentioned. It catalyses the hydrolysis of peptide linkages which involve the carboxyl functions of lysine or arginine. The specificity of the enzyme is further increased if the amino group of lysine is blocked with a benzyloxycarbonyl- or other amino-protecting group: under these conditions only arginyl bonds are attacked. Cysteinyl peptide bonds can be rendered susceptible to tryptic degradation by reaction with ethylenamine, to give an *S*-β-aminoethyl derivative.

Chymotrypsin cleaves on the carboxylic side of aromatic amino acids—for example, phenylalanine, tyrosine, tryptophan, etc. In the case of long peptide chains some attack at the carboxy groups of leucine, valine, asparagine and methionine may occur. *Pepsin* has a low side chain specificity, tryptophan, phenylalanine, degrading at tyrosine, methionine and leucine residues.

Subtilisin cleaves peptide bonds in the neighbourhood of serine, glycine and aromatic amino acids. *Elastase* is of lower specificty and hydrolyses adjacent to neutral amino acids. The main points attacked by *papain*, which also shows a low specificity, are arginine, lysine and glycine residues, but bonds involving acidic amino acids are not cleaved. *Thermolysin* from *Bacillus proteolyticus* catalyses cleavage at hydrophobic residues.

Generally, the shorter the time chosen for the hydrolysis the higher the specificity of the proteolytic cleavage. The purity of the enzyme preparation employed is very important and special precautions are sometimes necessary—for example, to remove the last traces of chymotrypsin from trypsin, specific inhibitors such as diphenylcarbamyl chloride or L-(1-tosylamido-2-phenyl)ethylchloromethyl ketone can be used.

The cleavage of polypeptide chains containing *disulphide bridges* usually leads to the formation of complex mixtures of lower peptides and the presence of disulphide bonds allows the occurrence of several undesirable side reactions. They are therefore decomposed before cleavage begins either by oxidation with performic acid or, better, by reduction with mercaptoethanol, dithiothreitol or sodium borohydride. The easy separation of cysteic acid peptides is an advantage of oxidation; a disadvantage is the destruction of tryptophan. After reduction, blocking of the free SH groups is required and this can usually be achieved by cyanoethylation using acrylonitrile. The localisation of disulphide bridges is a special problem of protein structure analysis: one solution to the problem is the elegant technique of *diagonal electrophoresis* of Hartley. According to this technique, the proteolytic fragments are separated by paper electrophoresis. Then, on oxidation with performic acid, the cystine bridges are broken and two new cysteic acid peptides are formed. The paper band of the first electrophoresis is fixed on a larger sheet and electrophoresis is repeated in a direction perpendicular to the first one. The unaffected peptides are located diagonally, while the acidic cysteic acid peptides, being outside the diagonal, can be identified easily.

In addition to the enzymatic methods for cleaving of peptide chains, specific chemical methods are also used[2,3]. Cyanogen bromide, for instance, splits peptide bonds involving the carboxyl group of methionine:

$$\text{--NH-CH(CH}_2\text{-CH}_2\text{-S-CH}_3\text{)-C(=O)-NH-CH(R)-CO--} \xrightarrow{\text{BrCN}} \left[\text{--NH-CH(CH}_2\text{-CH}_2\text{-}\overset{\oplus}{\text{S}}\text{(CN)-CH}_3\text{)-C(=O)-NH-CH(R)-CO--}\right]^{\oplus} \text{Br}^{\ominus}$$

$$\xrightarrow{-\text{CH}_3\text{SCN}} \left[\text{--NH-CH-C}(=\overset{\oplus}{\text{N}}\text{H-CHR-CO--})\text{ (ring: H}_2\text{C-CH}_2\text{-O)}\right]^{\oplus} \text{Br}^{\ominus} \xrightarrow{\text{H}_2\text{O}} \text{--NH-CH-C=O (ring: H}_2\text{C-CH}_2\text{-O)} + \left[\text{H}_3\overset{\oplus}{\text{N}}\text{-CH(R)-CO--}\right]^{\oplus} \text{Br}^{\ominus}$$

N-Bromosuccinimide (NBS) cleaves bonds of tyrosine and tryptophan:

$$\text{HO-C}_6\text{H}_4\text{-CH}_2\text{-CH(NH--)-CO-NH-CH(R)-CO--} \xrightarrow[\text{pH 4,5}]{\text{NBS}}$$

$$\left[\text{O=C}_6\text{H}_2\text{Br}_2\text{(Br)-CH}_2\text{-CH(NH--)-C(HO)=N-CH(R)-CO--}\right] \longrightarrow \text{O=C}_6\text{H}_2\text{Br}_2\text{(spiro: O-C(=O)-CH(NH--)-CH}_2\text{)} + \text{H}_2\text{N-CH(R)-CO--}$$

For the specific splitting of cystine peptide bonds, the cystine residue is treated with 2-nitro-5-thiocyanobenzoic acid to give *S*-cyanocysteine, whose amide bond can easily be hydrolysed without side reactions[4]. The peptide mixtures obtained by the different cleaving methods have to be separated and purified. Ion exchange chromatography, electrophoresis, paper or thin layer chromatography and liquid–liquid partition are the techniques available, and a combination of these methods usually enables homogenous material for sequantial analysis to be isolated.

3.6.1.2. *Sequence Analysis*[5,6]

The sequence of amino acids in di- and tripeptides can be determined with relative ease; it is sufficient to know the amino acid composition and the terminal amino acids. For larger peptides it is necessary to carry out stepwise degradation of the chain. The most important methods for the determination of terminal groups and for the stepwise degradation of peptide chains are discussed in the next section.

3.6.1.2.1. *Terminal Group Determination*

The chemical methods for terminal group determination are mostly based on the transformation or blocking of terminal amino acid functions. Following total hydrolysis, the terminal amino acid derivatives can then be separated from the unchanged amino acids and characterised.

The dinitrophenyl (DNP) method of SANGER[7] is of widespread application. This method involves the treatment of the peptide or protein with 2,4-dinitrofluorobenzene followed by hydrolysis and extraction of the Dnp-amino acids and their identification.

$$2,4\text{-}(NO_2)_2C_6H_3\text{-}F \; + \; H_2N\text{-}\underset{R_1}{CH}\text{-}CO\text{-}NH\text{-}\underset{R_2}{CH}\text{-}CO\text{-}NH\text{-}\underset{R_3}{CH}\text{-}CO\text{----}$$

$$\downarrow$$

$$2,4\text{-}(NO_2)_2C_6H_3\text{-}NH\text{-}\underset{R_1}{CH}\text{-}CO\text{-}NH\text{-}\underset{R_2}{CH}\text{-}CO\text{-}NH\text{-}\underset{R_3}{CH}\text{-}CO\text{----------}$$

$$\downarrow$$

$$2,4\text{-}(NO_2)_2C_6H_3\text{-}NH\text{-}\underset{R_1}{CH}\text{-}COOH \; + \; H_2N\text{-}\underset{R_2}{CH}\text{-}COOH \; + \; H_2N\text{-}\underset{R_3}{CH}\text{-}COOH$$

The dansyl (5-dimethylaminonaphthalenesulphonyl) method[8] is more sensitive; here, the N-terminal amino acid is converted into the corresponding dansyl derivative the identification of which is facilitated by its intense yellow fluorescence.

$$\text{Dansyl chloride (}(CH_3)_2N\text{-}C_{10}H_6\text{-}SO_2Cl) + H_2N-\underset{R_1}{CH}-CO-NH-\underset{R_2}{CH}-CO\cdots NH-\underset{R_n}{CH}-COOH$$

$$\downarrow$$

$$(CH_3)_2N\text{-}C_{10}H_6\text{-}SO_2-NH-\underset{R_1}{CH}-CO-NH-\underset{R_2}{CH}-CO\cdots NH-\underset{R_n}{CH}-COOH$$

$$\downarrow$$

$$(CH_3)_2N\text{-}C_{10}H_6\text{-}SO_2-NH-\underset{R_1}{CH}-COOH + H_2N-\underset{R_2}{CH}-COOH + H_2N-\underset{R_n}{CH}-COOH$$

Simple techniques are available for determining the dansyl amino acids by direct fluorescence scanning, photography and densitometry[9]. The method allows analysis at the picomole level and is accurate between 1.0 and 10×10^{-12} mol with a lower limit of detection around 10^{-14} mol.

There are a number of less important methods for blocking the N-terminal amino acids. These methods include arylsulphonation with naphthalene or benzenesulphonyl chloride, carbamylation with potassium cyanate and carboxymethylation with bromoacetic acid. N-Terminal residues can also be determined by converting the N-terminal amino acids into hydroxy acids, keto acids or nitriles followed by appropriate estimations performed on the hydrolysate.

Finally, leucine aminopeptidase can be used for N-terminal amino acid determination: it requires the presence of a free α-amino group and only splits the N-terminus from the polypeptide chain.

Carboxypeptidases A and B are used similarly for the determination of C-terminal amino acids in a polypeptide chain (J. Lens, 1949):

$$H_2N-AA-AA_n-NH-\overset{R}{CH}-COOH \xrightarrow{\text{carboxypeptidase}}$$

$$H_2N-AA-AA_{n-1}-AA-COOH + H_2N-\overset{R}{CH}-COOH$$

The two enzymes have different but complementary specificities and they are therefore usually used together.

The Akabori method[10] is one of the most important chemical methods for the determination of C-terminal groups. All amino acids, apart from the C-terminal ones, are converted into hydrazides on treatment with anhydrous hydrazine for 5 hours at 100 °C. The acid hydrazides which constitute the bulk of the reaction products are usually separated by treatment with isovaleraldehyde (or other aldehydes). It is also possible to treat the reaction mixture directly with dinitrofluorobenzene and to separate the C-terminal Dnp-amino acids from the acidic fraction:

$$\begin{array}{c} H_2N{-}\underset{\displaystyle R_1}{\overset{|}{CH}}{-}CO{-}NH{-}CH(R_2){-}CO \cdots\cdots NH{-}CH(R_n){-}COOH \\ \downarrow N_2H_4 \\ H_2N{-}CH(R_1){-}CO{-}NH{-}NH_2 + H_2N{-}CH(R_2){-}CO{-}NH{-}NH_2 + H_2N{-}CH(R_n){-}COOH \\ \downarrow R{-}CHO \\ R{-}CH{=}N{-}CH(R_1){-}CO{-}NH{-}N{=}CH{-}R \\ + R{-}CH{=}N{-}CH(R_2){-}CO{-}NH{-}N{=}CH{-}R + H_2N{-}CH(R_n){-}COOH \end{array}$$

C-terminal amino acids such as cysteine, glutamine, asparagine and tryptophan cannot be determined by this method and arginine may be partially converted to ornithine.

A less widely employed method involves the reduction of the C-terminal carboxyl group to the corresponding alcohol with $LiAlH_4$ (C. FROMAGEOT, 1950). It is an advantage to esterify the protein with diazomethane before reduction. Complete hydrolysis then yields the aminoalcohol of the C-terminal residue, which may be isolated and identified.

3.6.1.2.2. Stepwise Degradation of the Peptide Chain

3.6.1.2.2.1. Chemical Methods

EDMAN degradation (also known as the 'phenylthiohydantoin method') was developed by P. EDMAN[11], and is the most important chemical method for the stepwise degradation of a peptide from the N-terminal end. When a peptide is treated with phenylisothiocyanate at pH 9 and 40 °C, a thiourea derivative (I) is formed. In acid solution this is cleaved to give 2-anilinothiazolin-5-one(II) and the peptide chain with the N-terminal residue is removed. The thiazolone dericative is unsuitable for identification purposes because it is relatively unstable. It is therefore hydrolysed to a phenylthiocarbamic acid (PTC-derivative III), which cyclises to give a 3-phenyl-2-thiohydantoin (PTH-amino acid IV). On heating under reflux, direct isomerisation of thiazolone to thiohydantoin occurs.

$$C_6H_5{-}N{=}C{=}S \;+\; H_2N{-}\underset{R_1}{CH}{-}CO{-}NH{-}\underset{R_2}{CH}{-}CO\cdots NH{-}\underset{R_n}{CH}{-}COOH$$

$$\downarrow$$

$$C_6H_5{-}NH{-}\overset{S}{\overset{\|}{C}}{-}NH{-}\underset{R_1}{CH}{-}CO{-}NH{-}\underset{R_2}{CH}{-}CO\cdots NH{-}\underset{R_n}{CH}{-}COOH \qquad (I)$$

$$\downarrow + H^{\oplus}$$

(II) 2-Anilino-thiazolin-5-one $+\; H_2N{-}\underset{R_1}{CH}{-}CO\cdots NH{-}\underset{R_n}{CH}{-}COOH$

$$\xrightarrow{+H_2O}$$

$$C_6H_5{-}NH{-}\overset{S}{\overset{\|}{C}}{-}NH{-}\underset{R_1}{CH}{-}COOH \qquad (III)$$

$$(III) \xrightarrow[-H_2O]{+H^{\oplus}} (IV)$$

The PTH-amino acids formed are extracted and identified by paper, thin layer or gas–liquid chromatography. The remaining peptide is also isolated and kept for the next degradation cycle. Recently, it has been shown that mass spectrophotometry is a sensitive method for the unequivocal identification and analysis of both the PTH- and MTH(methylthiohydantoin)-amino acid derivatives resulting from Edman degradation[12–14]. This technique is particularly useful when the reaction mixture contains only one thiohydantoin. Identification of PTH-amino acids using high-pressure-liquid chromatography has been described by Frank and Strubert[15]. The technique is rapid and sensitive, takes about 7 minutes and permits the detection of a few nanomoles. A range of elution solvents has been described for the chromatography of hydrophobic and hydrophilic PTH-amino acids.

The *substractive Edman degradation* merely involves the separate determination of amino acid composition before and after degradation: this obviously permits identification of the residue lost.

Combinations of the Edman method and the dansyl technique[16–18] have sometimes proved useful. Other variants have also been tried—for example, a micro method on filter paper (H. Fraenkel-Conrat), thin layer chromatography on polyamide plates and the use of pentafluorophenylisothiocyanate as the reagent. Use of the latter reagent allows an electron-capture detector to be used in the gas–liquid chromatographic analysis of the thiohydantoin derivatives[19]. By continually repeating the reaction steps of the EDMAN degradation, a complete determination of the sequence is theoretically possible. In practice, sequence analyses can usually be performed over about 10 amino acid residues only, since the cleavage steps are rarely complete. Furthermore, additional non-specific splitting of the peptide chain leads to erroneous results. These problems were solved by EDMAN and BEGG[20], who succeeded in automating the degradation process and produced the so-called protein sequencer which is shown in Figure 42. In contrast to the manual technique, the unstable 2-anilino-thiazolin-5-one is separated in a fraction collector and converted into the PTH-amino acid outside the reaction container. The reaction vessel (1) is a cylindrical glass cup thermostated at 50 °C and mounted on the shaft of an electric motor (2). The cup spins at 1425 rev/min and all solutions entering the vessel are spread as thin films on the cup wall. The reagents and solvents are automatically admitted to the bottom of the glass cup through a valve assembly (3) and a special feed line (4). The extracting solvents climb to the groove, where they are removed and leave the cup through an effluent line (5). The glass cup is enclosed in a bell jar (6), so that the whole system can be evacuated by means of a vacuum pump (7). Reagents and solvents are stored in reservoirs (8) under a constant pressure of nitrogen (9 = nitrogen cylinder). The pressure difference between the reservoirs and reaction vessel allows continual transport of the required

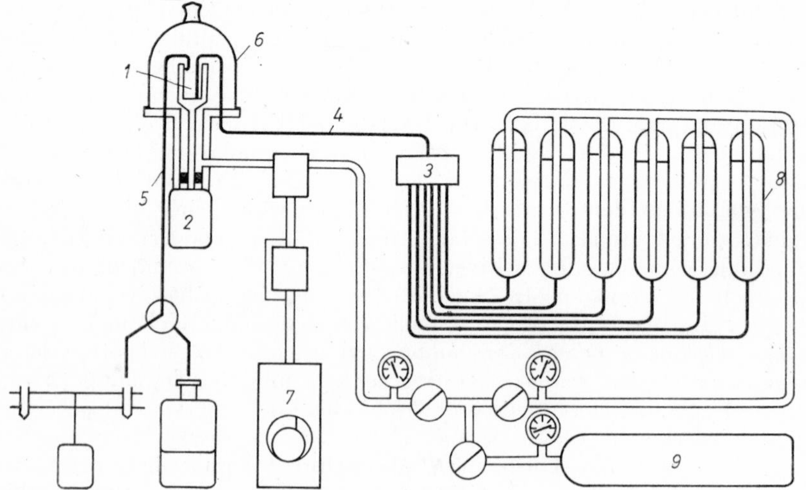

Figure 42. Diagram of the protein sequencer of P. EDMAN and G. BEGG (1967)

reagents and solvents. All operations—dissolution, concentration, drying under vacuum and extraction—are carried out in the same system. The shortened peptide remains in the reaction vessel. The reaction cycle necessary for splitting off one amino acid comprises 30 operations and is complete in 93.6 min. About 15 amino acids can therefore be degraded and identified in 24 hours. The yield is 97–98 per cent per cycle, allowing the stepwise degradation of about 100 amino acids under the conditions described. The classical sequencer is not suitable for the analysis of small peptides, since there is insufficient solubility difference between the peptides and the thiazolone for the selective extraction of the latter. BRAUNITZER and co-workers[21,22] have modified the method, and complete degradation to the end of the peptide chain is now possible. By the introduction of hydrophilic naphthalene sulphonic acid residues onto the ε-amino groups of lysine in peptides obtained by tryptic hydrolysis (so-called 'ε-labelling'), the smaller peptide derivatives are rendered more suitable for the working conditions of the sequencer.

The *solid-phase Edman degradation* is another alternative for the degradation of small peptides. In a sort of 'reverse-MERRIFIELD' technique, the peptide under investigation is covalently bound to a polymeric support. Degradation then takes place from the N-terminal amino acid. This method has been successfully employed by a number of research teams. Amino-alkylpolystyrene resins are preferred as supports. They are stable towards trifluoracetic acid and permit an amide linkage between peptide and resin component. The first sequencer based on the solid-phase degradation method was constructed in 1970–71 by R. LAURSEN[23,24]. It is cheaper than the Edman sequencer and suitable for the degradation of peptides comprising up to 30 amino acid residues. All degradation steps are carried out in a thermostated column. A mechanical disc-programmer controls the operations of the solenoid valves, pumps and fraction collectors.

Some problems remain to be resolved before the solid-phase EDMAN degradation can be considered to be generally applicable. For example, the side chain carboxyl groups of aspartic and glutamic acid may become bound to the resin or they may form cyclic imides, which prevent further degradations. This premature termination of the solid-phase degradation of peptides containing aminodicarboxylic acids has been studied in the case of glucagon[25]. Recently LAURSEN *et al.*[26] have described using *p*-phenyldiisothiocyanate to attach lysine containing tryptic peptides to aminopolystyrene via their α-amino groups, avoiding any bond with ω-carboxylic groups. Arginine peptides are first deguanidated to the corresponding ornithine peptides. As an alternative, LAURSEN suggests the degradation of aminodicarboxylic peptides with blocked side chains[27].

It has been found that an *N*,*N'*-disubstituted carbodiimide may affect the carboxyl groups of peptide chain in different ways depending on their enviroments. Initially all carboxyl groups react to give *O*-acyl urea derivatives.

In the case of the side chain carboxyl groups, rearrangement to the stable *N*-acylurea derivatives occurs. However, the C-terminal carboxyl group is activated as an oxazolinone which reacts with the amino groups of the support. The procedure is impossible in the case of a C-terminal proline peptide, which cannot form an oxazolinone.

NIALL and co-workers[28] have recently published a new automatic degradation method which can be operated on a nanomole scale. The method involves the use of synthetic polymers of the type H–(Nle–Arg_{27})–NH_2 as carriers.

Another solid-phase sequencer uses 3-aminopropyl glass (APG) as the support[29]. The introduction of amino groups into porous glass is achieved with 3-aminopropyltriethoxysilane. The peptide is joined to the support by carboxyl activating methods. Lysine-containing peptides are attacted by means of *p*-phenylenediisothiocyanate. The modified glass support has a constant bindung capacity and gives excellent flow rate properties in the reaction column. Th. WIELAND suggested the solid-phase degradation of peptides fixed on aluminium oxide by ionic adsorption. Isothiocyanatobenzene sulphonic acid is used as reagent instead of phenylisothiocyanate.

G. MANEKE and G. GÜNZEL have developed successfully a heterogeneous phase degradation using solid polymeric isothiocyanate as the reagent. The degradation method suggested by J. P. COLLMANN and D. A. BUCKINGHAM with *cis*-hydroxyaquoethylenetetramino cobalt complexes as intermediates has not found wide application. The resulting amino acid chelates are identified by chromatography or by IR and NMR spectroscopy.

A sequential degradation of peptides and proteins from the C-terminal residue has been reported by STARK[30]. Here ammonium isothiocyanate is used as the degradation reagent.

3.6.1.2.2.2. Enzymic Methods

Carboxypeptidase A and B, which are used for the determination of C-terminal amino acids, can also be used for the degradation of the whole peptide chain. The individual amino acids are liberated successively. Certain precautions must be taken: for instance, the enzyme used must be freed from other peptidases (inhibition with diisopropylfluorophosphate). Moreover, C-terminal D-amino acids, amide groups, proline and hydroxyproline interrupt the stepwise degradation. Some proteins may be degradated enzymically only after denaturation: and there are difficulties arising from differences in the splitting rates at different residues. The degradation of peptides starting from the N-terminal end can be carried out in an analogous manner with leucine aminopeptidase.

A new method of degrading peptide chains uses dipeptidyl aminopeptidase I (DAP-I, cathepsin C)[31]. The enzyme removes dipeptides sequentially from the N-terminal end. The dipeptides released are fractionated and identified. A second degradation of the same peptide chain after being shortened by one residue allows another set of dipeptides. In the case of small peptides the entire sequence can then be deduced by combining the results of dipeptide splitting with the results from complete hydrolysis of the intact molecule[32]. For longer peptide chains OVCHINNIKOV *et al.*[32] recommended the combined use of gas–liquid chromatography and mass spectrometry for identifying the dipeptides. An MS library of the 400 possible dipeptide sequences should allow automatic computer interpretation of the results.

CALLAHAN *et al.*[33] adopted a method in which the dipeptidase and peptide were arranged on one side of a special membrane. The liberated dipeptides gradually diffuse through the membrane and are identified by thin layer chromatography.

3.6.1.2.2.3. Physical Methods

Recently several different physical methods have been applied to sequence analysis. SHEINBLATT[34] determined the sequence of some di- and tripeptides by means of an NMR technique, based on changes in the spectra correlated with pH changes and sequence.

Mass spectrometry has proved the most useful physical method of sequence analysis[35]. The method is based on the observation of ions characteristic for the sequence and known as sequence peaks. These peaks are derived from splitting of C—CO or CO—N bonds. In simple cases the occurrence of amine (A) and aminoacyl (B) fragments allows the sequence of the amino acids to be deduced:

$$\begin{array}{l}
\qquad\quad R_1 \qquad\qquad\qquad R_2 \qquad\qquad\qquad R_3 \qquad\qquad\qquad\qquad\quad R_n \\
Y\text{—NH—CH—CO—NH—CH—CO—NH—CH—CO} \cdots\cdots \text{NH—CH—CO}\,OR' \\
\qquad\quad A_1 \qquad\qquad\qquad A_2 \qquad\qquad\qquad A_3 \qquad\qquad\qquad\qquad\quad A_n \\
\qquad\qquad\qquad B_1 \qquad\qquad\qquad B_1 \qquad\qquad\qquad B_3 \qquad\qquad\qquad\qquad\quad B_n
\end{array}$$

The interpretation of spectra from long-chain peptides and peptides containing polyfunctional amino acids is very complicated because the fragmentation types A and B become overlapped by several other modes of fragmentation.

The first method of mass spectrometric sequence analysis was developed by BIEMANN in 1959. It depends on the easy fission of the —C—C—bonds in —HN—CHR—CH_2—NH—groups in the mass spectrometer. Such groups are formed by reduction of the peptide bond with $LiAlH_4$. This reduction process also has the result of increasing the volatility of the peptide.

Mass spectrometric analysis allows the determination of sequences that could not be worked out by other methods. This is specially true for compounds with blocked amino terminals or unusual side chains. The structure of *fortuitine*, for instance, was elucidated by E. LEDERER (1965) using the MS technique.

$$CH_3-(CH_2)_n-CO\text{-Val-MeLeu-Val-Val-MeLeu}-\begin{pmatrix}\text{Ac}\\ | \\ \text{Thr}\end{pmatrix}_2\text{-Ala-Pro-OMe}$$

$$(n = 18 \text{ and } 20)$$

The research groups of K. BIEMANN, F. WEYGAND, M. SHEMYAKIN and E. LEDERER have pioneered the field of mass spectrometric sequence analysis. The automatic measurement of the spectra with high-resolution mass spectrometers equipped with computers for the calculation and interpretation of data will play an increasing important role in the future[36–38].

The number of publications on primary structures is continually increasing. Over 350 structures have been elucidated and reported. γG-Immunoglobulin[39] is one of the longest sequences determined so far (EDELMANN *et al.* 1969), with 1320 amino acid residues. Beef insulin was the first protein to have its primary structure elucidated, in over 10 years of work by SANGER *et al.*[40,41] in Cambridge.

The principle of stepwise degradation of peptide chains was introduced by P. EDMAN and the technique has now been automated. BRAUNITZER and co-workers[42] have successfully completed the first fully automatic analysis of a protein (β-lactoglobulin, 161 amino acid residues) using a modified EDMAN degradation.

References

1 HILL, R. L. (1965). *Advan. Protein Chem.*, **20**, **37**
2 WITKOP, B. (1961). *Advan. Protein Chem.*, **16**, **221**
3 WITKOP, B. (1968). *Science*, **162**, 318
4 STARK, G. R. (1973). *J. Biol. Chem.*, **248**, 6583
5 EDMAN, P. (1970). *Mol. Biol. Biochem. Biophys.*, **8**, 211
6 BLACKBURN, S. (1970). *Protein Sequence Determination Methods and Techniques*, Dekker, New York
7 SANGER, F. (1945). *Biochem. J.*, **39**, 507
8 GRAY, W. R. and HARTLEY, B. S. (1963). *Biochem. J.*, **89**, 59 p
9 VARGA, J. M. and RICHARDS, F. F. (1973). *Anal. Biochem.*, **53**, 397
10 AKABORI, S. *et al.* (1956). *Bull. Chem. Soc. Japan*, **29**, 507
11 EDMAN, P. (1950). *Acta Chem. Scand.*, **4**, 277
12 FAIRWELL, T., BARNESS, U. T. and LOVINS, R. F. (1970). *Biochemistry*, **9**, 2260
13 HAGENMAIR, H., EBBIGHAUSEN, W., NICHOLSON, G. and VÖTSCH, W. (1970). *Z. Naturforsch.*, **256**, 681
14 SUN, T. and LOVINS, R. E. (1972). *Anal. Biochem.*, **45**, 176
15 FRANK, G. and STRUBERT, W. (1973). *Chromatographia*, **6**, 522

16 GRAY, W. R. (1967). *Methods in Enzymology* (Ed. Hirs, C. H. W.), Vol. XI, pp. 139, 469, Academic Press, New York
17 GRAY, W. R. and SMITH, J. (1970). *Anal. Biochem.*, **33**, 36
18 GRAY, W. R. and HARTLEY, B. S. (1963). *Biochem. J.*, **89**, 379
19 LEQUIN, R. M. and NIALL, H. D. (1972). *Biochim. Biophys. Acta*, **257**, 76
20 EDMAN, P. and BEGG, C. (1967). *Europ. J. Biochemistry*, **1**, 80
21 BRAUNITZER, G., CHEN, R., SCHRANK, B. and STANGL, A. (1973). *Hoppe-Seyler's Z. Physiol. Chem.*, **354**, 867
22 BRAUNITZER, G., SCHRANK, B., PETERSEN, S. and PETERSEN, U. (1973). *Hoppe-Seyler's Z. Physiol. Chem.*, **354**, 1563
23 LAURSEN, R. A. and BONNER, A. G. (1970). *Chem. Engng News*, **48**, 52
24 LAURSEN, R. A. (1971). *Europ. J. Biochemistry*, **20**, 89
25 SCHELLENBERGER, A., JESCHKEIT, H., GRAUBAUM, H., MECH, C. and STERNKOPF, G. (1972). *Z. Chem.*, **12**, **63**
26 LAURSEN, R. A., HORN, M. J. and BONNER, A. G. (1972). *FEBS Letters*, **21**, 67
27 PREVIERO, A., DERANCOURT, J., COLETTI-PREVIERO, M. A. and LAURSEN, R. A. (1973). *FEBS Letters*, **33**, 135
28 NIALL, H. D., JACOBS, J. W., VAN RIETSCHOTEN, J. and TREGEAR, G. W. (1974). *FEBS Letters*, **41**, 62
29 WACHTER, E., MACHLEIDT, W., HOFNER, H. and OTTO, J. (1973). *FEBS Letters*, **35**, 97
30 STARK, G. R. (1972). *Meth. Enzymol.*, **25**, 369
31 LINDLEY, H. (1972). *Biochem. J.*, **126**, 683
32 OVCHINNIKOV, YU. A. and KIRYUSHKIN, A. A. (1972). *FEBS Letters*, **21**, 200
33 VALLAHAN, P. X., MCDONALD, J. K. and ELLIS, S. (1972). *Fed. Proc.*, **31**, **1105**
34 SHEINBLATT, M. (1966). *J. Amer. Chem. Soc.*, **88**, 2845
35 HEYNS, K. and GRÜTZMACHER, H. F. (1966). *Fortschr. Chem. Forsch.*, **6**, 536
36 BIEMANN, K., CONE, C. and WEBSTER, B. R. (1966). *J. Amer. Chem. Soc.*, **88**, 2597
37 SENN, M. and MCLAFFERTY, F. W. (1966). *Biochem. Biophys. Res. Comm.*, **23**, 381
38 BARBER, M., JOLLES, P., VILKAS, E. and LEDERER, E. (1965). *Biochem. Biophys. Res. Comm.*, **18**, 469
39 EDELMANN, G. M. *et al.* (1969). *Proc. Acad. Sci.* (*U.S.A.*), **63**, 78
40 SANGER, F. and TUPPY, H. (1951). *Biochem. J.*, **49**, 463
41 SANGER, F. and THOMPSON, E. O. P. (1953). *Biochem. J.*, **53**, 353
42 BRAUNITZER, G. *et al.* (1972). *Hoppe-Seyler's Z. Physiol. Chem.*, **353**, 832; (1973). *ibid.*, **354**, 867

3.6.2. Protein Conformation (*Secondary and Tertiary Structures*)

The chemical constitution of a protein as shown in the amino acid sequence is only one of several aspects of the structure. The linear sequence of a protein chain, which is genetically determined, controls the secondary and tertiary structures and therefore also its biological activity. No attempt will be made in the present discussion to draw a distinction between secondary and tertiary structures of proteins.

The native conformation of a protein is spontaneously adopted after its biosynthesis and is ultimately determined by the amino acid sequence, partly through the covalent peptide and disulphide bonds but also through stabilisation by interactions between side chain functions. The latter effect depends upon the type and character of the side chain of every individual amino acid in the peptide chain.

3.6.2.1. Conformation of Protein-backbone

The protein-backbone conformation is defined as the arrangement of the polypeptide chains without regard to the amino acid side chains. The pioneering work of L. PAULING and R. B. COREY enabled protein-backbone conformations to be determined. The bond distances as well as valency angles of amino acids, amides and simple peptides as examples were determined by X-ray diffraction see p. 65. It was shown that the peptide bond

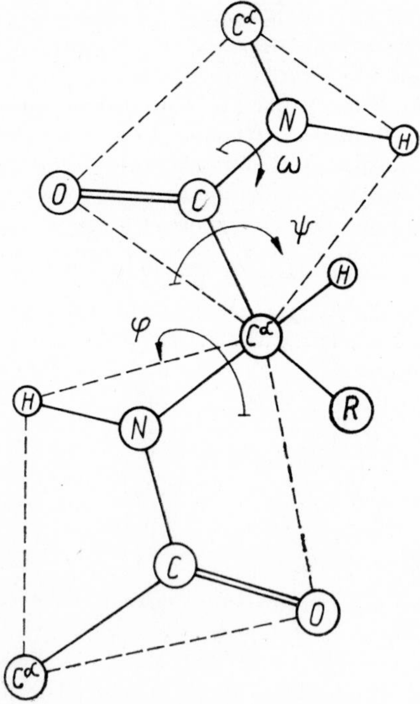

Figure 43. A pair of peptide units in a fully extended *trans* conformation ($\varphi = \psi = \omega = 0$)

possesses a high degree of double-bond character (about 50 per cent). The peptide bond is planar, and therefore the polypeptide chain has only two degrees of freedom: rotation about the C_α—N bond (dihedral angle φ) and rotation about the C_α—C bond (dihedral angle ψ). The three-dimensional structure of the chain is determined by φ, ψ values of all the amino acid residues.

The coplanarity of the peptide bond results in a dihedral angle $\omega = 0$, although a small departure from planarity has been shown in X-ray diffraction studies of some proteins. When rotation occurs in a clockwise manner with reference to the C_α-atom, φ and ψ are positive. It is important to note that not all possible combinations of the angles φ and ψ yield reasonable structures. G. N. Ramachandran *et al.*, using models and computers, examined all possibilities of angle combination of φ and ψ. The figures compiled in Table 25 represent the minimum distances for atoms not covalently connected.

The data in Table 25 formed the basis for deciding which steric conformations were possible and which were impossible. Ramachandran and his group have constructed plots of ψ against φ, indicating which areas are allowed on the basis of the forbidden interactions (Figure 44). In this Ramachandran map of the *trans*-peptide conformation, the allowed φ, ψ angle combinations are located in the shadowed area. In this case, the normal atomic radii (see Table 25) were taken into consideration. By applying the minimal contact distances the area of allowed conformations is extended to that outlined by the dotted line in the Ramachandran plot. Fully allowed conformations in the Ramachandran map represent conformations of maximum stability. Some pairs of angles provide stable structures while others do not because they bring non-bonded atoms too

Table 25. Minimum distances (Å) between non-covalently connected atoms calculated from van der Waals' radii. The values in parentheses represent minimum distances obtained by Ramachandran

	C	O	N	H
C	3.20 (3.00)	2.80 (2.70)	2.90 (2.80)	2.40 (2.20)
O		2.70 (2.60)	2.70 (2.60)	2.40 (2.20)
N			2.70 (2.60)	2.40 (2.20)
H				2.00 (1.90)

close together. In the case of $\psi = 180°$ and $\varphi = 0°$, the two adjacent carbonyl oxygen atoms overlap to such a degree that the structure is an impossible one. Similarly, if $\psi = 0°$ and $\varphi = 180°$, the amide hydrogen

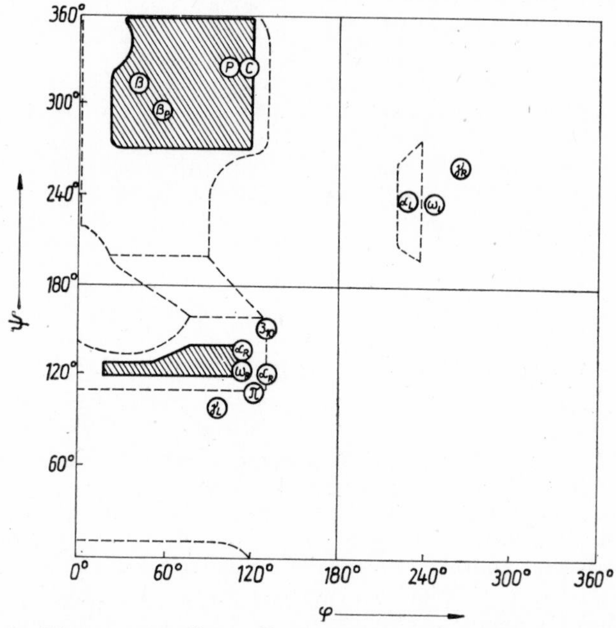

Figure 44. φ,ψ Diagram according to RAMACHANDRAN. Representative conformations are P = polyproline helix; C = collagen helix; β = anti-parallel pleated sheet; β_p = parallel pleated sheet; α_R and α_L = right-handed and left-handed α-helix, respectively; ω_R and ω_L = right-handed and left-handed ω-helix, respectively; γ_R and γ_L right-handed and left-handed γ-helix, respectively; 3_{10} = 3_{10}-helix; π = π-helix

atoms overlap so badly that the structure is impossible. The effect on the allowed structures of side chain functions bearing space-filling groups is significant. The calculation of the φ, ψ plot represented in Figure 44 is based on a methyl side chain (alanine). In the cases of valine and isoleucine, for example, the area of fully allowed conformations is smaller.

3.6.2.1.1. Helical Structures

As a result of identical sign of φ and ψ on every C_α-atom of a peptide chain, a helix is formed with a definite number of units per turn (n), a characteristic pitch of the helix (h), and a specific distance travelled parallel to the helix axis per residue (d). Assuming coplanarity of the peptide bond

($\omega = 0$), the values n and d are determined by the angles φ and ψ: the pitch of the helix (h) is the product of n and d ($h = n \cdot d$). An example of a helix

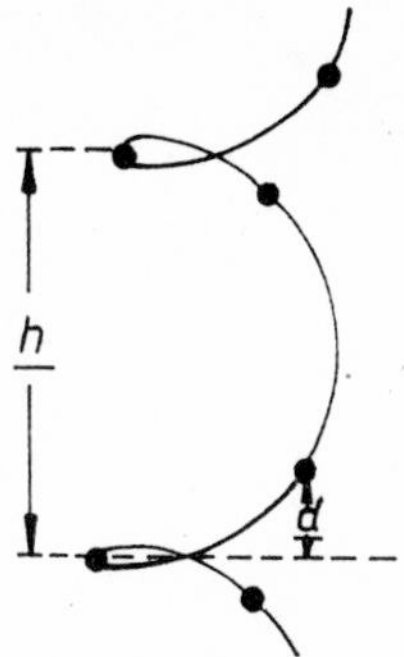

Figure 45. A helix with three units per turn (according to H. HAGENMAIR)

with 3 amino acid residues per turn ($n = 3$) is shown in Figure 45. Different helical structures are possible, based on the occurrence of hydrogen bonds between different peptide groups. The best-known example of these structures is the α helix (Figure 46) with the parameters $n = 3.6$, $d = 1.5$ Å and $h = 5.4$ Å. An α-helix is generated if φ is 132° and ψ is 123°. In the case of the α-helix, a ring of 13 atoms is formed by intramolecular hydrogen bonds. The correct definition is, therefore, $\alpha(3.6_{13})$-helix. Other backbone conformations of the helix type which have been described are the 3_{10}-helix (Figure 45), the $\pi(4.4_{16})$-helix and the $\gamma(5.1_{17})$-helix. Both right-handed and left-handed forms of the α-helix are known; the former occurs in natural compounds.

In the globular proteins, the α-helix is the dominant helical conformation, although the 3_{10}-helix has also been detected in them (in, for example, haemoglobin and lysozyme). In α-keratin the pitch of the helix is 5.1 Å, shorter than the corresponding identity period of the normal α-helix (5.4 Å). Keratin will be discussed further in Section 3.6.4.

3.6.2.1.2. Pleated Sheet Structures

Maximum formation of hydrogen bonds between two adjacent peptide chains leads to the so-called pleated sheet. The steric hindrance of the side chains necessitates a small rotation of the peptide groups opposite to each other. The resulting structure is also known as a β-structure.

Depending on whether the two peptide chains are arranged parallel or antiparallel to each other, it is possible to differentiate between *parallel pleated sheets* (a) and *anti-parallel pleated sheets* (b). The former is composed of a series of peptide chains that are parallel in the sense of having all their

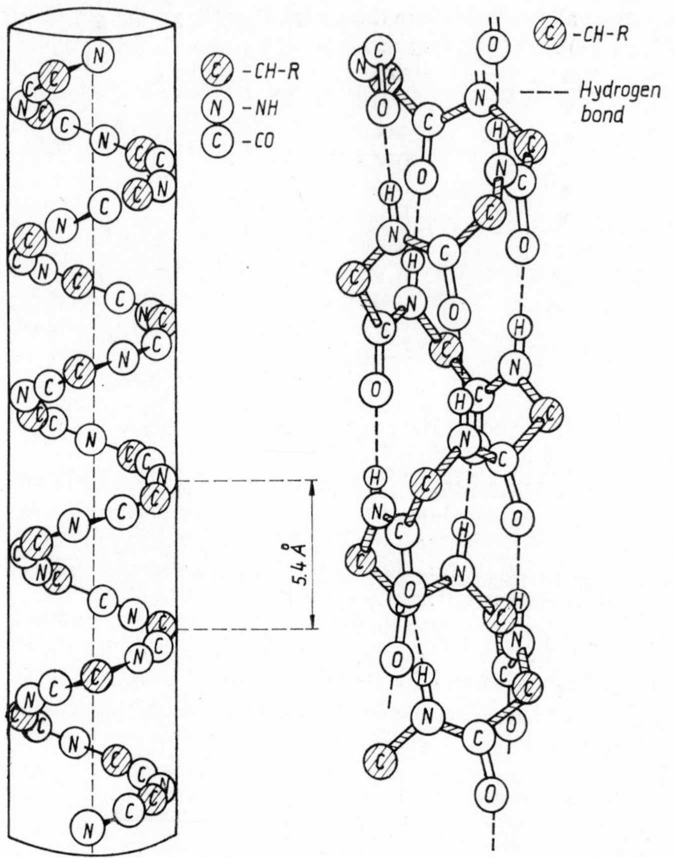

Figure 46. An α-helix (from S. M. RAPOPORT, *Medizinische Biochemie*, Verlag Volk und Gesundheit, Berlin, 1968)

N-termini at the same end, whereas the latter form has every other chain pointed in the opposite direction:

O=C   O=C
    N—H   N—H
RHC   RHC
    O=C   C=O
H—N   H—N
    CHR   CHR
O=C   O=C
    N—H   N—H

(a)

O=C   N—H
    N—H------O=C
RHC   CHR
    C=O------H—N
H—N   C=O
    CHR   RHC
O=C   N—H
    N—H------O=C

(b)

Figure 47 shows a diagrammatic representation of an anti-parallel pleated sheet structure, where successive side chains are located alternately over and under the pleated sheet. The corresponding pair of angles φ and ψ for parallel and anti-parallel pleated sheets can be deduced from Figure 44.

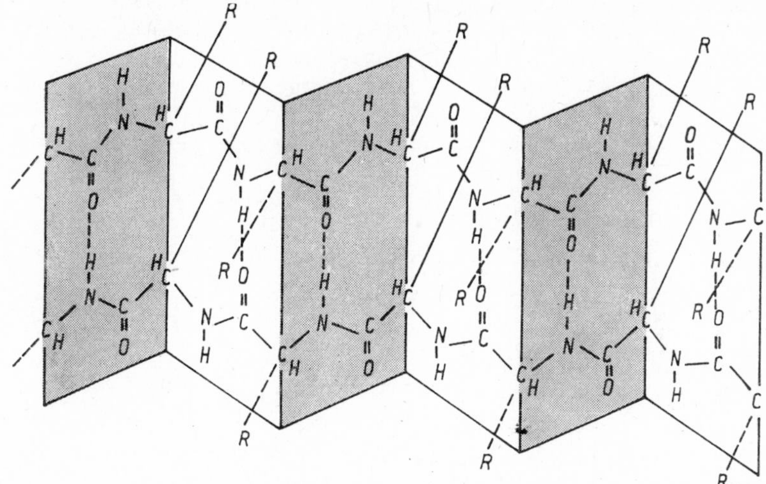

Figure 47. The anti-parallel pleated sheet structure

Proteins which contain extensive sheet structures are fibrous and generally insoluble in aqueous solvents. Fibrous proteins will be discussed later (see Section 3.6.4). β-Structure may also occur between different segments of polypeptide chains in globular proteins—for example, in carboxypeptidase A and in lysozyme.

3.6.2.1.3. Random Conformation

Conformations other than helical structures and pleated sheet structures are often described as random coils. This term is often used in connection with denaturation; one speaks, for example, of a helix–coil transition.

The angles φ and ψ of all amino acid residues in a helical structure, for example, occur at a fixed point of the Ramachandran plot (see Figure 44). However, in random conformations, the angle-pairs φ and ψ of the different amino acid residues occur at different points of the Ramachandran map. Accordingly, a great number of possible conformations may result, although the actual distribution is not completely random in the strict statistical sense. The molecules of globular proteins tend to be roughly spherical and devoid of the regular repeating units characteristic of the structural proteins (see Section 3.6.4). In addition, the folded structure adopted by globular proteins is highly specific, as is indicated by the extreme sensitivity of the biological activity to factors which influence it. In α-chymotrypsin, for example, only eight residues of the C-chain adopt an α-helical conformation and the remainder of the residues occur in practically fully extended chains, which run mostly parallel to each other, 5 Å apart.

There is a great deal of interchain hydrogen bonding between the three peptide chains of α -chymotrypsin, but the structure is not that of a regular pleated sheet. Also, the non-ordered regions in α-chymotrypsin cannot be compared with a random coil, because the φ,ψ angle-pairs have fixed values for each amino acid residue.

3.6.2.2. Methods for Elucidation of the Three-dimensional Structure of Proteins

X-ray diffraction analysis is one of the most powerful methods for the determination of the three-dimensional structure of proteins. X-ray diffraction patterns are obtained from protein crystals mounted in a quartz capillary tube on a stage in line with a beam of X-rays. The positions and intensities of the diffraction maxima are recorded with the aid of cameras or counting devices. The crystal is mounted so that it can be rotated about three perpendicular axes, and all possible orientations of the crystal with respect to the incident X-ray beam can be achieved. The whole process is nowadays highly automated and the operation of the diffractometer is computer-controlled[1,2].

The calculation of the distribution of the electron density is carried out using the following equation:

$$\varrho(x, y, z) = \frac{1}{V} \sum_{h=-\infty}^{+\infty} \sum_{k=-\infty}^{+\infty} \sum_{l=-\infty}^{+\infty} |F(hkl) \cos 2\pi| [hx + ky + lz - \alpha(hkl)]$$

The distribution of the electron density $\varrho(x, y, z)$ in the unit cell is calculated. For the characterisation of the reflections the MILLER indices hkl are required. $F(hkl)$ are the observable structure amplitudes of the crystal reflection hkl. V is the volume of the unit cell and $\alpha(hkl)$ the phase angle. The last must be determined using indirect methods, while the remaining values may be obtained from the X-ray diffraction patterns. The phase problem can be solved using the method of isomorphous replacement. For this purpose heavy atoms are introduced into the crystal structure of the protein in such a way that the size and symmetry of the unit cell remains unaltered.

Heavy atoms act as intense scattering centres and are relatively easy to locate in the unit cell from the differences between the diffraction patterns of the crystalline protein and the crystalline heavy-atom derivatives. After the location of the heavy atoms it is possible to determine the phase angles.

The electron densities calculated for every point are usually represented in an electron density diagram in which points relating to the same electron density are connected together by contours. Further, it is possible to construct three-dimensional electron density maps which indicate the electron density at that depth.

Carrying out an X-ray diffraction structure analysis is a very laborious task, even with the very high-speed electronic computers which are used to handle the data.

The resolving power of X-ray diffraction techniques is generally about 1 Å. Therefore, it is theoretically possible to identify each of the atoms in the protein crystal. Since the determination of electron density with a resolution of about 5 Å is not so tedious as a fine resolution, the former technique is usually preferred in the first stage of structural investigation of protein crystals. However, the fine structure (that is, the determination of the position of all individual atoms) is only visible at 1.5 Å resolution.

During the period 1952–1960 the groups of J. C. KENDREW and F. M. PERUTZ laid the foundations for research in this area. In 1968 HOPPE[3] reported briefly on current research, including finished and nearly finished structural analyses. In the following sections, the three-dimensional structures of myoglobin, lysozyme, ribonuclease, carboxypeptidase and haemoglobin will be discussed as examples.

In addition to these proteins, X-ray diffraction analyses of the following proteins have been carried out: papain[4], erythrocruorin[5], insulin[6], elastase[7], cytochrome c[8], pepsin[9], staphylococcal elastase[10], α-chymotrypsin[11], immunoglobulins[12], β-lactoglobulin, tobacco mosaic virus, etc.: see also the review by DICKERSON and GEIS[13].

The possibility of applying *electron beams*[14] or *high-resolution neutron diffraction*[15] to protein crystal structure analysis is under investigation.

Although the most important results so far obtained have come from the use of the methods discussed above and especially from X-ray diffraction studies, there are reasons for carrying out studies of conformation under actual physiological conditions and spectroscopic methods are best suited for this purpose.

Nuclear magnetic resonance (NMR) spectroscopy had been used for the investigation of protein conformation about 20 years ago. The sensitivity of the spectrometer used at that time and its resolution were both low and did not allow easy interpretation of the spectra. The use of high-frequency instruments with much greater resolving power and sensitivity has led to the development of high-resolution proton magnetic resonance spectroscopy[16, 17]. However, relatively little information can be obtained from proton resonance spectra of proteins even at 220 MHz. This is attributed to incomplete resolution, which results from the large number of proton resonances occurring within the spectral range of the resonances of individual amino acids. Furthermore, the high molecular weight of proteins tends to increase line widths. It has already been mentioned in the discussion of proton magnetic resonance spectra of amino acids (see pp. 21–28) that only a few characteristic absorption ranges are usually chosen for interpretation. Better results from the point of view of resolution are to be expected from unusual resonance positions outside the spectral range of the resonances of the bulk of the amino acid residues. Excellent information on the conformation of proteins is obtained by a direct comparison of the proton magnetic resonance spectra of individual amino acids with those of corresponding protein in its native and random coil conformation.

It is to be expected that further progress in protein conformation studies will be achieved in the future by applying the magnetic resonance properties of other nuclei—for example, ^{13}C, ^{19}F, ^{15}N—in combination with pulse FOURIER transformation techniques.

Infra-red spectroscopy can also be applied to the study of conformational changes in polypeptide chains (α-helix, pleated sheet, random coil conformation). The characteristic amide I and amide II bands change as the conformation changes and each of these bands exhibits two maxima in the ordered structures. It is possible to differentiate between α-helix and β-structures by *infra-red dichroism* studies. Infra-red dichroism is the dependence of the infra-red absorption on the orientation of the solid sample.

Ultra-violet difference spectroscopy is often employed in protein conformational studies. Information about the conformation of proteins in aqueous solution can be obtained from red or blue shifts in the spectrum, which are not necessarily related to intensity changes.

Although *optical rotatory dispersion* (*ORD*) and *circular dichroism* (*CD*) studies of several proteins have been carried out[18], the interpretation of the experimental data is relatively difficult. Both ORD and CD data have been employed in efforts to estimate the helical content of polypeptides, and trials have been carried out to try to correlate Cotton effects with results of the X-ray diffraction analysis. Applying the latter method, it was found that the α-helix content is high in myoglobin and lower in lysozyme (with β-structure content), carboxypeptidase A and papain. The helix contents of ribonuclease and chymotrypsin are very low. The X-ray diffraction results correlate well with both ORD and CD data in the cases of myoglobin and lysozyme (which are both proteins of a high helix content) but not so well in the case of chymotrypsin. It is probable in the case of non-helical proteins that the Cotton effects are strongly influenced by the presence of other chromophores. For example, the inherent asymmetry of disulphide bonds and aromatic residues in the asymmetric surroundings will produce a pronounced perturbation. In addition, the contribution to the optical rotation of the β-structure portions is much weaker than that of the α-helix. The situation is further complicated by the possible presence of structures other than α-helices, which can also induce Cotton effects. These and other studies[19] indicate that it may be unwise to assume too close an analogy between the ORD and CD spectra of simple synthetic polyamino acids and those of proteins.

In spite of these limitations, the ORD and CD studies of proteins are extremely important for the understanding of their conformations. Very useful results can be obtained by comparison of data from native and denatured proteins as well as by carrying out these investigations on chemically modified proteins.

The number and type of hydrogen bonds in a protein can be determined by *hydrogen–deuterium exchange.* It is a well-known fact that hydrogen bound to atoms containing unshared pairs of electrons can exchange with protons of the solvent. In the case of hydrogen bound to carbon, exchange is impossible. Thus exchange of the proton bound to the nitrogen atom of the peptide bond can be followed by observing the disappearance of the N–H deformation band at 1550 cm^{-1} in D_2O. The total exchangable hydrogen can be substituted by deuterium by prolonged treatment with D_2O. The kinetics of deuterium release in water can be determined by infra-red spectroscopy or by density measurement. The usefulness of the results obtained is generally limited by the purity of the protein studied.

It should be clear from the discussion above that any one of the various methods available for the study of polypeptides and proteins in solution gives only part of the over-all picture. To get a full picture, the results of several methods must be combined.

3.6.2.3. Structure of Myoglobin

The first three-dimensional X-ray structure analysis of a protein to be published was that of sperm-whale myoglobin. It was carried out by Kendrew *et al.*[20]. At the start of their work, the primary structure was unknown. Studies on the primary structure ran parallel with the X-ray diffraction studies. The complete sequence was published in 1965 by Edmundson[21].

Myoglobin consists of a single polypeptide chain of 153 amino acid residues with one haem group. It has a molecular weight of 17 816. Myoglobin is the oxygen storage protein of muscle. In the whale, myoglobin is the oxygen-transporting protein, while in other mammals haemoglobin (see p. 307) performs this function.

The elucidation of the three-dimensional structure of myoglobin carried out by the Kendrew group was performed in three stages. The early work involved the measurements on unmodified protein crystals and various mercury and gold derivatives. This yielded a structure at 6 Å resolution which showed that the polypeptide chain was folded in an intricate compact structure (Figure 48).

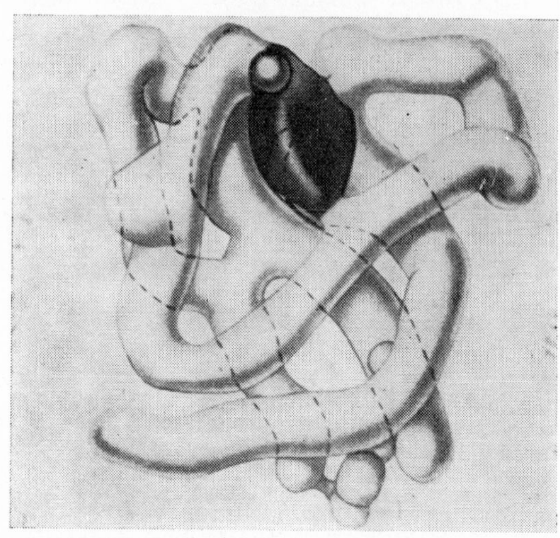

Figure 48. Sperm-whale myoglobin at 6 Å resolution (KENDREW *et al.* [20])

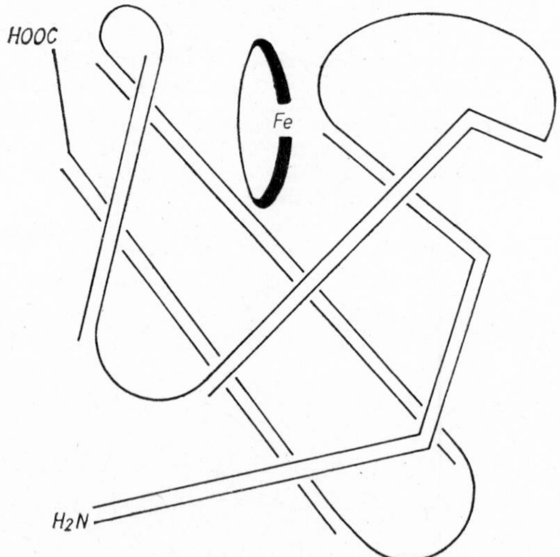

Figure 49. The polypeptide chain of myoglobin (the α-helical regions are represented by double lines)

To obtain a resolution of 2 Å, four isomorphous derivatives were examined, involving the determination of the amplitude and phase of 960 Fourier components. The polypeptide chain forms a right-handed helix whose characteristics closely resemble those of the α-helix in α-keratin. The content of α-helix is around 75–80 per cent. The regular spiral is, however, interrupted by loops and corners. Amino acids with hydrophobic side chains are oriented so that their side chains lie in the interior of the myoglobin molecule. Nearly all the polar groups lie on the surface of the molecule, exposed to the solvent. The possibility of contact with the surrounding water molecules leads to further stabilisation of the conformation. The general arrangment of the myoglobin molecule provides a compact enclosure for the haem group.

Figure 50. Electron density distribution of a part of the polypeptide chain of myoglobin (2 Å resolution) according to Kendrew *et al.* [20]

The existence of the right-handed helix was clearly shown by the 2 Å resolution. Figure 50 (a) shows the electron density distribution of the cylindrical projection of a part of the polypeptide chain (with the superimposed redrawn α-helix). Figure 50 (b) illustrates the arrangements of atoms in the α-helix, where the points β and β' indicate the alternatives for the arrangement of the β-carbon atom.

Finally, high-resolution studies (1.4 Å) were carried out on myoglobin, which permitted identification of the atomic coordinates of the polypeptide chain.

The tertiary structure of *erythrocruorin* obtained from the larvae of *Chryronomus* species[22] and elucidated by Huber *et al.*[5] is similar to that of myoglobin. Larvae of this species are only able to exist in oxygen-poor pools and waters because of the ability of erythrocruorin to store oxygen.

3.6.2.4. Structure of Lysozyme

Lysozyme was the first enzyme whose three-dimensional structure was elucidated by X-ray diffraction studies.

Lysozyme degrades the polysaccharides of bacterial cell walls. One of the best substrates is an alternating copolymer of *N*-acetyl-D-glucosamine and *N*-acetyl muramic acid (see below).

Although several groups have studied the amino acid sequence of lysozyme, the sequence published by Canfield and Liu[23] is generally considered to be the correct one (Figure 51). The results of the sequence studies,

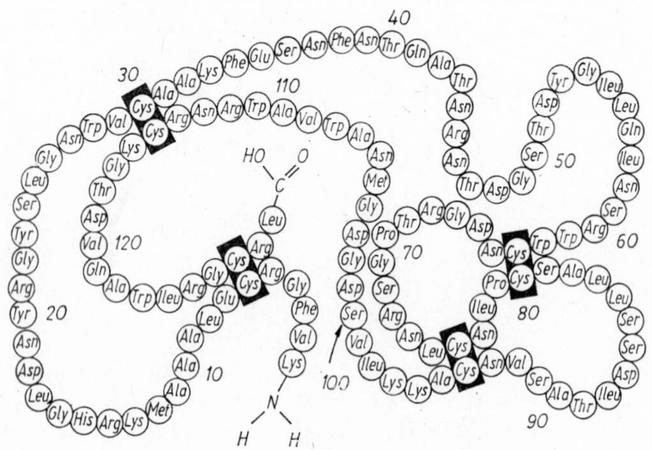

Figure 51. Primary structure of lysozyme according Canfield and Liu[23]

which were crucial to the interpretation of the X-ray diffraction data obtained by Phillips and co-workers[24], showed that lysozyme consists of a single polypeptide chain with 129 amino acid residues and four intramolecular disulphide bridges and a molecular weight of 14600.

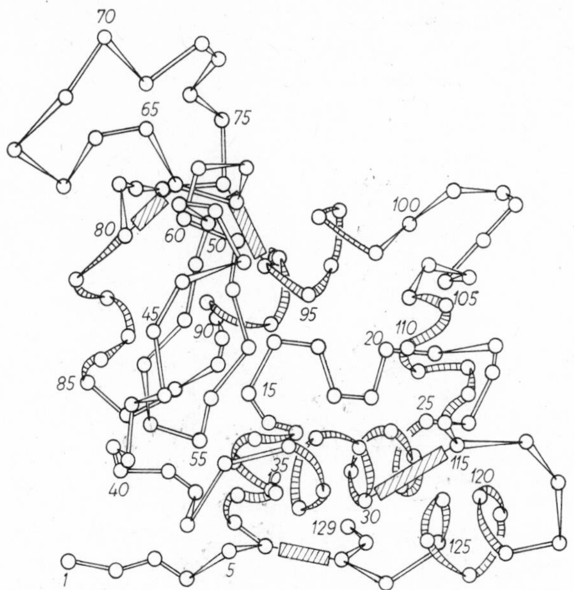

Figure 52. Lysozyme at 2 Å resolution (Phillips *et al.* [24])

The three-dimensional structure of lysozyme at a resolution of 2 Å (Figure 52) is rather more complicated than that of myoglobin. The lysozyme molecule is ellipsoidal with dimensions of 45 Å × 30 Å × 30 Å. Only about 25 per cent of the 129 amino acid residues occur in α-helical regions: the sequence 41–54 is arranged in an anti-parallel pleated sheet structure.

The first information about the active size of the enzyme was obtained from X-ray diffraction studies of lysozyme–inhibitor complexes[25]. It was shown that the groove on the surface of the molecule is large enough to accommodate six monosaccharide units of the substrate. In addition, model building based on the known location of inhibitors gave an indication how the enzymatic cleavage of the glycosidic bond might occur. The model hexasaccharide substrate is held in a cleft in the lysozyme molecule. In addition to apolar bonds, there are well-defined hydrogen bonds responsible for locating the substrate in such a way as to interact favourably with the lysozyme. Figure 53 shows schematically the way in which the catalytic action of lysozyme operates. Lysozyme catalyses the cleavage of the

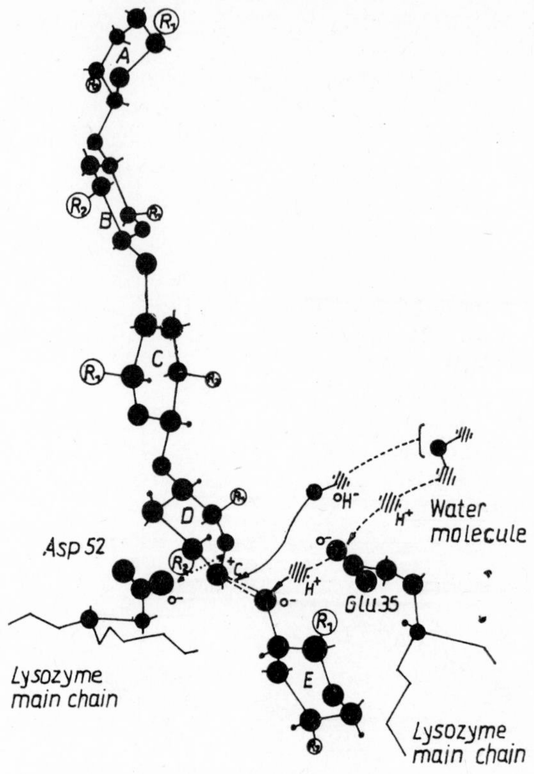

Figure 53. Mechanism of action of lysozyme as proposed by PHILLIPS [26]

C_1—O bond between the hexose unit D and the hexose residue E. To achieve this it is necessary to distort ring D from its normal chair conformation into half-chair conformation so that the oxygen atom and the carbon atoms 1, 2 and 5 lie in the same plane. The side chains of the two carboxylic amino acids Glu^{35} and Asp^{52} of the lysozyme chain participate in the catalysis and lie in the vicinity of the glycosidic bond to be broken. The glycosidic oxygen atom is protonated by the glutamic side chain, which is normally undissociated in its hydrophobic environment and can therefore act as a general acid catalyst. After cleavage of the C_1—O bond a carbonium ion is formed at C_1 which is stabilised by the planarity of the ring D as well as through electrostatic interaction with the side chain of the aspartic acid residue which is present (because of its hydrophilic environment) as a carboxylate ion. The catalytic process is completed by hydroxylation of the carbonium ion.

A polypeptide with lysozyme activity was synthesised using the MERRI-technique by SHARP *et al.* in 1973. The specific activity of the crude product was 0.5–1 per cent that of native lysozyme (or 9–25 per cent that of native lysozyme which had been similarly treated with HF).

3.6.2.5. Structure of Ribonuclease

Ribonuclease A, isolated from bovine pancreas, consists of a single polypeptide chain with 124 amino acid residues and four intramolecular disulphide bonds (M.W. = 13700).

Ribonuclease A (RNase A) is an endonuclease, capable of catalysing the hydrolysis of the phosphodiester bonds of ribonucleic acids.

R. DUBOS was the first to isolate the enzyme from bovine pancreas: it was crystallised in 1940 by M. KUNITZ. SMYTH, STEIN and MOORE[27] reported the correct primary structure (Figure 54) in 1963. F. H. WHITE and C. B.

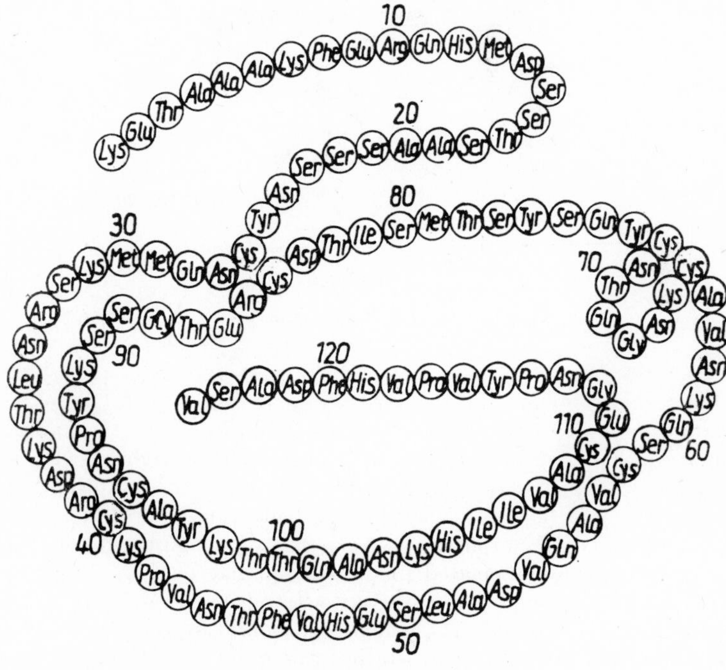

Figure 54. Primary structure of ribonuclease A (SMITH, STEIN and MOORE [27])

ANFINSEN (1962) showed that reduction of the disulphide bonds of RNase A yielded a biologically inert product, but that most of the enzymatic activity was recovered on oxidation. RICHARDS[28] discovered that ribonuclease A can be cleaved by the enzyme subtilisin at the peptide bond between Ala^{20} and Ser^{21} to yield two fragments, the S-peptide and the S-protein. An equimolar mixture of S-peptide and S-protein possesses all the enzymic activity of the parent enzyme and is called ribonuclease S (RNase S). On separation of the two products by gel filtration or other techniques, the S-peptide and S-protein are (separately) quite inactive, but mixing stoichiometric amounts of the two generates a fully active enzyme complex (ribonuclease S′) even though the covalent bond has not been reformed (Figure 55).

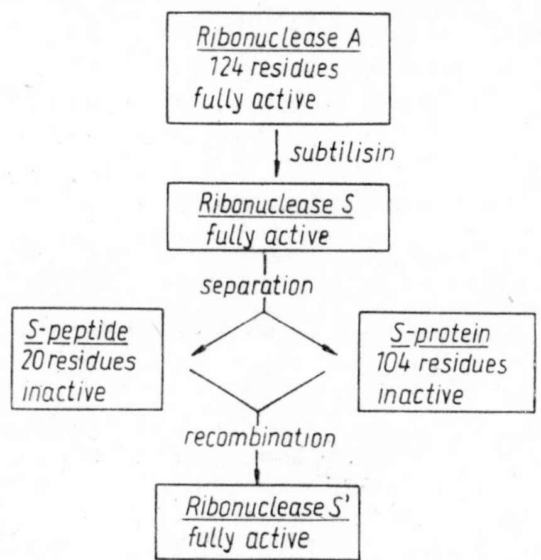

Figure 56. The enzymatic cleavage of ribonuclease A and recombination of the fragments

HOFMANN *et al.*[29] synthesised the S-peptide and mixed their synthetic product with an equimolar amount of S-protein. The mixture possessed all of the activity of the enzyme. Many analogues of the S-peptide have been synthesised in efforts to understand the relation between primary structure and activity. It has been shown that the sequence 8–12 is essential to the catalytic function of the enzyme. Further, Glu^2 and the sequence 13–15 (–Met–Asp–Ser) are not essential for activity but participate in the binding of S-peptide to S-protein. His^{12}, His^{119} and Lys^{41} take part in the catalytic process and are located in the binding site of the enzyme. The great distance between these residues in terms of position in the chain suggests that a special chain conformation is necessary for activity, and this was borne out by the results of a three-dimensional structural analysis. The complete structure of ribonuclease A at a resolution of 2 Å has been determined by KARTHA, BELLO and HARKER[30], whose results indicated that the active site did include indeed these amino acid residues. The active site is located in a depression in the side of the kidney-shaped molecule, which measures 38 × 28 × 22 Å. It was shown that the sequences 6–12 (–Ala–Lys–Phe–Glu–Arg–Gln–*His*–) and 117–120 (–Pro–Val–*His*–Phe–) are located on one side of the depression, while the sequence 41–48 (–*Lys*–Pro–Val–Asn–Thr–Phe–Val–His–) lies on the opposite side (Figure 56). K. HOFMANN's results ob-

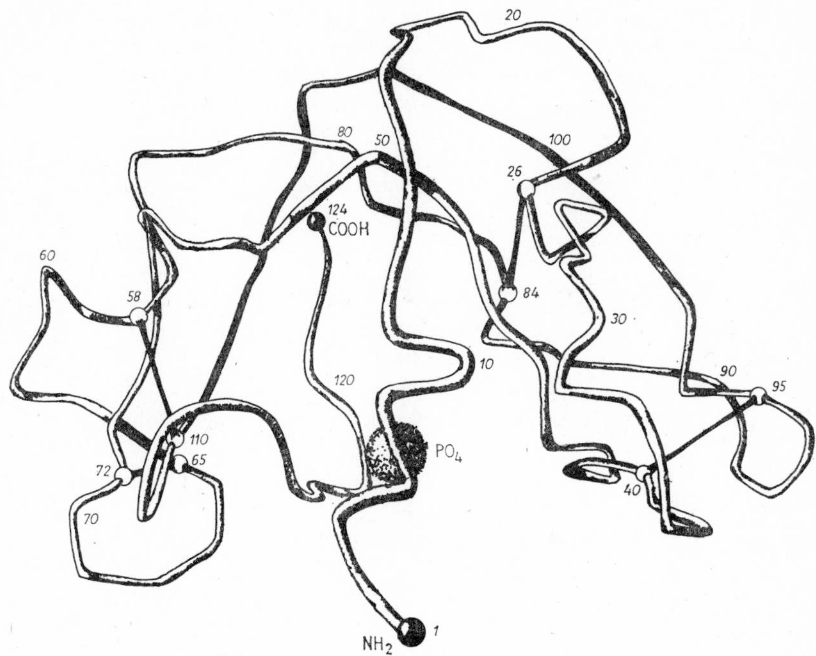

Figure 55. Ribonuclease A (KARTHA *et al.* [30])

tained from studies of the S-peptide agree well with the results of the X-ray diffraction studies. RICHARDS, WYCKOFF *et al.*[31] have described a three-dimensional structure analysis of ribonuclease S at 3.5 Å resolution. In general, the conformations of the two enzymes are quite similar. The sequences 2–13, 26–33 and 50–58 are α-helical. In addition, a large proportion of the polypeptide chain is in an anti-parallel β-pleated sheet structure involving residues 42–29, 71–92, 80–86, 94–110 and 116–124. The charged side chains of the individual amino acids are located on the outer surface with the exceptions of Asp^{14} and His^{48}. In contrast, the interior of the ribonuclease molecule is strongly hydrophobic, as in the case of lysozyme. SCOFFONE *et al.*[32] have synthesised many analogues of the S-peptide and their conformational studies showed that, in aqueous solution, the S-peptide forms a random coil. This is in contrast to the situation in native enzyme, where more than 50 per cent of amino acid residues of the S-peptide sequence are involved in helical structures. It has been suggested that the formation of the enzyme–substrate complex involves interaction of the phosphate group of the nucleotide with the ε-amino group of Lys^{41}, with hydrogen bonding of the pyrimidine via a water molecule to one of the histidine residues. A conformational change brings the other histidine residue within hydrogen-bonding distance of the 2-hydroxyl group of the ribose ring. This arrangement facilitates the formation of 2,3-cyclic phosphate and, through this intermediate, the cleavage of the substrate.

Three groups have tried to synthesise ribonuclease as an example of the total synthesis of an enzyme. R. B. MERRIFIELD and B. GUTTE (1969) synthesised the polypeptide chain of 124 amino acid residues by the solid-phase method (see p. 149). The synthetic product showed 78 per cent of the activity of the native enzyme after extensive purification. The MERCK group (see p. 180) synthesised the S-protein of ribonuclease, since the S-peptide had been synthesised by K. HOFMANN *et al.* some time ago. Combination of synthetic S-protein with the S-peptide gave a product which showed about 30 per cent of the specific activity of ribonucleases.

K. HOFMANN *et al.* (see p. 180) are at work on the synthesis of ribonuclease T_1, the sequence of which was reported by K. TAKAHASHI in 1965. Their synthetic approach involves the synthesis of six fragments each with C-terminal-protected hydrazide groups and a fragment which forms the sequence of the C-terminal tetracosapeptide amide. N. IZUMIYA *et al.*[33] are also working on the synthesis of ribonuclease T_1.

3.6.2.6. Structure of Carboxypeptidase A

Carboxypeptidase A (CPA) is a zinc-containing enzyme, which catalyses the hydrolysis of C-terminal peptide bonds and which can degrade a peptide chain in a stepwise fashion (see p. 280). It is secreted in the pancreas as an inactive zymogen, procarboxypeptidase A. Tryptic activation of the precursor gives rise to four forms of CPA: CPA_α, CPA_β, CPA_γ, CPA_δ. During this process about 64 amino acid residues are released from the N-terminus part of procarboxypeptidase A. W. N. LIPSCOMB *et al.* showed by X-ray diffraction studies that CPA α contains 307 amino acid residues in 1967, and H. NEURATH *et al.* established the complete primary structure of CPA α 2 years later. The single polypeptide chain consists of 307 residues (M.W. 34 472): there are two forms, CPA_α^{Val} having Ile ^{179}, Ala^{228}, Val^{305} and CPA_α^{Leu} having Val^{179}, Glu^{228}, and Leu^{305}. These investigations provide an interesting example of how X-ray studies and chemical sequence analysis can be used to complement each other. Comparing the results of the chemical sequence determination with the X-ray diffraction results showed that only 60–85 per cent of the side chains had been correctly identified in the X-ray studies.

Between 1966 and 1968, W. N. LIPSCOMB's group at Harvard[34, 35] carried out a study of the three-dimensional structure of CPA using the isomorphous replacement method. They achieved resolutions of 6 Å, 2.8 Å, and 2 Å (Figure 57), and showed that CPA is an ellipsoidal molecule, having dimensions of 50 by 42 by 38 Å. It has about 17 per cent of β-structure, with an α-helical content of about 38 per cent. The zinc atom is located in a depression in the surface which is adjacent to a pocket which extends into the interior of the molecule. The zinc ligands are His[69], Glu[72], His[196] and one molecule of water. The FOURIER difference technique has been used to study the bonding of substrates and inhibitors to carboxypeptidase A. In particular the complex of Gly–Tyr with CPA has been studied at 2 Å, and the

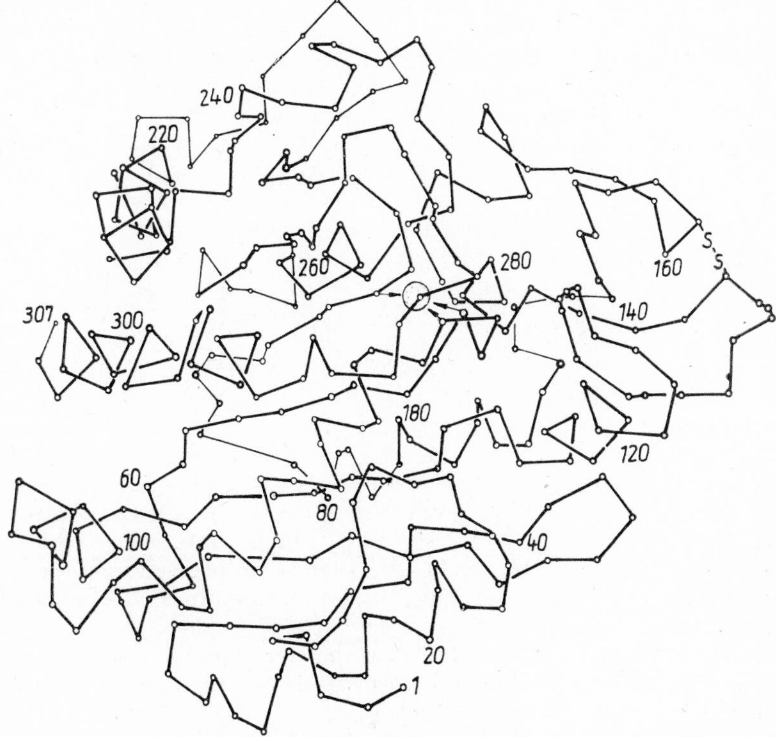

Figure 57. The folding of the polypeptide chain in carboxypeptidase A (LIPSCOMB *et al.* [35])

complex of Phe–Gly–Phe–Gly at 2.8 Å resolution. In this manner LIPSCOMB's group have identified the active site and revealed some of the details of the way in which substrates are bound. It appears that the C-terminal side chain of the substrate fits into a so-called 'dead-end' pocket, and the C-terminal carboxylate group forms an ionic link with Arg^{145}. The carbonyl oxygen of the peptide bond which is to be cleaved binds to the zinc atom, displacing a water molecule. As the substrate is bound, considerable conformational change takes place. The Arg^{145} is changed in a manner which allows Tyr^{248} to donate a proton to the NH group of the peptide bond. In addition, the glutamic acid residue at position 270 is brought into an orientation such that it can attack the carbonyl carbon atom of the peptide bond. The zinc atom plays a fundamental role both in controlling the over all stereochemistry and in polarising the peptide bond which is to be cleaved.

References

1 HOLMES, K. C. and BLOW, D. M. (1966). *The Use of X-ray Diffraction in the Study of Protein and Nucleic Acid Structure*, Wiley/Interscience, New York

2 FRUBERG, S. (1967). *Naturw. Rundsch.*, **20**, 185

3 HOPPE, W. (1968). *Naturwissenschaften*, **55**, 65

4 DRENTH, J. *et al.* (1968). *Nature*, **218**, 929; (1971). *Advan. Protein Chem.*, **25**, 79

5 HUBER, H. *et al.* (1969). *Naturwissenschaften*, **56**, 262

6 HODGKIN, D. *et al.* (1969). *Nature*, **224**, 491

7 WATSON, H. C. *et al.* (1970). *Nature*, **225**, 806, 811

8 DICKERSON, R. E. *et al.* (1970). International Congress of Biochemistry, abstracts, p. 13

9 ANDREEVA, N. S. *et al.* (1970). International Congress of Biochemistry, abstracts, p. 45

10 ARNONE, A. *et al.* (1970) International Congress of Biochemistry, abstracts, p. 48

11 MATTHEWS, B. W., SIGLER, P. B., HENDERSON, R. and BLOW, D. M. (1967). *Nature*, **214**, 652; see also KRAUT, J. *et al.* (1967). *Proc. Nat. Acad. Sci.* (*U.S.A.*), **58**, 304

12 POLJAK, R. J. (1970). International Congress of Biochemistry, abstracts, p. 294

13 DICKERSON, R. E. and GEIS, J. (1969). *The Structure and Action of Proteins*, Harper and Row, New York

14 SCHOENBORN, B. P. (1970). International Congress of Biochemistry, abstracts, p. 3

15 HOPPE, W., LANGER, R., KNESCH, G. and POPPE, Ch. (1968). *Naturwissenschaften*, **55**, 333

16 WÜTHRICH, K. (1970). *Chimia*, **24**, 409

17 BOVEY, F. A. (1969). *Nuclear Magnetic Resonance Spectroscopy*, Academic Press, New York

18 JIRGENSONS, B. (1969). *Optical Rotatory Dispersion of Proteins and other Macromolecules*, Springer Verlag, Berlin, Heidelberg, New York; BEYSCHOK, S. (1966). *Science*, **154**, 1288

19 JIRGENSONS, B. (1970). *Biochim. Biophys. Acta,* **200**, 9

20 KENDREW, J. C., DICKERSON, R. E., STRANDBERG, B. E., HART, R. G., DAVIES, D. R., PHILLIPS, D. C. and SHORE, V. S. (1960). *Nature,* **185**, 422

21 EDMUNDSON, A. E. (1965). *Nature,* **205**, 883

22 see reference 5

23 CANFIELD, R. E. and LIU, A. K. (1965). *J. Biol. Chem.,* **240**, 1997

24 BLAKE, C. C. F., KOENIG, D. F., MAIR, G. A., NORTH, A. C. T., PHILLIPS, D. C. and SARMA, V. R. (1965). *Nature,* **206**, 757

25 JOHNSON, L. N. and PHILLIPS, D. C. (1965). *Nature,* **206**, 762

26 PHILLIPS, D. C. (1966). *Sci. Amer.,* **215**, 78

27 SMYTH, D. G., STEIN, W. H. and MOORE, S. (1963). *J. Biol. Chem.,* **238**, 227

28 RICHARDS, F. M. (1958). *Proc. Nat. Acad. Sci. (U.S.A.),* **44**, 162

29 HOFMANN, K., SMITHERS, M. J. and FINN, F. M. (1966). *J. Amer. Chem. Soc.,* **88**, 4107

30 KARTHA, G., BELLO, J. and HARKER, D. (1967). *Nature,* **213**, 862; KARTHA, G. (1967). *Nature,* **214**, 234

31 WYCKOFF, H. W., HARDMAN, K. D., ALLEWELL, N. M., INAGAMI, T., JOHNSON, L. N. and RICHARDS, F. M. (1967). *J. Biol. Chem.,* **242**, 3984

32 SCOFFONE, E. *et al.* (1967). *Chem. Comm.,* 1273

33 IZUMIYA, N. *et al.* (1968). *Bull. Chem. Soc. Japan,* **41**, 2480

34 LIPSCOMP, W. N. *et al.* (1966). *J. Mol. Biol.,* **19**, 423

35 QUIOCHO, F. A. and LIPSCOMP, W. N. (1971). *Advan. Protein Chem.,* **25**, 1

3.6.3. Quaternary Structure[1]

The quaternary structure of a protein is the term used to describe the organisation of the components when a protein comprises several protein sub-units. The association or aggregation of two or more sub-units occurs through interactions between the polar, ionisable and non-polar side chains by apolar or hydrophobic bonds, ionic bonds, side chain hydrogen bonds, interpeptide hydrogen bonds, etc. In a few exceptional cases, disulphide bonds also take part in the stabilisation of quaternary structure. Quaternary structures show both characteristic stoichiometry and geometry. Each of the polypeptide chains making up the whole is characterised by its own individual secondary and tertiary structure, and it is the interactions between the exposed groups of the folded sub-units which are responsible for holding together the whole complex. Each non-covalently bonded polypeptide chain making up a protein which has quaternary structure is defined as a sub-unit. The quaternary structure may be homogeneous or heterogeneous. In the case of *homogeneous quaternary structure,* identical sub-units are associated to form a functional whole, whereas differring sub-units constitute the building blocks of proteins which show *heterogeneous quaternary structure.* In general, it seems that globular proteins with molecular weights above 50 000 are composed of sub-units. Up to 1974 about 650 proteins with quaternary structure had been recognised, of which about 500 were enzymes. The sub-unit structure of proteins is interesting for many reasons. The quaternary structure appears to be the basis of regulation of the activity of enzymes by substrates or products of the reactions which they catalyse. In general, regulatory enzymes consist of sub-units, while enzymes which have no regulatory function are usually simpler molecules. In allosteric enzymes non-identical sub-units may perform the regulatory and catalytic functions. A sub-unit which contains an intact active centre is known as a *monomer,* while the smallest identical sub-unit is known as a *protomer.* The terms 'monomer' and 'protomer' are thus intermediate between the definitions 'sub-unit' and *oligomer* (protein with quaternary structure).

Proteins with quaternary structure have other advantages compared with single-chain proteins. These include some protection against synthetic faults during translation, economy of genetic material, optimisation of turnover, etc.

Quaternary structure differences form the basis for some cases of *isozymes* (*isoenzymes*). These are enzymes which perform the same function, but which are present in more than one molecular form within the same species or tissue. Lactic dehydrogenase is a well-known example. Five different lactic dehydrogenases detected in various species (Th. Wieland and G. Pfleiderer, 1957). Lactic dehydrogenase is made up of two principal sub-units: one type of sub-unit designated as H (heart or α) and the other M (muscle or β). Since lactic dehydrogenase is a tetramer, five structurally distinct lactic dehydrogenases may be formed: H_4 (α_4), and H_3M ($\alpha_3\beta$), H_2M_2 ($\alpha_2\beta_2$), HM_3 ($\alpha\beta_3$), M_4 (β_4).

Finally, the formation of *multienzyme complexes* is closely tied up with sub-unit structure. For example, the pyruvate dehydrogenase of *E. coli*, which has a molecular weight of 4 000 000, is composed of three types of enzymes: pyruvate decarboxylase, dihydrolipoamide dehydrogenase and dihydrolipoate transacetylase. The latter is composed of 24 sub-units. Electron micrography shows that the pyruvate dehydrogenase complex has a distinct polyhedral appearance with a diameter of 350 ± 50 Å and a height of 225 ± 25 Å.

The quaternary structure of proteins plays an important part in the regulation of enzymes by the feedback inhibition. It has frequently observed in multistep metabolic pathways that the enzyme catalysing the first step is inhibited by the final product of the synthesis. J. Monod coined the term 'allosteric regulation' for this phenomenon. According to Monod's theory, the activity of enzymes can be controlled by allosteric inhibitors not only directly, at or close to the active site, but also indirectly, at so-called allosteric sites well distant from the active site. Such an inhibitor can be a metabolic end product, which inhibits the key enzyme in the metabolic sequences. For example, during the synthesis of isoleucine from threonine, the end product isoleucine inhibits the enzyme which catalysed the conversion of threonine into α-keto glutaric acid. Similarly, in the reaction of carbamyl phosphate with aspartic acid to give carbamyl aspartate, which then leads to cytidine 5′-triphosphate in an enzyme-catalysed four-step process the enzyme aspartate transcarbamylase which catalyses the first step is inhibited by the end product. The native aspartate transcarbamylase (molecular weight 310 000) can be separated into two catalytic sub-units (molecular weight 100 000) and three regulatory sub-units (molecular weight 30 000). The catalytic sub-unit interacts with aspartate and the regulatory one interacts with cytidine *5′-triphosphate*. The interaction of the inhibitor with the allosteric site of an enzyme is frequently associated with changes in the quaternary structure of the protein.

The *quaternary structure* of a protein can be investigated directly by electron microscopy and X-ray diffraction analysis. However, other methods are available which enable the dissociation of oligomeric proteins and subsequent characterisation of the sub-units. Dissociation can be brought about using 1 per cent sodium dodecylsulphate, 6 M guanidine hydrochloride or 8 M urea solutions. It is often possible to cause dissociation by changing the ionic strength, pH, protein concentration or temperature, or by the addition or removal of cofactors, or by chemical modification. The most useful techniques for structural elucidation and molecular weight determination of the sub-units are: ultracentrifuge, polyacrylamide electrophoresis (with sodium dodecylsulphate), gelfiltration and the other methods of molecular weight determination which have already been discussed (see p. 257). It should of course be mentioned that some conditions for dissociation cause simultaneous denaturation with loss of both regulatory and catalytic activity.

Examples of proteins with quaternary structure are given in Table 26.

X-Ray studies are the most powerful tool for the investigation of subunit arrangement: examples include lactic dehydrogenase, malic dehydrogenase, alcohol dehydrogenase, glyceraldehyde-3-phosphate dehydrogenase, five other enzymes of the glycolysis pathway and haemoglobin. Haemoglobin will now be discussed in detail as being representative of proteins with quaternary structure.

3.6.3.1. Haemoglobin

Haemoglobin is the most important respiratory protein in vertebrates. The blood of a average human being contains about 950 g haemoglobin, and the 3.5 g of iron in it represents 80 per cent of the total bodily iron.

Haemoglobin is made up of two pairs of identical sub-units all of which have a haem group. The different polypeptide chains are referred to as

Table 26. Molecular weights and constitution of some proteins exhibiting quaternary structure

Protein	Occurrence	Mol. wt.	Sub-units	
			Number	Mol. wt.
Haemoglobin	Blood of mammals	64250	4 ($\alpha_2\beta_2$)	α: 15130 β: 15870
Tobacco mosaic virus	Infected leaves	39400000	2130*	17530
Concanavalin A	Jack beans	55000	2*	27000
Lactoglobulin	Cow	36750	2*	18375
Enterotoxin	Cholera vibrios	84000	6*	14000
Ceruloplasmin	Plasma	125000	4 ($\alpha_2\beta_2$)	α: 16000 β: 53000
Nerve growth factor	Mouse	26520	2*	13260
Lactose repressor	*E. coli*	150000	4*	39000
Catalase	Liver of cow	240000	4*	60000
Alcohol dehydrogenase	Yeast	141000	4*	35000
Luciferase	*Renilla*	34000	3*	12000
Malic dehydrogenase	*Neurospora*	54000	4 ($\alpha_2\beta_2$)	13500
Aspartate kinase	*B. polymyxa*	116000	4 ($\alpha_2\beta_2$)	α: 17000 β: 43000
Leucine aminopeptidase	Eye lens	330000	6*	58000
	Porcine kidneys	255000	4*	63500
Glutamine synthetase	Porcine brain	370000	8*	46000
	Porcine kidneys	370000	4*	90000
Phosphofructokinase	Yeast	770000	6*	130000

* identical subunits

the α- and β-chains and the symmetrical construction of the molecule is usually abbreviated to $\alpha_2\beta_2$. The adult human has mainly haemoglobin $A_1(\alpha_2\beta_2)$, but also $A_2(\alpha_2\delta_2)$ to the extent of 1.5–3.5 per cent. Two other haemoglobins which are characterised by their high oxygen affinity are found during ontogenesis. Embryonic haemoglobin (H. E, $\alpha_2\varepsilon_2$) or pre-haemoglobin is formed in the first week after conception. Foetal haemoglobin (H. E, $\alpha_2\gamma_2$) is formed in the third month. Besides the α-chains, there are also some developmental modifications which take place in the other identical chain pairs. The β-, γ- and δ-chains each contain 146 amino acid residues, but the α-chain contains only 141 amino acid residues. The primary structures of the α-, β- and γ-chains and the polypeptide chain of myoglobin (153 residues) show numerous homologous regions. These regions are shown in Figure 58. The same haemoglobin is found in all the human races

α	Val		Leu	Ser	Pro	Ala	Asp	Lys	Thr	Asn	Val	Lys	Ala	Ala	Try	Gly	Lys	Val	Gly	Ala	His	Ala	Gly	Glu	Tyr	Gly	Ala	Glu	Ala	Leu
β	Val	His	Leu	Thr	Pro	Glu	Glu	Lys	Ser	Ala	Val	Thr	Ala	Leu	Try	Gly	Lys	Val	Asn			Val	Asp	Glu	Val	Gly	Gly	Glu	Ala	Leu
γ	Gly	His	Phe	Thr	Glu	Glu	Asp	Lys	Ala	Thr	Ile	Thr	Ser	Leu	Try	Gly	Lys	Val	Asn			Val	Glu	Asp	Ala	Gly	Gly	Glu	Thr	Leu
M	Gly		Leu	Ser	Glu	Gly	Glu	Try	Gln	Leu	Val	Leu	His	Val	Try	Ala	Lys	Val	Glu	Pro	Asp	Val	Ala	Gly	His	Gly	Asp	Asp	Ile	Leu
										10										20										30

α	Glu	Arg	Met	Phe	Leu	Ser	Phe	Pro	Thr	Thr	Lys	Thr	Tyr	Phe	Pro	His	Phe		Asp	Leu	Ser	His						Gly	Ser	Ala
β	Glu	Arg	Leu	Leu	Val	Val	Tyr	Pro	Try	Thr	Gln	Arg	Phe	Phe	Glu	Ser	Phe	Gly	Asp	Leu	Ser	Thr	Pro	Asp	Ala	Val	Met	Gly	Asn	Pro
γ	Gly	Arg	Leu	Leu	Val	Val	Tyr	Pro	Try	Thr	Gln	Arg	Phe	Phe	Asp	Ser	Phe	Gly	Asn	Leu	Ser	Ser	Ala	Ser	Ala	Ile	Met	Gly	Asn	Pro
M	Ile	Arg	Leu	Phe	Lys	Gly	His	Pro	Glu	Thr	Leu	Glu	Lys	Phe	Asp	Lys	Phe	Lys	His	Leu	Lys	Ser	Glu	Asp	Glu	Met	Lys	Ala	Ser	Glu
										40										50										60

α	Gln	Val	Lys	Gly	His	Gly	Lys	Lys	Val	Ala	Asp	Ala	Leu	Thr	Asn	Ala	Val	Ala	His	Val	Asp	Asp	Met	Pro	Asn	Ala	Leu	Ser	Ala	Leu
β	Lys	Val	Lys	Ala	His	Gly	Lys	Lys	Val	Leu	Gly	Ala	Phe	Ser	Asp	Gly	Leu	Ala	His	Leu	Asp	Asn	Leu	Lys	Gly	Thr	Phe	Ala	Thr	Leu
γ	Lys	Val	Lys	Ala	His	Gly	Lys	Lys	Val	Leu	Thr	Ser	Leu	Gly	Asp	Ala	Ile	Lys	His	Leu	Asp	Asp	Leu	Lys	Gly	Thr	Phe	Ala	Gln	Leu
M	Asp	Leu	Lys	Lys	His	Gly	Val	Thr	Val	Leu	Thr	Ala	Leu	Gly	Ala	Ile	Leu	Lys	Lys	Lys	Gly	His	His	Glu	Ile	Glu	Leu	Lys	Pro	Leu
										70										80										90

α	Ser	Asp	Leu	His	Ala	His	Lys	Leu	Arg	Val	Asp	Pro	Val	Asn	Phe	Lys	Leu	Leu	Ser	His	Cys	Leu	Leu	Val	Thr	Leu	Ala	Ala	His	Leu	Pro	Ala
β	Ser	Glu	Leu	His	Cys	Asp	Lys	Leu	His	Val	Asp	Pro	Glu	Asn	Phe	Arg	Leu	Leu	Gly	Asn	Val	Leu	Val	Cys	Val	Leu	Ala	His	His	Phe	Gly	Lys
γ	Ser	Glu	Leu	His	Cys	Asp	Lys	Leu	His	Val	Asp	Pro	Glu	Asn	Phe	Lys	Leu	Leu	Gly	Asn	Val	Leu	Val	Thr	Val	Leu	Ala	Ile	His	Phe	Gly	Lys
M	Ala	Gln	Ser	His	Ala	Thr	Lys	His	Lys	Ile	Pro	Ile	Lys	Tyr	Leu	Glu	Phe	Gln	Ser	Glu	Ala	Ile	Ile	Ser	Val	Leu	Asn	Ser	Lys	His	Pro	Gly
										100										110										120		

α	Glu	Phe	Thr	Pro	Ala	Val	His	Ala	Ser	Leu	Asp	Lys	Phe	Leu	Ala	Ser	Val	Ser	Thr	Val	Leu	Thr	Ser	Lys	Tyr	Arg						
β	Glu	Phe	Thr	Pro	Pro	Val	Gln	Ala	Ala	Tyr	Gln	Lys	Val	Val	Ala	Gly	Val	Ala	Asn	Ala	Leu	Ala	His	Lys	Tyr	His						
γ	Glu	Phe	Thr	Pro	Glu	Val	Gln	Ala	Ser	Try	Gln	Lys	Met	Val	Thr	Gly	Val	Ala	Ser	Ala	Leu	Ser	Ser	Arg	Tyr	His						
M	Asn	Phe	Gly	Ala	Asp	Ala	Gln	Gly	Ala	Met	Asn	Lys	Ala	Leu	Glu	Leu	Phe	Arg	Lys	Asp	Met	Ala	Ser	Asp	Tyr	Lys	Glu	Leu	Gly	Tyr	Gln	Gly
								130											140									150				

and in chimpanzees. Point mutations lead to pathological haemoglobins. More than 150 of these pathological haemoglobins have been identified to date. The most common of the pathological haemoglobins is that which occurs in sickle-cell anaemia (H. S). Sickle-cell haemoglobin and normal haemoglobin differ only in the substitution of a valine residue for a glutamic acid residue in position 6 of the β-chain. Details of the amino acid sequences of the various haemoglobins as well as their evolution (and the evolution of proteins generally) will not be dealt with here. The subject has been reviewed by ACHER[2,3] and BRAUNITZER[4].

The three-dimensional structure of the oxyhaemoglobin of the horse was elucidated by PERUTZ and co-workers[5,6] by X-ray crystal structure analysis. Although the exact location of individual amino acids was not possible at a resolution of 5.5 Å, the tetrameric structure and the position of the four haem groups were clearly defined. The four chains are arranged in a roughly spherical structure, each having conformation rather like that of myoglobin. A model of horse oxyhaemoglobin is shown in Figure 59; the folding of the α- and β-chains is shown in Figure 60. PERUTZ and his co-workers[7] have also elucidated the structure of human haemoglobin at a re-

Figure 59. Model of horse oxyhaemoglobin (5.5 Å resolution) according to PERUTZ *et al.*[5] The backbone of the α-chain (white unit) and β-chain (black unit), respectively, is outlined. One α-chain is omitted

solution of 5.5 Å. PERUTZ published the three-dimensional FOURIER synthesis of horse oxyhaemoglobin at 2.8 Å resolution in 1968[8]: virtually all the positions of the amino acid residues and some details of the haem groups were provided by this study. At 2 Å resolution, the haemoglobin model could be refined, and the positions of its ca. 10000 atoms were fixed with an accuracy of ± 1 Å.

From the structural data now available, it is possible to explain the physiological function of haemoglobin. Haemoglobin performs the essential function of transporting oxygen from the lungs to the various tissues in which oxidative metabolism occurs, by combining reversibly with molecular oxygen to give oxyhaemoglobin, in which oxygen becomes the sixth ligand of the haem iron atom, without oxidation of the iron atom. In contrast, methhaemoglobin, in which the iron atom is oxidised, is incapable of combining with oxygen.

The X-ray diffraction studies have shown that haemoglobin is composed of four polypeptide chains, which occur in pairs ($_2\beta_2$), together with four haem groups bound to each of the chains through weak forces. A fifth coordination site of every haem iron atom is occupied by the imidazole

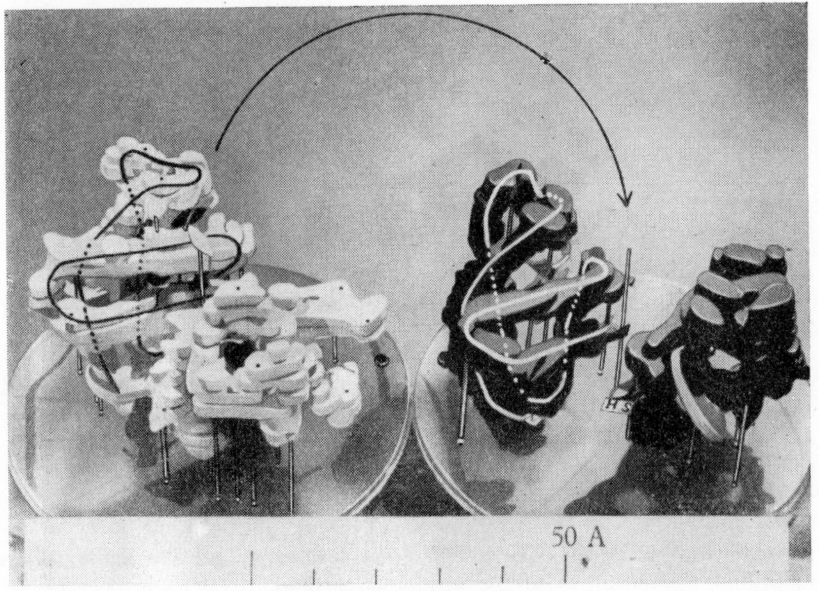

Figure 60. Models of the four polypeptide chains of haemoglobin according to PERUTZ *et al*[5]. In onec ase of an α-chain (white unit) and a β-chain (black unit), respectively, the course of the main chain is outlined. Joining together the four chains the completed model of haemoglobin results

group of a histidine residue. In the oxygen-free form haemoglobin forms a high-spin Fe^{2+} complex of exceedingly low potential in which the sixth ligand site is vacant. When the π-bonding O_2 molecule enters the sixth co-ordination position, a structural rearrangement to a low-spin state takes place. It is known that the arrangement of the four chains is such that the interactions between like chains are weak and those between unlike chains are strong. Therefore, as a result of one $\alpha\beta$ half sliding past the other, the contact regions for the reversible binding of the four oxygen molecules are formed. No detectable change in the conformations of the individual polypeptide chains is observed. The sigmoid oxygen binding curve is attributed to this allosteric effect. The X-ray studies have enabled the precise stereochemistry of this co-operative effect to be elucidated[9].

References

1 SUND, H. and K. WEBER (1966). *Angew. Chem.*, **78**, 217

2 ACHER, R. (1966). *Angew. Chem.*, **78**, 356

3 ACHER, R. (1974). *Angew. Chem.*, **86**, 209

4 BRAUMITZER, G. (1967). *Naturwissenschaften*, **54**, 407

5 PERUTZ, M. F. *et al.* (1960). *Nature*, **185**, 416

6 PERUTZ, M. F. *et al.* (1961). *Proc. Roy. Soc.* (*London*), A **265**, 161

7 PERUTZ, M. F. *et al.* (1967). *J. Mol. Biol.*, **28**, 117

8 PERUTZ, M. F. *et al.* (1968). *Nature*, **219**, 29, 131

9 PERUTZ, M. F. (1970). 8th International Congress of Biochemistry, abstracts, p. 5

3.6.4. Structural Proteins

The term 'structural proteins' includes those fibrous proteins which fulfil protective and skeletal functions in animals. They are also known as scleroproteins or fibrous proteins. They generally have a characteristic amino acid composition, are generally insoluble in water and are not attacked by common proteolytic enzymes. They are unsuitable as food because of their indigestibility and low content of essential amino acids. On the basis of chain conformation three types of structure can be distinguished; the *α-helical type, the pleated sheet type* and the *triple helix type*. The most important representatives of these types are: keratin, silk proteins and collagen. In structural proteins, specific physical properties can be achieved by three-dimensional cross-linking of chains through covalent bridges (cross-links). *Resilin* is the protein component that coats the chitin lamellae of insects and gives rise to the flexibility of the exo-skeleton. The cross-linking takes place through three tyrosine residues in the chains. *Elastin* has a characteristically high content of hydrophobic amino acids. It is the structural protein of the elastic fibre components of ligaments, bands and vessels. Its hardness and elasticity are attributed to crosslinking through the pyridine amino acid *desmosine*.

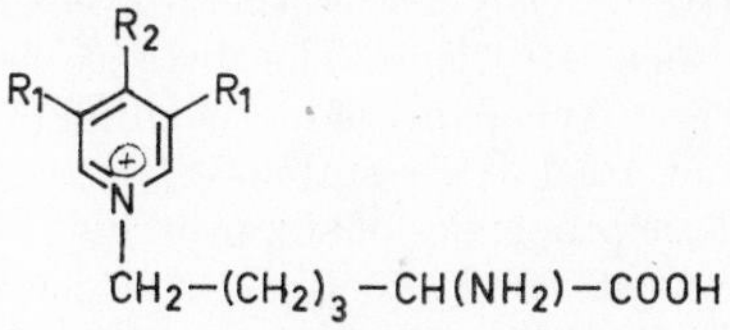

$$\text{pyridinium ring (N}^{+}\text{) with } R_1 \text{ at positions 3 and 5, } R_2 \text{ at position 4; N-substituent: } CH_2-(CH_2)_3-CH(NH_2)-COOH$$

$$R = -CH_2-(CH_2)_n-CH(NH_2)-COOH$$

$$R_1: n = 1; \quad R_2: n = 2$$

Amino acid side chains of lysine take part in the biosynthesis of desmosine cross-links. Desmosine and isodesmosine were discovered in the acid hydrolysates of elastin. (In isodesmosine, R_2 is in position 2 of the pyridine ring instead of position 4.)

3.6.4.1. Keratins[1]

The basic structure of the keratin of wool and hair is the α-helix stabilised by hydrogen bonds. The helices are linked to one another through disulphide bonds. The amino acid composition is high in hydrophobic amino acids and in cysteine (11 per cent). Three right-directed α-helices are twisted to form *protofibrils* with a diameter about 20 Å. The twisting process reduces the repeating distance of the normal α-helix from 5.4 Å to 5.1 Å. Eleven of the protofibrils twist together to form a cable-like structure with a diameter of 80 Å. These so-called *microfibrils* can be observed with the electron microscope. Several hundred microfibrils form '*macrofibrils*' of 2000 Å diameter, which are embedded in a cysteine-rich protein matrix which hardens the structure. The macrofibrils are stacked parallel to the fibre axis in the dead cells of the wool fibre, which reaches a diameter of 200000 Å in its over-all structure. The special and valuable characteristics of wool fibres are their ability to be stretched and their elasticity. Stretching wool fibres causes stretching of the α-helices due to breakage of hydrogen bonds. Their elasticity arises from the cross-linking of the helices by disulphide bonds. As well as the α-helix structure, some complex β-structures have been identified in keratins of skin, feathers, claws and beaks. The precise details of these β-structures have not yet been fully elucidated.

3.6.4.2. Silk Proteins

Eighty-seven per cent of the amino acid content of the fibroin of natural silk of the commercially exploited silkworm *Bombyx mori* consists of glycine, alanine and serine. Other amino acids occur in small concentrations, but cysteine and methionine are absent. In the extended polypeptide chains, which run anti-parallel to one another, the sequence Gly–Ser–Gly–Ala–Gly–Ala occurs repeatedly. The chains themselves are stabilised in an β-anti-parallel pleated sheet structure. The distance between the sheets is 3.5 and 5.7 Å alternately. The pairs of polypeptide chains are bound to protein complexes which are further stabilised by *sericin*, a water-soluble silk protein. Characteristic properties of silk proteins are a lack of elasticity, due to the strong covalent bonding of the stretched chains, and great flexibility, which is due to the weak VAN DER WAALS forces between the pleated sheets.

3.6.4.3. Collagens[2–6]

Collagens occur widely in animals and constitute 25–30 per cent of their total proteins. They are components of skin, cartilage, arterial walls and connective tissues. They have characteristically high proline (12 per cent) and hydroxyproline (9 per cent) contents, and glycine (33 per cent) is the main amino acid component. Cysteine and methionine occur only in the collagens of vertebrates. Collagens also contain hydroxylysine and, depending on the source, a little carbohydrate. The basic structure of collagen consists of *tropocollagen* molecules with molecular weights of 300000. These occur in the form of stick-like structures, 30000 Å in length and 15 Å in diameter. Tropocollagen is formed from three similar large polypeptide chains of which two are usually identical and termed the α_1-chains. The collagen of cartilage contains three identical α_1-chains, while the chains of tuna fish-skin collagen are completely different from each other. In 1973 it was shown by sequence analysis that the α_1-chains consist of 1012 amino acids and have the molecular weight 100000. The chains occur as left-handed helices with three amino acids per twist. The sequences $(\text{–Gly–X–Pro–})_n$ and $(\text{Gly–X–Hypro–})_n$ are required for the structure and affect the twisting together of the individual chains to form the so-called *super-* or *triple helix*, which is right-handed. The helical structure is stabilised by hydrogen bridges between the peptide bonds of the individual chains and can be destroyed by heating in aqueous medium. On cooling, the solution begins to 'gelatinise'.

The triple helical molecules combine to form fibrils up to 5000 Å in diameter. In these fibrils the individual molecules are arranged parallel to one another and are shifted about one-fourth of their length to form the typical transverse marking of collagen fibrils. The fibrils can be observed with the electron microscope by staining with heavy metal salts or by small-angle X-ray diffraction. On treatment with phosphomolybdic acid, uranyl or chromic salts the polar regions of collagen units become cross-linked by multivalent bonds through the acidic and basic amino acid

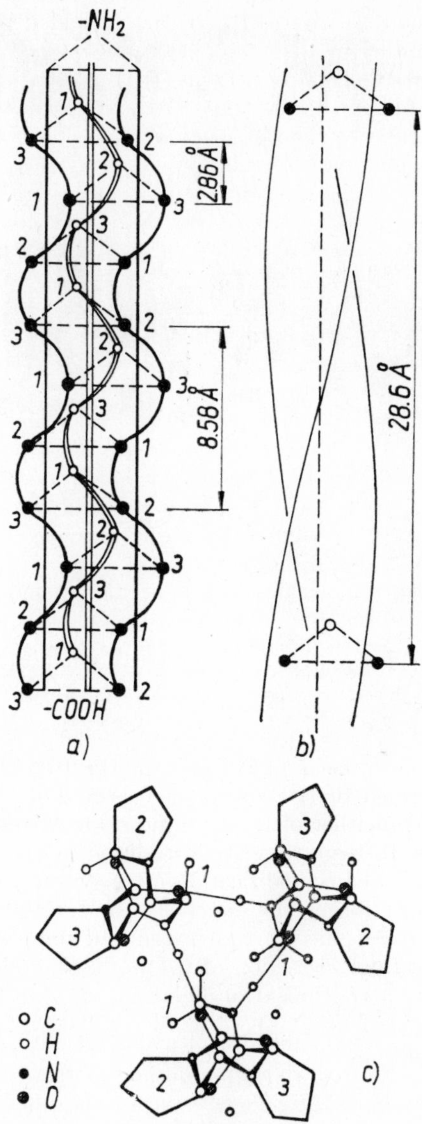

Figure 61. Structure of collagen as proposed by A. Rich and F. H. Crick (from Kühn, K., *Naturwissenschaften*, **54**, 101, 1967)

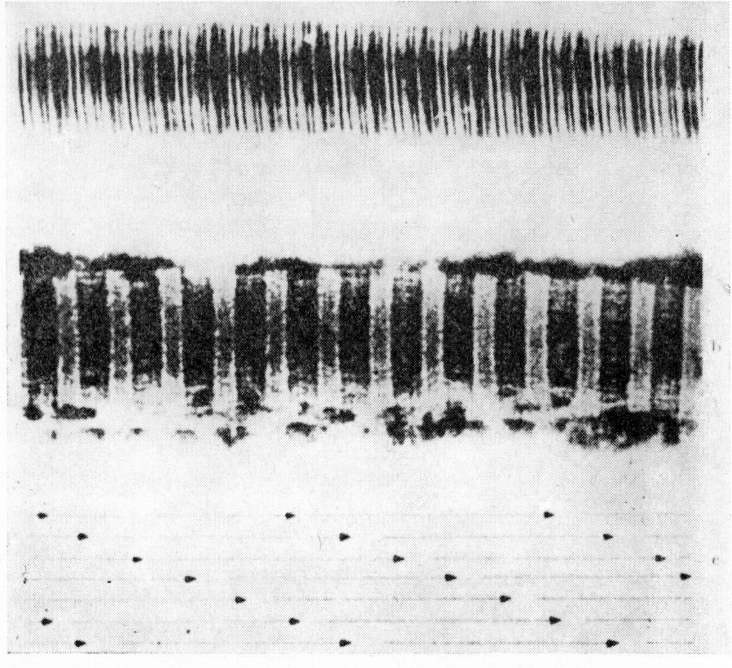

Figure 62. Electron microscope photograph of collagen fibrils (from KÜHN, K., *Naturwissenschaften*, **54**, 101, 1967)

residues. In the biosynthesis of collagen in the fibroblasts water-soluble protocollagen is formed first. Protocollagen is free of hydroxyproline and hydroxylysine, since neither of these amino acids is contained in the code of the nucleic acids. Instead, they are introduced by specific protocollagen-hydroxylases by side chain modification after synthesis of the chain. The triple-helix structure forms spontaneously and then the carbohydrate components (galactose, glucose) are introduced onto the OH groups of hydroxylysine. The final formation of the insoluble collagen fibrils occurs when the precursor enters the extracellular region.

References

1 BRADBURY, J. H. (1973). *Advan. Protein Chem.*, **27**, 111
2 RICH, A. and CRICK, F. H. (1961). *J. Mol. Biol.*, **3**, 483
3 GRASSMANN, W. (1965). *Fortschr. Chem. Organ. Naturstoffe*, **23**, 196
4 NEMETSCHEK, Th. (1969). *Chem. Labor Betrieb*, **20**, 433

5 REICH, G. (1967). *Kollagen, Eine Einführung in Methoden, Ergebnisse und Probleme der Kollagenforschung*, Theodor Steinkopff Verlag, Dresden
6 TRAUB, W. and PIEZ, K. A. (1971). *Advan. Protein Chem.*, **25**

3.7. Biosynthesis of Proteins[1,2]

Although the fine details of the *in vivo* synthesis of proteins have not yet been completely elucidated, the principles of protein biosynthesis are well established. The mechanism of protein biosynthesis is a stepwise process and will now be discussed.

3.7.1. Activation of Amino Acids

An amino acid reacts with adenosine triphosphate (ATP) with elimination of pyrophosphate to yield a mixed anhydride of the amino acid and phosphoric acid:

$$\text{Ⓟ}\sim\text{Ⓟ}\sim\text{Ⓟ}-O-\underset{OH\quad OH}{[\text{O-ring}]}-\text{Ⓐ} + H_2N-\underset{R}{CH}-COOH \underset{\text{Synthetase}}{\rightleftharpoons} H_2N-\underset{R}{CH}-\underset{O}{\overset{\|}{C}}-O-\text{Ⓟ}-O-\underset{OH\quad OH}{[\text{O-ring}]}-\text{Ⓐ} + \text{Ⓟ}\sim\text{Ⓟ}$$

This reaction is catalysed by a specific synthetase. Each enzyme appears to be highly specific for its individual amino acid and also catalyses the formation of aminoacyl-transfer ribonucleic acid (aminoacyl-tRNA). The mixed anhydride, which is sometimes called aminoacyl adenylate, is tightly bound to the enzyme protein.

3.7.2. Formation of Aminoacyl-tRNA Molecules

In the next step, the activated aminoacyl group is transferred to a low-molecular weight RNA, known as transfer RNA (tRNA) and adenosine monophosphate is lost. The product, amino-acyl tRNA, is an active ester with a high group-transfer potential.

For every amino acid there is at least one specific tRNA and one aminoacyl-tRNA synthetase. As already mentioned, the aminoacyl-tRNA synthetase catalyses both the activation of the amino acid and the formation of aminoacyl-tRNA molecules. All amino acids are activated directly by the two steps described above, with exception of glutamine. In this case the formation of Gln-tRNA^{Gln} appears to be indirect and requires the conversion of Glu-tRNA^{Gln} into Gln-tRNA^{Gln}, which is brought about by an amino transferase in an additional step.

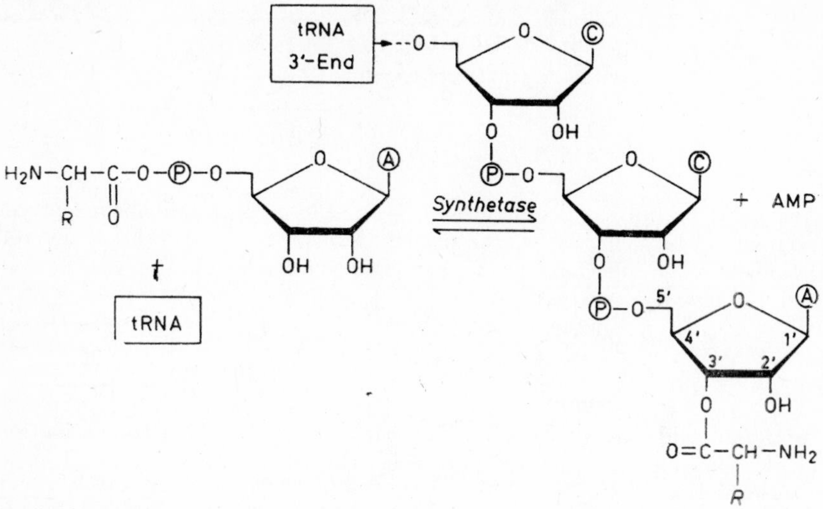

The tRNAs provide the link between the nucleotide and the amino acid codes embodied in DNA. The groups of R. A. Holley, J. T. Madison and H. C. Zachau[3,4] have elucidated the primary structure of a number of tRNAs. In all tRNAs, the acceptor group of the amino acids consists of the triplete: cytidine phosphate–cytidine phosphate–adenosine. The amino acid is bound via an ester bond to the 3-hydroxyl group of the ribose part of the terminal adenosine. The other terminal group of a tRNA consists of guanosine phosphate. The various tRNAs differ in the composition and sequence of their nucleotides. The length of the tRNAs of different species appears to vary over a rather narrow range, from 76 to 85 residues. Besides the four common nucleosides adenosine (A), guanosine (G), cytosine (C) and uridine (U), they contain a variety of unusual nucleosides, such as inosine (I), pseudouridine (ψ), dihydrouridine (H_2U), mono- and dimethyl derivatives of adenosine (A^{Me}), guanosine (G^{Me}) and cytosine (C^{Me}). The tRNAs exhibit a high degree of internal sequence complementarity and of organised secondary and of tertiary structure. Mg^{2+} ions are essential for their biological activity. A typical secondary structure is the cloverleaf arrangement shown in Figure 63. The cloverleaf structure of $tRNA^{Phe}_{yeast}$ has been refined using X-ray diffraction analysis to a resolution of 3–4 Å. The three-dimensional structure corresponds to a twisted L, where the branches orginate from each of the two cloverleaf stalks. The terminal points of the L are marked by the 'anticodon loop' and the binding site for the amino acid.

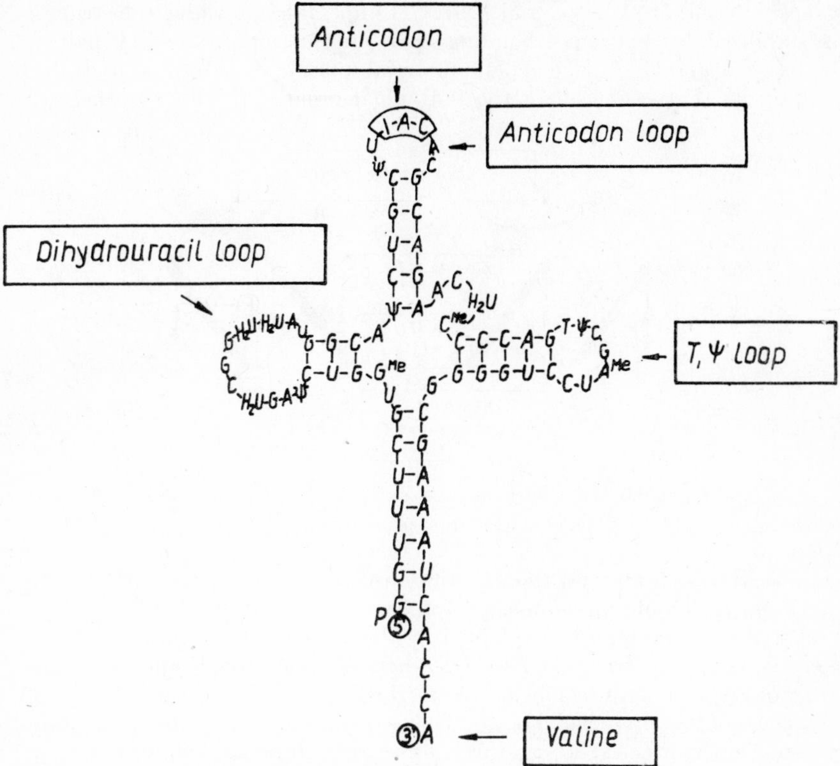

Figure 63. 'Cloverleaf' structure of $tRNA^{Val}_{Yeast}$ [2]

3.7.3. Binding of the First Aminoacyl-tRNA to the Complex between Ribosomes and mRNA and Formation of Peptide Bonds

The characteristic amino acid sequence of a protein is essential for its specific biological action. The information required for the the biosynthesis of a protein is contained in a gene. The genes are arranged linearly on the chromosomes, which exercise control over the development and daily metabolism of the organism. The synthesis of the right enzyme at the appropriate time is essential to the biosynthesis of all other natural products. Desoxyribonucleic acid (DNA) is responsible for this process. The DNA, as the name indicates, contains 2-desoxyribose instead of ribose in the case of RNA.

DNA is a polydesoxyribonucleotide containing the heterocyclic bases adenine (A), guanine (G), cytosine (C) and thymine (T) and forms a regular *double helix* (J. D. Watson and F. H. C. Crick, 1953) in which the structure is stabilised by hydrogen bonding between the complementary pairs of bases (A–T and G–C). The genetic material must be able to reproduce itself exactly so that each cell in the individual organism will have the same

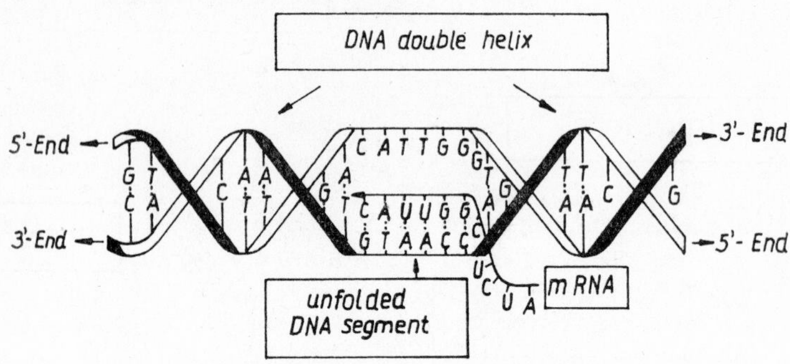

Figure 64. Enzymatic mRNA synthesis on the DNA matrix (asymmetric transcription of DNA)

genetic composition, and the genetic information can be passed on to the next generation of individuals. The process of duplication of DNA, which is called *replication,* may be explained by the Watson–Crick model (M. Meselson and F. W. Stahl, 1958). DNA acts like a punched-tape program to control its own replication. DNA carries its genetic information in the sequence of its heterocyclic bases. This sequence also contains the program for protein synthesis: every three bases determine one amino acid (the triplet code). However, DNA does not conduct protein biosynthesis directly, but does so via RNA intermediaries. Owing to the asymmetric transcription of DNA, the interpretation of the sequence of the genetic code is done by messenger RNA. There is no overlapping in the genetic code[5–8] embodied in the DNA strand; every nucleotide triplet codifies for one amino acid and one only. Since the biosynthesis of proteins takes place on the ribosomes, the genetic information must be transferred from the chromosomes to the ribosomes. This is accomplished by the messenger RNA (abbreviated mRNA), a single-stranded ribonucleic acid, which contains uracil (U) in place of the thymine found in DNA. The mRNA carries the information which determines the amino acid sequence of the polypeptide chain. The synthesis of the mRNA is programmed by the hydrogen bonding of ribonucleotide triphosphates to the bases of the DNA. This process is called *transcription* (see Figure 64). During transcription the genetic information of one or more genes is copied. The mRNA then goes into the cytoplasma, combines with the ribosome and forms the matrix for the synthesis of a polypeptide chain. The step in which protein is synthesised under the control of mRNA is called *translation*[9]. The aminoacyl tRNA is attached to the ribosomes as well to the mRNA. The mRNA selects the appropriate aminoacyl tRNA by a base-pairing mechanism which involves a specific code in form of three bases in the mRNA sequence and three bases in the aminoacyl tRNA (the anticodon). In this manner, the 'correct' amino acid is incorporated into the growing polypeptide chain. Once the binding has been accomplished, the mRNA together with the aminoacyl tRNA moves further into the ribosome and the next code of the mRNA recognises and binds the next aminoacyl tRNA by the same mechanism. The formation of the actual peptide bonds within the complex occurs by aminolysis of the ester bond of the first aminoacyl RNA by the amino group of the second amino acyl RNA (see Fig. 65). Then the complex moves relative to the ribosome and the procedure is repeated and so-on down the chain. The sequence of the base in the mRNA is read from the 5-hydroxyl to the 3'-hydroxyl ends. It is normal for several ribosomes to be combined with the mRNA and engaged in synthesising polypeptide chains at the same time. The resulting structures are known as polysomes.

The process of chain elongation (*propagation*) takes place in the presence of guanosine triphosphate (GTP) and adenosine triphosphate (ATP) as well as divalent (Mg^{2+}) and monovalent (NH_4+, K^+) ions, several soluble proteins (elongation factors, transfer factors, transferases), peptide synthetase and reactive thiol groups.

The formation of the polypeptide chain begins at the N-terminal amino acid. The chain elongation is a stepwise process in which the polypeptide is built up one amino acid at a time (see Figure 65).

If the mRNA contains several cistrons each of which genetically determine a specific polypeptide chain, the question of identifying the beginning and the end of each cistron arises. The start of the synthesis in *E. coli,* for example, was shown to be determined by the triplete AUG (incorporation of *N*-formyl methionine). SMITH and MARCKER[10] have proposed that this is also the chain-initiating signal for the biosynthesis of mammalian proteins. S. BRENNER *et al.* found that the tripletes UAG and UAA (which, like UGA, do not code for amino acids and have been called 'nonsense codons') determine the end of the chain.

3.7.4. Release of the Synthesised Polypeptide Chain from the Ribosomes and Formation of Native Conformation

In the final step of biosynthesis, the newly formed polypeptide chain is released from the ribosomes. Before this can happen a specific signal for chain termination is required: this is provided by the 'nonsense codons' mentioned above. Then an enzyme system is needed to cleave the *C*-terminal ester bond in order to detach the completed polypeptide chain from the

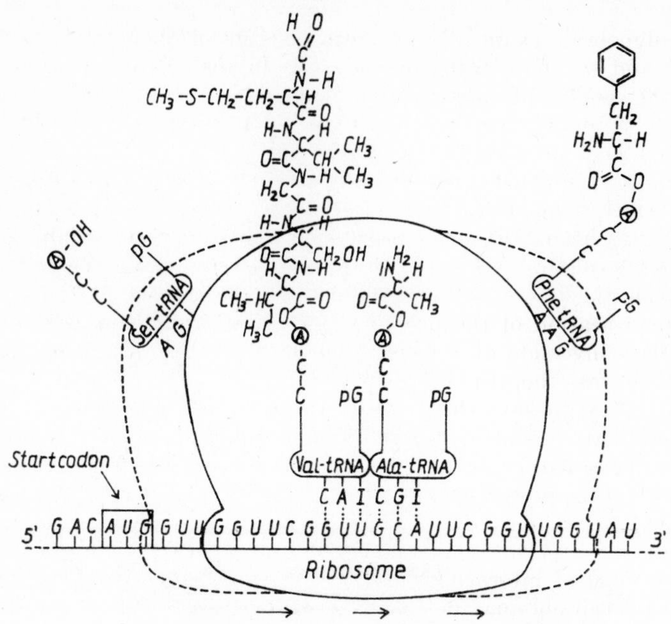

Figure 65. Schematic representation of translation

ribosome. Further, a specific protein is necessary to detach the ribosomal subunits from the complex and a number of additional release factors are necessary to release the polypeptide chain from the ribosomes. The over-all mechanism of chain termination is a multistep process whose details have not yet been fully elucidated.

The biosynthesis of proteins is a very rapid process. The formation of each peptide bond is carried out in about one second. A large protein molecule can therefore be formed in a few minutes. There is now little doubt that the native conformation of a protein forms spontaneously. However, it is unclear as to when the necessary folding reaction takes place, and what factors determine the adoption of one particular conformation out of several possibilities.

Quaternary structure, the building up of large proteins from numbers of sub-units, is extremely economical in terms of genetic information since only the sequence of each type of sub-unit needs to be encoded. For instance, the protein of tobacco mosaic virus (refer to Table 26) contains 2130 identical polypeptide sub-units, and apoferritin has 20 identical sub-units. Immediately after biosynthesis the sub-units associate to form the over-all protein complexes.

The polypeptide chain of apoferritin consists of about 200 amino acid residues, and so about 600 pairs of bases in the DNA are necessary to contain the genetic information for its biosynthesis. If apoferritin consisted of a single chain 20 times as long as the sub-unit chain, 12000 base pairs would have been necessary.

It is quite common for the primary product of translation to undergo enzyme-catalysed modifications before it can assume its final biologically active conformation. One example of such a modification is the sequential removal of formate, methionine, alanine and serine from the N-terminus N–For–Met–Ala–Ser– of most bacterial and bacteriophage proteins. A more radical modification of the primary product of translation occurs in the case of the conversion of zymogens, enzyme precursors, proenzymes and prohormones into their final forms.

Every cell synthesises the proteins and enzymes necessary for growth. Cell differentiation controls the individual requirements of any particular cell, and the regulation mechanism controls not only the synthesis but also the activity of the proteins synthesised. Regulation is possible at the mRNA synthesis stage (transcription), during translation or even during the enzymic catalysis.

A great deal of information about the regulation of the transcription process has been obtained from studies with bacteria. JACOB and MONOD[11] elucidated the mechanism of enzyme induction from their studies of the lactose system in *E. coli*. When *E. coli* is allowed to grow on lactose as a source of carbon, the sugar acts as an *inducer* and stimulates the synthesis of three enzymes (permease, β-galactosidase, transacetylase) which allow the utilisation of this 'unusual' nutritive source. The three genes which determine the sequences of the three enzymes are called *structure genes*. In the absence of lactose these enzymes are not synthesised by the cell, and this repression occurs on the *lac operon*, which is a cluster of genes of the bacterial chromosome or genome. The lac operon consists of three structural genes and two regulatory regions, one for the operator and one for the promoter. In the course of repression an allosteric protein, known as the *repressor*, is bound by the operator and thus inactivates the whole operon. In addition to induction and repression, there are other factors involved in the process of regulation at the level of transcription. These factors are directly related to RNA polymerase and to the velocity of release of mRNA. The finer details of the regulation of protein biosynthesis will not be discussed here. For more details see references 2 and 12.

The biological aspects of the biosynthesis of proteins have been only briefly discussed here owing to lack of space. In spite of an enormous increase of excellent experimental data, numerous problems in this area remain to be answered. Recent investigations have shown that the generally accepted dogma of molecular genetics is, in special cases, questionable. It has been found that the RNA of phage can act as a matrix for DNA synthesis after infection. RNA-dependent polymerase, previously detected in different virus RNA causing cancer has been isolated.

Lipmann *et al.*[13] have studied the biosynthesis of peptide antibiotics and found that it does not occur by the usual mechanism of protein biosynthesis. The mechanism underlying the biosynthesis of the peptide antibiotics is similar to that of fatty acids, in which activated thioesters play an important role.

Finally, there is the possibility that the biosynthetic process described above may possibly be used in the future for the industrial production of special proteins, since it is possible to carry out all the individual steps of protein biosynthesis in cell-free preparations.

References

1 Campbell, P. N. and Sargent, J. R. (1967). *Techniques in Protein Biosynthesis*, Academic Press, London
2 Träger, L. (1969). *Einführung in die Molekularbiologie*, VEB G. Fischer Verlag, Jena
3 Zachau, H. G. (1969). *Angew. Chem.*, **81**, 645
4 Holley, R. W. (1969). *Angew. Chem.*, **81**, 1039
5 Wittman, G. and Jokusch, H. (1967). *Molekularbiologie — Bausteine des Lebendigen* (ed. Wieland, Th. and Pfleiderer, G.), p. 49, Umschau-Verlag, Frankfurt/M.
6 Ochoa, S. (1966). *Naturwiss. Rundsch.*, **19**, 483
7 Mirenberg, M. (1969). *Angew. Chem.*, **81**, 1017
8 Khorana, H. G. (1969). *Angew. Chem.*, **81**, 1027
9 Ochoa, S. (1968). *Naturwissenschaften*, **55**, 506
10 Smith, A. E. and Marcker, K. A. (1970). *Nature*, **226**, 607
11 Jacob, F. and Monod, J. (1961). *J. Mol. Biol.*, **3**, 318
12 Wallenfels, K. and Weil, R. (1967). In reference 5, p. 67
13 Lipmann, F. *et al.* (1971). *Arch. Biochem. Biophys.*, **143**, 485

Subject Index